Climate Change and Agricultural Water Management in Developing Countries

CABI Climate Change Series

Climate change is a major environmental challenge to the world today, with significant threats to ecosystems, food security, water resources and economic stability overall. In order to understand and research ways to alleviate the effects of climate change, scientists need access to information that not only provides an overview of and background to the field, but also keeps them up to date with the latest research findings.

This series addresses many topics relating to climate change, including strategies to develop sustainable systems that minimize impact on climate and/or mitigate the effects of human activity on climate change. Coverage will encompass all areas of environmental and agricultural sciences. Aimed at researchers, upper level students and policy makers, titles in the series provide international coverage of topics related to climate change, including both a synthesis of facts and discussions of future research perspectives and possible solutions.

Titles Available

1. Climate Change and Crop Production
Edited by Matthew P. Reynolds

2. Crop Stress Management and Climate Change
Edited by José L. Araus and Gustavo A. Slafer

3. Temperature Adaptation in a Changing Climate: Nature at Risk
Edited by Kenneth Storey and Karen Tanino

4. Plant Genetic Resources and Climate Change
Edited by Michael Jackson, Brian Ford-Lloyd and Martin Parry

5. Climate Change Impact and Adaptation in Agricultural Systems
Edited by Jürg Fuhrer and Peter J. Gregory

6. Livestock Production and Climate Change
Edited by Pradeep K. Malik, Raghavendra Bhatta, Junichi Takahashi, Richard A. Kohn and Cadaba S. Prasad

7. Climate Change and Insect Pests
Edited by Christer Björkman and Pekka Niemelä

8. Climate Change and Agricultural Water Management in Developing Countries
Edited by Chu Thai Hoanh, Robyn Johnston and Vladimir Smakhtin

9. Climate Change Challenges and Adaptations at Farm-level
Edited by Naveen Prakash Singh, Cynthia Bantilan, Kattarkandi Byjesh and Swamikannu Nedumaran

Climate Change and Agricultural Water Management in Developing Countries

Edited by

Chu Thai Hoanh

Robyn Johnston

Vladimir Smakhtin

International Water Management Institute (IWMI), Colombo, Sri Lanka

www.cabi.org

in association with
International Water Management Institute (IWMI)

CABI is a trading name of CAB International

CABI
Nosworthy Way
Wallingford
Oxfordshire OX10 8DE
UK

Tel: +44 (0)1491 832111
Fax: +44 (0)1491 833508
E-mail: info@cabi.org
Website: www.cabi.org

CABI
745 Atlantic Avenue
8th Floor
Boston, MA 02111
USA

T: +1 (617)682-9015
E-mail: cabi-nao@cabi.org

A catalogue record for this book is available from the British Library, London, UK.

Library of Congress Cataloging-in-Publication Data

Names: Chu, Thai Hoanh, 1949– | Johnston, Robyn. | Smakhtin, Vladimir.
Title: Climate change and agricultural water management in developing countries / edited by Chu Thai Hoanh, Robyn Johnston, and Vladimir Smakhtin, International Water Management Institute (IWMI), Colombo, Sri Lanka.
Description: Boston, MA: CABI, 2015. | Series: CABI climate change series | Includes bibliographical references and index.
Identifiers: LCCN 2015027427 | ISBN 9781780643663 (hbk : alk. paper)
Subjects: LCSH: Water-supply, Agricultural--Management--Developing countries. | Crops and climate--Developing countries. | Climatic changes--Developing countries.

Classification: LCC S494.5.W3 C55 2015 | DDC 631.709172/4--dc23 LC record available at http://lccn.loc.gov/2015027427

ISBN-13: 978 1 78064 366 3

Commissioning editor: Nicki Dennis and Ward Cooper
Editorial assistant: Emma McCann
Production editor: Lauren Povey

Typeset by AMA DataSet Ltd, Preston, UK.
Printed and bound in the UK by CPI Group (UK) Ltd, Croydon, CR0 4YY.

Contents

Contributors

Pramod K. Aggarwal, International Water Management Institute, Hyderabad, India. E-mail: p.k.aggarwal@cgiar.org

Mukand S. Babel, Asian Institute of Technology, Water Engineering and Management (WEM), PO Box 4, Klong Luang, Pathumthani 12120, Thailand. E-mail: msbabel@ait.asia

Dinesh Bhatt, UNESCO-IHE Institute for Water Education, PO Box 3015, 2601 DA, Delft, the Netherlands and Ministry of Irrigation, Department of Irrigation, Jawalakhel, Lalitpur, Nepal. E-mail: dinesh_bhatta@yahoo.com

Amparo C. Dela Cruz, University of the Philippines Los Baños, Los Baños, Laguna, Philippines. E-mail: ampie.delacruz@gmail.com

Tran Duc Dong, Southern Institute for Water Resource Planning, Ho Chi Minh City, Vietnam. E-mail: tranducdongvn@yahoo.com.vn

Mark Giordano, Georgetown's School of Foreign Service, Georgetown University, Washington, DC. E-mail: mg1382@georgetown.edu

Weiping Hao, Institute of Environment and Sustainable Development in Agriculture, Chinese Academy of Agricultural Sciences, Beijing 100081 and Key Laboratory of Dryland Agriculture, Ministry of Agriculture, Beijing 100081, China.

Nguyen Xuan Hien, Southern Institute for Water Resources Planning, Ho Chi Minh City, Vietnam. E-mail: nxhien2009@yahoo.com

Chu Thai Hoanh, International Water Management Institute, Colombo, Sri Lanka. E-mail: c.t.hoanh@cgiar.org

Robyn Johnston, International Water Management Institute, Colombo, Sri Lanka. E-mail: r.johnston@cgiar.org

Hui Ju, Institute of Environment and Sustainable Development in Agriculture, Chinese Academy of Agricultural Sciences, Beijing 100081 and Key Laboratory of Agricultural Environment, Ministry of Agriculture, Beijing 100081, China.

Krishna Reddy Kakumanu, International Water Management Institute, Hyderabad, India. E-mail: k.krishnareddy@cgiar.org

Suan Pheng Kam, The WorldFish Center, Penang, Malaysia. E-mail: s.kam@cgiar.org

Nguyen Huy Khoi, Southern Institute for Water Resource Planning, Ho Chi Minh City, Vietnam. E-mail: nhkhoihl@yahoo.com

Felino P. Lansigan, University of the Philippines Los Baños, Los Baños, Laguna, Philippines. E-mail: fplansigan@up.edu.ph
Qin Liu, Institute of Environment and Sustainable Development in Agriculture, Chinese Academy of Agricultural Sciences, Beijing 100081 and Key Laboratory of Dryland Agriculture, Ministry of Agriculture, Beijing 100081, China. E-mail: liuqin02@caas.cn
Xinping Luo, Key Laboratory of Water Cycle and Related Land Surface Processes, Chinese Academy of Sciences, Beijing 100101, China. E-mail: lxp_830520@163.com
Shreedhar Maskey, UNESCO-IHE Institute for Water Education, PO Box 3015, 2601 DA, Delft, the Netherlands. E-mail: s.maskey@unesco-ihe.org
Xu-rong Mei, Institute of Environment and Sustainable Development in Agriculture, Chinese Academy of Agricultural Sciences, Beijing 100081 and Key Laboratory of Dryland Agriculture, Ministry of Agriculture, Beijing 100081, China. E-mail: meixr@ieda.org.cn
Xingguo Mo, Key Laboratory of Water Cycle and Related Land Surface Processes, Chinese Academy of Sciences, Beijing 100101, China. E-mail: moxg@igsnrr.ac.cn
Udaya Sekhar Nagothu, Norwegian Institute for Agricultural and Environmental Research-Biforsk, Ås, Norway. E-mail: nagothu.udayasekhar@nibio.no
Nguyen Trung Nam, Southern Institute for Water Resource Planning, Ho Chi Minh City, Vietnam. E-mail: nguyentrungnam47v@gmail.com
Nguyen van Ngoc, Southern Institute for Water Resource Planning, Ho Chi Minh City, Vietnam. E-mail: nvngoc221151@yahoo.com
Tran Nhuong, The WorldFish Center, Penang, Malaysia. E-mail: n.tran@cgiar.org
Kuppannan Palanisami, International Water Management Institute, Hyderabad, India. E-mail: k.palanisami@cgiar.org
Ngo Dang Phong, International Rice Research Institute (IRRI), Los Baños, Laguna, Philippines and Nong Lam University, Ho Chi Minh City, Vietnam. E-mail: n.phong@irri.org
Krishna C. Prasad, Chandra Engineering Consultants, Mills Area, Janakpur-1, Post Code 45600, Nepal. E-mail: krishna.prasad@colorado.edu
Jana Prosinger, School of Oriental and African Studies, London. E-mail: janaprosinger@gmail.com
Coimbatore Ramarao Ranganathan, Tamil Nadu Agricultural University, Coimbatore, India.
Bjoern Ole Sander, Crop and Environmental Sciences Division, International Rice Research Institute (IRRI), Los Baños, Laguna, Philippines. E-mail: b.sander@irri.org
Oxana Savoskul, International Water Management Institute (IWMI), PO Box 2075, Colombo, Sri Lanka. E-mail: xenia.savoskul@gmail.com
Elena Shevnina, Finnish Meteorological Institute, PO Box 503, FI-00101, Helsinki, Finland. E-mail: eshevnina@gmail.com
Joel D.L.C. Siopongco, Crop and Environmental Sciences Division, International Rice Research Institute, Los Baños, Laguna, Philippines. E-mail: jsiopongco@irrialumni.org
Vladimir Smakhtin, International Water Management Institute, Colombo, Sri Lanka. E-mail: v.smakhtin@cgiar.org
Aditya Sood, International Water Management Institute, Colombo, Sri Lanka. E-mail: a.sood@cgiar.org
Diana Suhardiman, International Water Management Institute, South-east Asia Regional Office, Vientiane, Lao PDR. E-mail: d.suhardiman@cgiar.org
Tran Quang Tho, Southern Institute for Water Resource Planning, Ho Chi Minh City, Vietnam. E-mail: tranquangtho41c@gmail.com
To Phuc Tuong, International Rice Research Institute (IRRI), Los Baños, Laguna, Philippines. E-mail: t.tuong@irri.org

Stefan Uhlenbrook, UNESCO-IHE Institute for Water Education, PO Box 3015, 2601 DA, Delft, the Netherlands. E-mail: s.uhlenbrook@unesco-ihe.org
Karen Villholth, International Water Management Institute, Regional Office for Southern Africa, Pretoria, South Africa. E-mail: k.villholth@cgiar.org
Jinxia Wang, Center for Chinese Agricultural Policy, Institute of Geographical Sciences and Natural Resources Research, Chinese Academy of Sciences, Beijing 100101, China. E-mail: jxwang.ccap@igsnrr.ac.cn
Reiner Wassmann, Crop and Environmental Sciences Division, International Rice Research Institute, Los Baños, Laguna, Philippines. E-mail: r.wassmann@cgiar.org
Jun Xia, State Key Laboratory of Water Resources and Hydropower Engineering Science, Wuhan University, China and Key Laboratory of Water Cycle and Related Land Surface Processes, Chinese Academy of Sciences, Beijing 100101, China. E-mail: xiaj@igsnrr.ac.cn
Changrong Yan, Institute of Environment and Sustainable Development in Agriculture, Chinese Academy of Agricultural Sciences, Beijing 100081 and Key Laboratory of Dryland Agriculture, Ministry of Agriculture, Beijing 100081, China. E-mail: yanchangrong@caas.cn
Jianying Yang, Institute of Environment and Sustainable Development in Agriculture, Chinese Academy of Agricultural Sciences, Beijing 100081 and Key Laboratory of Dryland Agriculture, Ministry of Agriculture, Beijing, China. E-mail: yangjy@ieda.org.cn

Preface

Water used for agriculture is the largest water consumer globally, about 70% of all freshwater withdrawals. Climate change impacts on water resources – through changes in precipitation, snowfall, soil moisture, river flow and groundwater recharge – translate into impacts on agricultural production. With current trends in population growth, rising incomes and changing diets, the demand for food will double over the next 50–80 years. Significant improvements are necessary in agricultural water management now to reduce the vulnerability of poor people in developing countries to climate-induced changes in precipitation and water availability.

The objective of this book is to provide experiences from studies on agricultural water management under climate change as references for agriculture and irrigation planners, decision makers, researchers and students. Chapters in this book present an overview on global assessment of climate change impacts and water requirement for future agriculture, detailed crop water requirements in case studies in developing countries, irrigation management under sea-level rise in deltas, and agricultural adaptation options to climate change such as water-saving techniques and groundwater exploitation, and related policy settings. Findings and conclusions from the studies presented in this book may help in identifying subjects for further research and actions in management to filling the information and knowledge gaps in agricultural water management.

This book was edited by a team of scientists based at the International Water Management Institute (IWMI). It is produced as part of the CGIAR Research Program on Climate Change, Agriculture and Food Security (CCAFS), which is a strategic partnership of CGIAR and Future Earth. The views expressed in this book cannot be taken to reflect the official opinions of CGIAR or Future Earth.

The Editors

1 Climate Change and Agricultural Development: A Challenge for Water Management

Chu Thai Hoanh,* Robyn Johnston and Vladimir Smakhtin

International Water Management Institute (IWMI), Colombo, Sri Lanka

Abstract

Freshwater-related risks of climate change increase significantly with increasing global temperatures. Globally, the negative impacts of future climate change on freshwater systems are expected to outweigh the benefits, and agriculture and irrigation, as the largest consumers of water globally, are most at risk. This book analyses the potential impacts of climate change on water for agriculture, and the adaptation strategies in water management to deal with these impacts, drawing on global assessments and regional studies.

This chapter introduces the book, sets the scene for research on climate change in agricultural water management, and synthesizes the issues, methodologies and findings in the chapters to follow. Chapters 2 and 3 provide an overview of global assessment of climate change impacts and water requirement for future agriculture. Chapters 4–7 provide analyses of crop water requirements in four case studies in developing countries. Chapters 8 and 9 are studies of irrigation management under sea-level rise in Vietnam's Mekong Delta. Chapters 10–12 discuss examples of adaptation alternatives such as water-saving techniques and groundwater exploitation, and related policy settings. The last chapter links the dominant approach of uncertainty presented in the climate change discourse with policy discussions on climate adaptation strategies.

1.1 Climate Change and Agricultural Water Management

Observational records and climate projections provide abundant evidence that freshwater resources will be strongly impacted by climate change, with wide-ranging consequences for human societies and ecosystems. Observed global warming over several decades has been linked to changes in the large-scale hydrological cycle such as: increasing atmospheric water vapour content; changing precipitation patterns, intensity and extremes; reduced snow cover and widespread melting of ice; and changes in soil moisture and runoff (Bates *et al.*, 2008). Over the 20th century, precipitation changes show substantial spatial and inter-decadal variability. Since the 1970s, increases in precipitation have been observed over land in high northern latitudes, while decreases have dominated from 10°S to 30°N. Climate model simulations for the 21st century are consistent in projecting trends of increasing annual average river runoff and water availability at high latitudes and in some wet tropical areas, and decrease over some dry regions at mid-latitudes and in the dry tropics (IPCC, 2007b). These trends are reconfirmed in the Fifth Assessment Report by IPCC (2013).

Freshwater-related risks of climate change increase significantly with rising

* Corresponding author, e-mail: c.t.hoanh@cgiar.org

temperatures associated with increasing greenhouse gas (GHG) concentrations. Globally, the negative impacts of future climate change on freshwater systems are expected to outweigh the benefits. By the 2050s, the area of land subject to increasing water stress due to climate change is projected to be more than double that with decreasing water stress (Bates *et al.*, 2008). In particular, frequency of short hydrological droughts is likely to increase in the presently dry regions (IPCC, 2014).

Rainfed agriculture is vulnerable to climate change through the direct impacts of changed rainfall and temperature conditions; irrigated agriculture bears the additional risk of changes in availability of surface and groundwater. Irrigation is the largest water consumer globally: about 70% of all freshwater withdrawals go to irrigated agriculture (UN Water, 2009). Increasing water scarcity may limit food production in rainfed systems, putting pressure on food prices and increasing countries' dependence on food imports or investments into irrigation. Shifting from rainfed to irrigated agriculture is a front-line adaptation strategy to hotter, drier conditions, but will in turn increase demand for water. According to FAO (Food and Agriculture Organization for the United Nations) projections, developing countries, with 75% of the global irrigated area, will need to expand their irrigated areas by 0.6% per year until 2030 to meet food demand (Bruinsma, 2003), although a smaller expansion of irrigated area is assumed under all four scenarios of the Millennium Ecosystem Assessment, with global growth rates of only 0–0.18% per year until 2050 (Millennium Ecosystem Assessment, 2005).

Regarding crop water requirement, the physiological effect of CO_2 is associated with an increased intrinsic water use efficiency of plants, which means that less water is transpired per unit of carbon assimilated (IPCC, 2014). For C3 plant species, including most food crops, the CO_2 effect may be relatively greater for crops that are under moisture stress compared to well-irrigated crops. The large-scale implications of CO_2–water interactions (i.e. at canopy, field and regional level) are highly uncertain. In general, the positive effects of elevated CO_2 on plant–water relations are expected to be offset by increased evaporative demand under warmer temperatures (IPCC, 2007b); therefore the impact of elevated CO_2 on water demand is usually not considered in studies of agricultural water management under climate change.

Higher water temperatures and projected changes in water extremes, including floods and droughts, are projected to affect water quality and exacerbate many forms of water pollution such as sediments, nutrients and dissolved organic carbon, with possible negative impacts on agricultural production, although the extent of these changes is highly uncertain and depends on rainfall seasonality, land cover and soil management practices (IPCC, 2014). Mullan and Barrett-Lennard (2010) suggested that climate change is expected to reduce water availability in general, making the use of low quality water resources more common.

In low-lying coastal areas, sea-level rise is projected to increase flooding depth, resulting in shorter duration for crops, and to extend salinity intrusion into the surface and groundwater systems in estuaries, causing a decrease of freshwater availability for agricultural production. Without adaptation, it is projected that hundreds of millions of people will be affected by coastal flooding and displaced due to land loss by the year 2100; the majority of those affected are from East, South-east and South Asia. Protecting against flooding and erosion is considered economically rational for most developed coastlines in many countries under all socio-economic and sea-level rise scenarios analysed (IPCC, 2014).

Locally, irrigated agriculture may face new problems linked to the spatial and temporal distribution of streamflow. For instance, at low latitudes, especially in South-east Asia, early snowmelt may cause spring flooding and lead to a summer irrigation water shortage (IPCC, 2007b). In most dry subtropical regions, climate change is projected to reduce renewable surface water and groundwater resources significantly. This will intensify competition for water among agriculture and other sectors, and affect regional water supply, energy generation and food security. In contrast, water resources are projected to increase at high latitudes. Proportional changes are typically one to three times greater for runoff than for precipitation. The effects on water resources and irrigation requirements of changes in vegetation due to increasing GHG concentrations and climate change remain uncertain (IPCC, 2014). There is high confidence that irrigation demand will increase significantly in many areas (by more than 40% across Europe, USA and parts of Asia). Other regions – including major irrigated areas in India, Pakistan, and south-eastern China – might experience a slight decrease in irrigation demand, due for example to higher precipitation, but only under some climate change scenarios.

However, region-to-region variations in different studies were very heterogeneous. For example, using seven global hydrological models with a limited set of projections, Wada *et al.* (2013) estimated a global increase in irrigation demand by the 2080s (ensemble average 7–21% depending on emissions scenario). By contrast, based on projections from two general circulation models (GCMs) and two emissions scenarios, Zhang and Cai (2013) suggested a slight global decrease in crop water deficits in both irrigated and rainfed areas by the 2080s, which can be explained partly by a smaller difference between daily maximum and minimum temperatures.

With the changes in temperature and precipitation under climate change, shifting of crops from present locations to new locations will occur as one option in adaptation to climate change. In a study assessing the impact of climate change on agricultural land for 16 crops by comparing the global spread of farmland between the 'current' period (1981–2010) and a future period (2071–2100), Zabel *et al.* (2014) calculate a potential global net increase in land climatically suitable for crops of 4.7 million km^2, from 54.2 million km^2 to 58.9 million km^2. These figures include land that is both rainfed and irrigated, and exclude protected areas and dense forest because of the importance of conserving these to protect ecosystem services, such as carbon storage. However, the authors emphasize a downward shift in land quality: 3.9 million km^2 is projected to be 'highly suitable' for crops, compared with 4.6 million km^2 for the current period. In turn, more land will be classified as 'marginally suitable' or 'moderately suitable', with increases of 3.8 million km^2 and 1.6 million km^2, respectively, under these categories. In this study, irrigation areas are assumed not to change, but in practice, it is likely that irrigation systems will be expanded to the areas highly suitable for crops in terms of soil, terrain and climate.

IPCC (2007a, 2014) warn that increased demand for irrigation, from both surface and groundwater sources, will result both from changed climatic conditions and from increased demand for food by a growing population. Globally, the impacts of climate change on increasing irrigation water requirements could be nearly as large as the changes projected from socio-economic development in this century (Turral *et al.*, 2011). Agricultural production systems and water resources will be critically shaped in the coming decades by the interaction of climate with non-climatic drivers: demographic, socio-economic, technological and lifestyle changes.

IPCC (2013) stress that current water management practices may not be robust enough to cope with the impacts of climate change on agriculture and other water use sectors. Adaptive water management techniques, from field-scale water-saving techniques, to basin-level scenario planning, learning-based approaches, and flexible and low-regret management solutions, can help create resilience to uncertain hydrological changes and impacts due to climate change. Much of the adaptation cost in water management will be needed in developing countries, which face barriers including lack of human and institutional capacity, financial resources, awareness and communication.

1.2 Global Assessment on Agricultural Water Management

Maskey *et al.* (Chapter 2, this volume) review the global literature on impacts of climate change on agriculture and prospects for adaptation. They find that sensitivity of agriculture to climate change varies across the globe, and in general, developing countries, where more than 800 million people are already undernourished, will be hardest hit. There has been considerable progress in approaches for assessing the impact of climate change on agriculture and irrigation and in evaluation of adaptation measures, but critical challenges and constraints associated with climate change impact and adaptation remain. Limited understanding of the interactions between, and relative importance of, factors like elevated ozone and CO_2 levels, extreme weather conditions, weed variety, socio-economic changes and adaptation responses mean that diverse adaptation options from farm to policy level are essential. Most published studies on adaption focus on modification of existing management practices to improve crop yield, using process-based models, but do not account for trade-offs between crop production and resource availability, which influence the farmer's decision-making and profitability. The authors propose that more effort is required to incorporate social, financial, institutional and technical constraints and the limitations of resources and adaptive capacity into adaptation frameworks and planning.

Sood (Chapter 3, this volume) warns that, with world population increasing to about 9 billion and the per capita GDP rising, the future demand for water for domestic and industrial sectors will compete with the agricultural water demand. Results from a water accounting and global food trade model, the Water, Agriculture, Technology, Environmental and Resource Simulation Model (WATERSIM), indicate that the proportion of consumptive water demand for agriculture will decrease from around 72% of the total demand in 2010, to 37% in case of business-as-usual, 27% in case of an optimistic socio-economic scenario and 50% in case of a pessimistic socio-economic scenario. The greatest increase in demand for consumptive water comes from domestic and industrial sectors, although to maintain optimum yields, around 16–19% more irrigation water will be required on average due to higher evaporative demand. Sood concludes that with improved agricultural water management and liberalized global food trade, there are enough resources globally not only to meet the future demand but also to reduce malnutrition.

1.3 Crop Water Requirements under Climate Change

Analysing past climate data is one way to assess potential impacts of climate factors on crop water requirement. Liu *et al.* (Chapter 4, this volume) evaluate the effects of climate in the Huang-Huai-Hai Plain (3H Plain) in the northern China using observed climate data for the period 1981–2009 to estimate reference evapotranspiration (ET_0), then analyse the sensitivity of crop water requirement (ET_c) of major climatic variables and regional responses of precipitation deficit in different crop growth stages. The results show that temperature was the most sensitive variable in general for the 3H Plain, followed by solar radiation, wind speed and relativity humidity. Relative humidity was the factor most closely correlated with precipitation deficit.

Working in the same region in China, where water is scarce and climate change is likely to exacerbate water stress, Xia *et al.* (Chapter 5, this volume) explore regional crop responses to climate change, using the wheat–maize double-cropping system as an example. Their results indicate that under likely climate change scenarios, production of winter wheat will increase (with slightly intensified evapotranspiration), but in contrast, summer maize production will slightly decline (with a significant increase of evapotranspiration). The results also indicate that wheat is more resilient to climate change than maize. To mitigate the impacts of climate change on agricultural water use, they propose a range of agricultural water management measures and policies, including improving the performance of participatory irrigation management reform, establishing a water rights system, reforming agricultural water price, and promoting the adoption of agricultural water-saving technology.

Lansigan and Dela Cruz (Chapter 6, this volume) also report local observational evidence of changing climate based on available historical weather data sets in the Philippines. The authors select plausible climate scenarios and downscaled climate projections, both dynamic and statistical, for analysing crop yields using a calibrated crop simulation model for a standard rice variety (IR-64) and a local maize variety (IPB-911). Their results indicate a reduction from 8 to 14% in crop yields for every 1°C temperature increase depending on season and location. For adaptation strategies in agricultural water management, the authors propose location-specific measures based on best practices including adjusting the planting calendar, improving water use efficiency and irrigation water management, water impoundment, planting stress-tolerant varieties, and weather index-based insurance for crop production.

Since the agriculture sector is highly vulnerable to climate change in many parts of the world, there is an increasing concern among farmers, researchers and policy makers about the potential impacts of climate change on food security and livelihoods. Kakumanu *et al.* (Chapter 7, this volume) review the current state of understanding of climate change impacts on irrigation water in South Asia and specifically on the crop yield and relevant adaptation measures in three major river basins, the Godavari, Krishna and Cauvery in India. An optimization model was used to evaluate the different adaptation practices and their potential to maximize rice production and income, and minimize water use for the mid- and end-century climate change scenarios. The authors conclude that adaptation practices at farm level, such as system of rice intensification, machine transplantation, alternate wetting and drying and direct seeding, could reduce the water and labour use by 10–15% and stabilize rice production in the long term. They suggest that to adapt to the impacts of climate change on agricultural water management, technology upscaling is an alternative but should be backed up with well-planned capacity-building programmes for farmers.

1.4 Agricultural Water Management under Sea-level Rise

The Mekong River Delta is one of the regions most seriously impacted by sea-level rise (World Bank, 2007), which will aggravate inundation and salinity intrusion and hence strongly influence agricultural production. Phong *et al.* (Chapter 8, this volume) focus on the impacts of sea-level rise in Bac Lieu province, a low-lying coastal province in the Mekong River Delta of Vietnam. The province receives fresh water from the Mekong mainstream on the east side, and dyke and sluices for protection from salinity intrusion were built at the south-west. The authors use a hydraulic and salinity model for simulating water level, flow and salinity in the canal network for years of low, average and high water volume from upstream, and different levels of projected sea-level rise from 12 to 75 cm. Their results indicate that flooding depth increases directly with sea-level rise but that for sea-level rise less than 30 cm, salinity slightly decreases, due to increasing freshwater inflows from the mainstream. This contradicts many previous studies, which generally conclude that the area of salinity intrusion is always larger under sea-level rise. However, when sea-level rise is higher than 30 cm, saline water intrudes into the mainstream and freshwater intake canals, and additional structures (dykes and sluices) will be needed for salinity control.

Most of the Mekong Delta's aquaculture production occurs in the floodplains and coastal areas that are highly exposed and vulnerable to climate change impacts and sea-level rise. Adaptation in water management for cropping systems, such as upgrading dykes to reduce flooding and salinity intrusion, will benefit other production systems, particularly aquaculture. Kam *et al.* (Chapter 9, this volume) present an example of economic evaluation of autonomous adaptation by shrimp and catfish farms in the Mekong Delta of Vietnam. The study shows that planned adaptation measures can help defray catfish farmers' escalating costs of raising pond dykes in response to increased flooding in the delta, if government policy and public investment in adaptation to climate change, particularly for water management, take account of socio-economic development targets of the aquaculture industry. Because of the high level of uncertainty surrounding commodity prices and changes in production technologies, a 'no-regrets' strategy of reducing the high dependence on shrimp and catfish culture and diversifying into more ecologically oriented production systems is recommended, to hedge the aquaculture industry against the increasing risks and uncertainties brought about by climate change.

1.5 Adaptive Agricultural Water Management

Groundwater resources have been exploited in many regions to respond to the increase in water demand for agriculture under climate variability and climate change. Villholth (Chapter 10, this volume) considers groundwater resources in transboundary aquifers as significant water reserves for sustainable agricultural and socio-economic development in the context of Africa, Asia, the Middle East and Latin America. The author highlights that integrated surface and groundwater measures are important for local, smaller-scale, no-regrets adaptation, as well as for larger-scale, strategic adaptation. Socio-economic and institutional aspects, encompassing international law and specific adapted international agreements as well as bottom-up participatory processes, are critical for attaining success on the ground.

Although climate change challenges the traditional assumption that past hydrological experience provides a good guide to future conditions (IPCC, 2007b), adaptation to climate change is still based on past experiences. An approach in adaptation is to compare future water demand with lessons from similar conditions in the past, to select possible alternatives. In an example of this approach, Savoskul and Shevnina (Chapter 11, this volume) model the future inflow to the irrigation scheme of the Syr Darya Basin in Central Asia under two climate scenarios based on IPCC core models. Simulating water allocation in the basin in 2070–2099 shows that 14–21% of water demands in the agriculture sector in a normal hydrological year and 28–51% in a dry year are likely to be unmet. The challenges expected from future climate change can be paralleled to those resulting from the political change due to the collapse of the USSR, which left 18% (normal year) and 46% (dry year) of agricultural water demands unmet in 1992–2001. Therefore the authors suggest that the adaptation measures employed in the post-Soviet transitional period can serve as a basis for future climate change adaptation strategies.

Water-saving techniques in agriculture are considered as potential measures to adapt to climate change impacts. These techniques do not only improve water use efficiency but can also contribute to mitigation of long-term climate change. The effects of mitigation on irrigation water requirements could be significant in the coming decades, with large overall water savings, both globally and regionally, although Turral *et al.* (2011) suggest that production in some regions could initially be negatively affected by mitigation actions. Flooded rice fields are a large anthropogenic source of the greenhouse gas (GHG) methane (CH_4). Aeration of the paddy field can reduce methane emissions and at the same time save water. Sander *et al.* (Chapter 12, this volume) analyse the effects of different water-saving techniques such as alternate wetting and drying (AWD) on revenues of rice farmers and climate-change mitigation in the Philippines. Results from field experiments show that methane emissions can be reduced by an average of 37% with a single drainage and by 43% with multiple aerations; nitrous oxide emissions increase but not sufficiently to offset the reduction in CH_4 emissions. To improve uptake, the authors propose initial promotion of a simple, single midseason drainage that only requires water control during approximately 1 week. Uptake is highly dependent on provision of incentives for farmers, which could be coordinated through the Clean Development Mechanism. Other indirect benefits from AWD, such as less crop lodging, reduced pest damage and better soil conditions, need to be scientifically validated.

Policy settings for adaptation to climate change have to deal with the uncertainty inherent in projections. Prosinger *et al.* (Chapter 13, this volume) link the dominant approach of uncertainty presented in the climate-change discourse with policy discussions on climate adaptation strategies in Lao PDR as a case study. While the different perceptions and interpretations of climate-change uncertainty by different policy makers might lead to multiple problem framings, they also reflect structural impediments and institutional barriers in the overall formulation process of climate-change policy and adaptation strategies. The authors emphasize that understanding of these different notions of uncertainty is crucial to increase the actual significance of climate change policy, in particular for sectors that are strongly affected by climate change such as agriculture and water management, and that policy and governance responses to climate change need to be formulated based on a more nuanced, sophisticated understanding of how various policy actors and stakeholders perceive and experience uncertainty.

1.6 Uncertainty and Knowledge Gaps

Despite significant efforts in quantifying future changes in hydrological variables and their impacts on agricultural water management, responses to climate change are limited by uncertainty at all stages of the assessment process. Uncertainty derives from the range of socio-economic development scenarios, the range of climate model projections for a given scenario, the downscaling of climate effects to local/regional scales, impact assessments, and feedbacks from adaptation and mitigation activities (Bates *et al.*, 2008). Approaches and tools are needed to facilitate the appraisal of adaptation and mitigation options across multiple water-dependent sectors (IPCC, 2007b).

A number of key gaps in information and knowledge for agricultural water management, particular in developing countries, is indicated in the studies in this book and other reports (IPCC, 2007b, 2014; Bates *et al.*, 2008):

- Records of climate parameters such as solar radiation, relative humidity and wind speed, and of hydrological parameters such as flow and water quality, including sediment, in river and irrigation systems are often very short, and available for only a few regions.
- Knowledge is lacking on plant evapotranspiration responses to the combined effects of rising atmospheric CO_2, rising temperature and rising atmospheric water vapour concentration, as well as soil moisture changes due to reduction in surface water availability.
- Uncertainty in modelling climate variability, in particular precipitation, remains high. Projections vary widely between models, in particular when downscaling from large-scale climatic models to catchment. The approaches of ensemble of climate projections using climate models and observational constraints (Stott and Forest, 2007) or probabilistic approaches (Frieler *et al.*, 2012) do not assure removal of uncertainty.
- Feedbacks between land use and climate change (including vegetation change and anthropogenic activity such as irrigation and reservoir construction) have not been analysed extensively, in particular if impacts on local climate are considered.
- Information on groundwater is lacking in many regions.

- Climate-change impacts on water quality are still poorly understood; so far only salinity impacts have been studied in any detail.

1.7 Building Resilience Through Agricultural Water Management

Nicol and Kaur (2009) emphasize that adaptation options designed to ensure water supply require integrated demand-side as well as supply-side strategies. Demand management, which aims to regulate withdrawals at sustainable levels through such measures as the promotion of sustainable use, pricing mechanisms and water-saving techniques, will become increasingly important in areas where relative scarcity and competition between water-dependent sectors is increasing. Supply management, through increased storage capacity, abstraction from water course, rainwater harvesting and recharge activities and/or introducing incentives for water conservation, will become a priority where inter-annual resource availability is likely to change significantly. One novel and promising alternative to deal with extreme events under climate change is the harvesting of floods for later use in agriculture by underground taming of floods (Pavelic *et al.*, 2012; Smakhtin *et al.*, 2014).

Over the last decade, the International Water Management Institute has formulated strategies to help communities in developing countries to reduce risks and build resilience through better water management (see, for example Comprehensive Assessment of Water Management in Agriculture, 2007; Johnston *et al.*, 2010; Giordano *et al.*, 2012; McCornick *et al.*, 2013):

- Think more creatively about water storage to overcome short and long-term dry spells, employing a range of approaches along the storage continuum, from small ponds to large reservoirs, groundwater recharge, water harvesting and soil-water conservation.
- Tailor water management strategies to meet changing local needs, as climate change drives new cropping and land-use patterns, ensuring that water resources are developed and managed fairly so that vulnerable groups are not disproportionately burdened by the impacts of variability.
- Increase water productivity through higher yields, crop diversification, and integrating livestock and fisheries. At the global scale, improved productivity helps reduce GHG emissions by curbing the need to convert land for agricultural purposes.
- Improve basin water management and allocation to deal with increased variability in flows, by engaging social and institutional governance mechanisms, as well as technical approaches.
- A shift from 'drought response' to 'drought risk mitigation', with new approaches to early warning and insurance to prepare farmers for climate variability.
- Use our understanding of existing climate variability as the basis for dealing with future change, and improve understanding of the impacts of climate change on variability.
- Improve the understanding of the role of natural ecosystems in buffering variability.

Adapting water management to climatic variability cannot be done in isolation, but is best addressed in the context of sustainable development. Improving water management to deal with climate variability can help vulnerable rural communities to build resilience, diversify their livelihoods and reduce risk. Building resilience now will bring benefits regardless of how and when climate change plays out on the ground (McCornick *et al.*, 2013).

References

Bates, B.C., Kundzewicz, Z.W., Wu, S. and Palutikof, J.P. (eds) (2008) *Climate Change and Water*. Technical Paper of the Intergovernmental Panel on Climate Change. IPCC Secretariat, Geneva, Switzerland, 210 pp.

Bruinsma, J. (2003) *World Agriculture: Towards 2015/2030. An FAO Perspective.* Earthscan, London, 444 pp.

Comprehensive Assessment of Water Management in Agriculture (2007) *Water for Food, Water for Life: A Comprehensive Assessment of Water Management in Agriculture.* Earthscan, London, and International Water Management Institute (IWMI), Colombo, Sri Lanka, 643 pp.

Frieler, K., Meinshausen, M., Mengel, M., Braun, N. and Hare, W. (2012) A scaling approach to probabilistic assessment of regional climate change. *Journal of Climate* 25(9), 3117–3144.

Giordano, M., de Fraiture, C., Weight, E. and van der Bliek, J. (eds) (2012) *Water for Wealth and Food Security: Supporting Farmer-Driven Investments in Agricultural Water Management.* Synthesis Report of the AgWater Solutions Project, International Water Management Institute (IWMI), Colombo, Sri Lanka, 48 pp.

IPCC (2007a) Climate Change 2007: Synthesis Report. Contribution of Working Groups I, II and III to the Fourth Assessment Report of the Intergovernmental Panel on Climate Change (Core Writing Team, Pachauri, R.K and Reisinger, A. (eds)). IPCC, Geneva, Switzerland, 104 pp.

IPCC (2007b) Climate Change 2007: Impacts, Adaptation and Vulnerability. In: Parry, M.L., Canziani, O.F., Palutikof, J.P., van der Linden, P.J. and Hanson, C.E. (eds) Contribution of Working Group II to the Fourth Assessment Report of the Intergovernmental Panel on Climate Change. Cambridge University Press, Cambridge, UK, 976 pp.

IPCC (2013) Climate Change 2013: The Physical Science Basis. In: Stocker, T.F., Qin, D., Plattner, G.-K., Tignor, M., Allen, S.K., Boschung, J., Nauels, A., Xia, Y., Bex, V. and Midgley, P.M. (eds) Contribution of Working Group I to the Fifth Assessment Report of the Intergovernmental Panel on Climate Change. Cambridge University Press, Cambridge, UK and New York, 1535 pp.

IPCC (2014) Climate Change 2014: Impacts, Adaptation, and Vulnerability. Part A: Global and Sectoral Aspects. In: Field, C.B., Barros, V.R., Dokken, D.J., Mach, K.J., Mastrandrea, M.D., Bilir, T.E., Chatterjee, M., Ebi, K.L., Estrada, Y.O., Genova, R.C., Girma, B., Kissel, E.S., Levy, A.N., MacCracken, S., Mastrandrea, P.R. and White, L.L. (eds) Contribution of Working Group II to the Fifth Assessment Report of the Intergovernmental Panel on Climate Change. Cambridge University Press, Cambridge, UK and New York, 1132 pp.

Johnston, R., Lacombe, G., Hoanh, C.T., Noble, A., Pavelic, P., Smakhtin, V., Suhardiman, D., Kam, S.P. and Choo, P.S. (2010) *Climate Change, Water and Agriculture in the Greater Mekong Subregion.* IWMI Research Report 136. International Water Management Institute (IWMI), Colombo, Sri Lanka, 60 pp.

McCornick, P., Smakhtin, V., Bharati, L., Johnston, R., McCartney, M., Sugden, F., Clement, F. and McIntyre, B. (2013) *Tackling Change: Future-Proofing Water, Agriculture, and Food Security in an Era of Climate Uncertainty.* International Water Management Institute (IWMI), Colombo, Sri Lanka, 36 pp.

Millennium Ecosystem Assessment (2005) *Ecosystems and Human Well-being: Synthesis.* Island Press, Washington, DC.

Mullan, D.J. and Barrett-Lennard, E.G. (2010) Breeding crops for tolerance to salinity, waterlogging and inundation. In: Reynolds, M.P. (ed.) *Climate Change and Crop Production.* CABI Series in Climate Change. CAB International, Wallingford, UK.

Nicol, A. and Kaur, N. (2009) *Adapting to Climate Change in the Water Sector.* Background Note. Overseas Development Institute, London.

Pavelic, P., Srisuk, K., Saraphirom, P., Nadee, S., Pholkern, K., Chusanathas, S., Munyou, S., Tangsutthinon, T., Intarasut, T. and Smakhtin, V. (2012) Balancing-out floods and droughts: opportunities to utilize floodwater harvesting and groundwater storage for agricultural development in Thailand. *Journal of Hydrology* 470–471, 55–64.

Smakhtin, V., Pavelic, P., Amarnath, G. and McCartney, M. (2014) Managing water variability, from floods to droughts. In: van der Bliek, J., McCornick, M. and Clarke, J. (eds) *On Target for People and Planet: Setting and Achieving Water-Related Sustainable Development Goals.* International Water Management Institute (IWMI), Colombo, Sri Lanka, 52 pp.

Stott, P.A. and Forest, C.E. (2007) Ensemble climate predictions using climate models and observational constraints. *Philosophical Transactions of the Royal Society A – Mathematical, Physical and Engineering Sciences* 365, 2029–2052.

Turral, H., Burke, J. and Faures, J.-M. (2011) *Climate Change, Water and Food Security FAO Water Report 36.* Food and Agriculture Organization for the United Nations (FAO), Rome.

UN Water (2009) *Water in a Changing World.* The United Nations World Water Development Report 3. UN Water, Geneva, Switzerland.

Wada, Y., Wisser, D., Eisner, S., Flörke, M., Gerten, D., Haddeland, I., Hanasaki, N., Masaki, Y., Portmann, F.T., Stacke, T., Tessler, Z. and

Schewe, J. (2013) Multi-model projections and uncertainties of irrigation water demand under climate change. *Geophysical Research Letters* 40(17), 4626–4632.

World Bank (2007) *The Impact of Sea Level Rise on Developing Countries: A Comparative Analysis.* World Bank Policy Research Working Paper (WPS4136), February 2007. The World Bank Group, Washington, DC.

Zabel, F., Putzenlechner, B. and Mauser, W. (2014) Global agricultural land resources – a high resolution suitability evaluation and its perspectives until 2100 under climate change conditions. *PLoS ONE* 9(9), e107522. DOI: 10.1371/journal.pone.0107522.

Zhang, X. and Cai, X. (2013) Climate change impacts on global agricultural water deficit. *Geophysical Research Letters* 40(6), 1111–1117.

2 Adaptation to Climate Change Impacts on Agriculture and Agricultural Water Management – A Review

Shreedhar Maskey,[1*] Dinesh Bhatt,[1,2] Stefan Uhlenbrook,[1] Krishna C. Prasad[3] and Mukand S. Babel[4]

[1]UNESCO-IHE Institute for Water Education, Delft, the Netherlands; [2]Ministry of Irrigation, Department of Irrigation, Lalitpur, Nepal; [3]Chandra Engineering Consultants, Mills Area, Janakpur, Nepal; [4]Asian Institute of Technology, Pathumthani, Thailand

Abstract

This chapter reviews the global literature on impacts of climate change on agriculture and prospects for adaptation. Sensitivity of agriculture to climate change varies across the globe. Developing countries, where more than 800 million people are already undernourished, will be hardest hit. We review approaches for assessing the impact of climate change on agriculture and irrigation water requirements, and present recent progress in the assessment of adaptation measures. The challenges and constraints associated with climate change impact and adaptation research are critically discussed.

The review leads to the conclusion that warmer temperatures will tend to reduce the crop yields in many regions, mainly due to reduction of crop duration associated with water stress during the critical stages of crop development. Although efforts have been made to understand better the climate–crop relationships, there is still limited understanding of the interactions between and relative importance of factors such as elevated ozone and CO_2 levels, extreme weather conditions, weed variety, socio-economic changes and adaptation responses.

Evaluation of diverse adaptation options from farm to policy level, and covering a range of scales and issues, including availability of resources, constraints and associated uncertainties, are essential to address adequately the impacts of climate and other changes on agriculture. Most of the published studies on adaption focus on modification of existing management practices to improve crop yield, using process-based models. Trade-offs between crop production and resource availability, which influence the farmer's decision making and profitability, have not received substantial attention. More effort is required to incorporate constraints (such as social, financial, institutional, technical and resources) and adaptive responses into the model frameworks that most studies used.

2.1 Introduction

Global climate change is expected to have direct impacts on agricultural and food systems (Brown and Funk, 2008). Most staple crops are likely to experience yield reductions under various climate change scenarios, and the estimated reductions are generally larger in the developing countries (Nelson *et al.*, 2009). Increasing population

* Corresponding author, e-mail: s.maskey@unesco-ihe.org

and high rates of natural resource degradation will further increase the rates of poverty and food insecurity for Asia, sub-Saharan Africa and Latin America (Fischer *et al.*, 2002). As a consequence, the large population dependent on agriculture and living in the developing world, where more than 800 million people are already undernourished (UN Millennium Project, 2005), will live under increased food insecurity.

Agricultural systems have constituted one of the main subjects of analysis undertaken to understand the impact of both climate variability and climate change, as crop performance is strongly linked to the meteorological conditions of the growing season (Meza and Silva, 2009). Moreover, climatic conditions at critical stages of crop development, such as flowering and yield formation stages, have a pronounced impact on yield (Porter and Semenov, 2005). However, it is only partly understood to what extent the changed climate and variability will impact on agriculture.

This chapter reviews the current literature on impacts of climate change on crop production and possible adaptation measures to cope with the changing climate. There exist recent reviews on climate–crop modelling (e.g. Hansen *et al.*, 2006), ecosystem–hydrology–climate interaction (Betts *et al.*, 2006), agricultural contaminant fate (Boxall *et al.*, 2009), and impact of future hydrological changes on agricultural mitigation and adaptation options (Fallon and Betts, 2010). In this chapter, we particularly focus on the impacts of climate and socio-economic changes on agriculture and irrigation water requirement and the assessment of adaptation options, considering the limitations as well as challenges. Section 2.2 of this chapter presents some of the key issues related to the impact of variation in temperature, precipitation and CO_2 concentration on crop production. This section also presents the impact of climate and socio-economic developments on irrigation water requirements. In Section 2.3, we summarize some of the possible adaptation measures in agriculture and agricultural water managements that are deemed to be essential to offset the adverse impact of climate change. In Section 2.4, we present model-based evaluations of a range of adaptation measures across various geographical regions. In Section 2.5, some of the foremost challenges and constraints associated with the research on climate change impacts and adaptation are discussed, followed by some specific conclusions of the review in Section 2.6. While our review focuses on the developing countries, we make reference to other regions where appropriate.

2.2 Impact of Climate Change on Agriculture and Irrigation Water Requirement

2.2.1 Effects of elevated carbon dioxide

Plant development and crop production respond to rising atmospheric CO_2 concentration (one of the key indicators of human-induced global warming), higher temperature, altered precipitation regimes, increased frequency of extreme temperature and precipitation events (IPCC, 2007) as well as local factors, such as changes in water availability, agricultural practices and methods. However, the relative importance of these factors is a major topic for research.

A wide range of studies conducted in the last few decades have established that an increase in CO_2 concentration level enhances water-use efficiency and this tends to increase the plant biomass and yield for most agricultural plants (Tubiello *et al.*, 2007). Many experiments in controlled environments illustrate that the crop growth and biomass production increase up to 33±6% for C3 crops (such as rice, wheat, soybean) under doubled CO_2 condition (e.g. Kimball, 1983; Porter, 1992; Ewert *et al.*, 1999; Hsiao and Jackson, 1999; Amthor, 2001), while for C4 (such as maize, sugarcane) crops the increase is in the range of 0–10% (e.g. Long *et al.*, 2004; Ainsworth and Long, 2005). Similarly, free air CO_2 enrichment (FACE) experiments in well-managed fields have confirmed these results (Kimball *et al.*, 2002). Overall, the sensitivity to atmospheric CO_2 and surface ozone is relatively higher for C3 crops, such as rice, wheat, soybean, than C4 crops, such as maize and sugarcane (e.g. Brown and Rosenberg, 1999; Gifford, 2004; Long *et al.*, 2004; Ainsworth and Long, 2005; Slingo *et al.*, 2005).

However, it is still uncertain whether effects of CO_2 fertilization observed in controlled and FACE environments will be seen in the farmers' fields in the future (e.g. Tubiello and Ewert, 2002). On the other hand, the estimated benefits of elevated CO_2 may not be fully achieved due to many limiting factors such as increase in surface ozone level (Long *et al.*, 2005), water and nitrogen (Erda *et al.*, 2005), pests, weeds and air quality (Ainsworth and Long, 2005; Tubiello *et al.*, 2007), which are neither well understood nor well represented in the simulation models. Similarly, otherwise positive CO_2 effects on yield may be lowered by high temperature during the critical period of a crop (Caldwell *et al.*, 2005) and increased temperature during the growing season (e.g. Xiao *et al.*, 2005). Moreover, crop management practices such as irrigation and fertilization significantly influence the crop production under climate change (Tubiello *et al.*, 2002). For instance, water limitation enhances the positive benefits of CO_2 fertilization (Tubiello and Ewert, 2002). Further studies are required to understand the net effects of these interactions on crop production. Furthermore, as illustrated in Fig. 2.1, the possibility of more severe climate and saturated effect of CO_2 on plants after 2050 is likely to decrease significantly the yield and consequently the agricultural GDP (with reference to 1990 prices) in developing countries. This discrepancy between developed and developing countries is because of the dominancy of agriculture in the economy of the developing countries where a large fraction of population is employed on the farm. For example, in 2000 the GDP share of agriculture in developed countries was only 2.1% compared to 16% in the developing countries

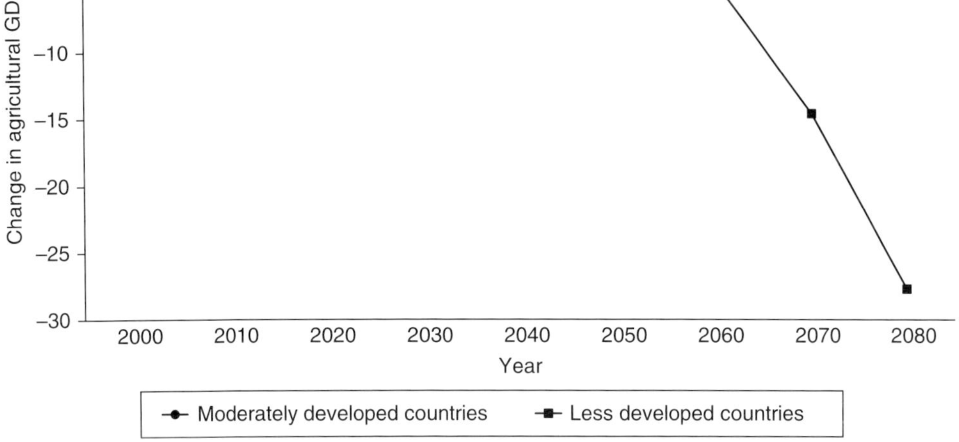

Fig. 2.1. Projected variation in agricultural GDP (billion US$, reference to 1990 price) due to climate change under A2r Hadley climate scenario. A2r is a revised Special Report on Emissions Scenarios (SRES), A2 scenario, with a revised population projection (see Riahi *et al.*, 2006) (from Tubiello and Fischer, 2007).

(Tubiello and Fischer, 2007). Additionally, lack of capital and technology for adaptation, warmer baseline climate, higher exposure to extreme events (Parry *et al.*, 2001) and reliance on weather-dependent natural resources make the developing countries more vulnerable to climate change than the developed countries.

2.2.2 Impact of temperature and precipitation

Temperature and precipitation are the major climatic variables in determining the crop yield. For instance, Kutcher *et al.* (2010) found that the number of days with maximum temperature (greater than 30°C) showed the strongest correlation with canola yield followed by the growing season total precipitation for Canada. Precipitation influences plant growth through altering soil moisture, humidity levels and general cloudiness (altering evaporation and surface level photosynthetically active radiation). The impact of warming on crop yield depends on the region and type of crops. In temperate regions, crop yields are expected to benefit slightly from moderate to medium increases in mean temperature (1–3°C) considering the effect of CO_2 fertilization and changing rainfall patterns (IPCC, 2007); yet large uncertainties remain (Easterling *et al.*, 2007). On the other hand, in semi-arid and tropical regions, this would decrease crop yield. Modelling studies have indicated that in low latitude regions a moderate temperature increase (1–2°C) is likely to have negative yield impacts for major cereals. Hence, for main cereal crops, climate change is expected to have negative impacts on crop productivity and yields in the tropics, while there may be some beneficial effects at high latitudes. This pattern is expected to be more pronounced as time progresses. However, the projected warming for the end of the 21st century is likely to have a negative impact on crop yield in all the regions (Tubiello *et al.*, 2007). Furthermore, increased evapotranspiration due to change in temperature could intensify drought stress (Tao *et al.*, 2003).

Existing literature also indicates the disparity in the climate-change-driven impacts on crop yield between the developed and developing countries, with mostly positive impacts in developed countries and negative impacts in developing countries. This discrepancy is estimated to be more pronounced for A1 and A2 Special Report on Emission Scenarios (SRES), based on the Basic Linked System (BLS) simulation for wheat, rice, maize and soybean, considering the beneficial effect of CO_2 fertilization (Fig. 2.2). The CO_2 level considered is maximum (810 ppm) for the scenario A1F1 (IPCC, 2000) and minimum (498 ppm) for the scenario S550 (see Arnell *et al.*, 2002). As crops are subjected to multiple stresses, the analysis of climate change alone provides only a partial view of the likely future yields. For a more vigorous assessment of impacts of climate change on agriculture, a range of drivers needs to be considered. However, if the climate-change effects dominate, crop yields are likely to be more negatively affected. Thus, we need to be prepared for the range of possible agriculture futures and search for ways to adapt to a more uncertain world in the coming decades (Parry *et al.*, 2004).

2.2.3 Impact of climate change on irrigation water requirement

Climate change also impacts agriculture and irrigation water requirements through the changes in local hydrology. Warmer temperature and change in precipitation (pattern and event characteristics) can cause significant changes in hydrological responses, e.g. evaporation, surface runoff, soil moisture, infiltration, percolation, base flow and groundwater recharge/discharge (Uhlenbrook, 2009). Existing studies have indicated an increase in irrigation water requirements both on global and regional scales, irrespective of the beneficial impact of increased CO_2 on crop water use efficiency. Considering the direct effect (without considering CO_2 effects) of climate change on crop evaporative demand, Döll

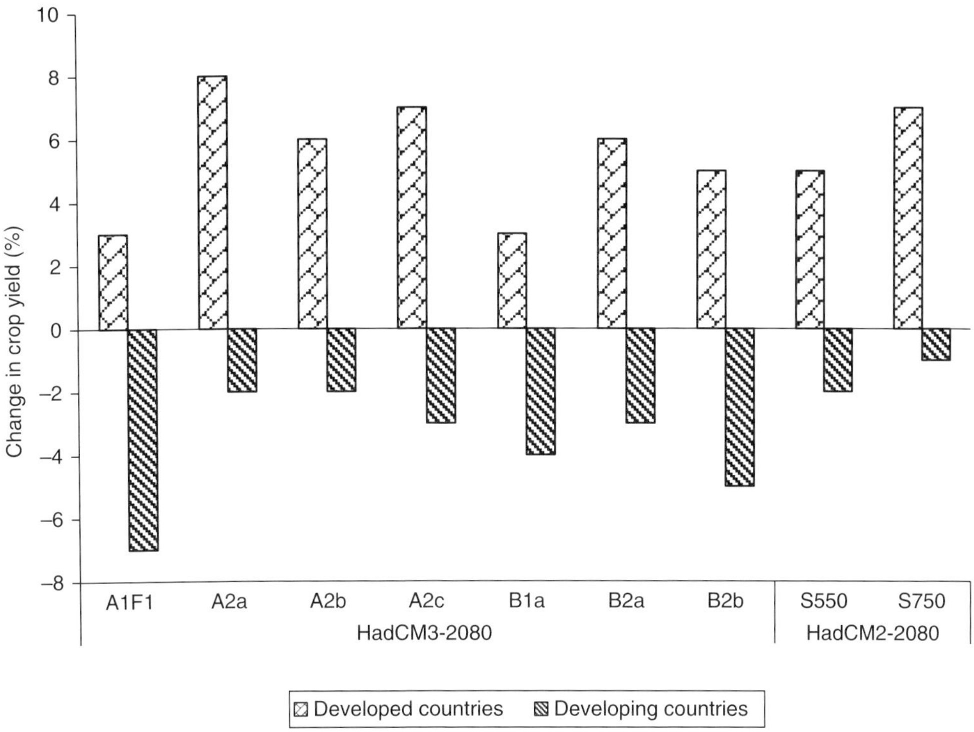

Fig. 2.2. Projected changes in crop yield (%) from baseline (1990) for various emission scenarios (Special Report on Emissions Scenarios) from HadCM3 and HadCM2 models in developed and developing countries (from Parry *et al.*, 2004).

(2002) estimated an increase of net crop irrigation requirements by 3–5% until the 2020s and by 5–8% until the 2070s, with a large regional variation, e.g. +70% in South-east Asia by the 2070s (Fig. 2.3). The increase in crop water requirement can be attributed to both direct (changes in temperature and precipitation) and indirect (changes in cropping pattern and growing season) impacts of climate change. Döll (2002) applied a raster-based Global Irrigation Model (Döll and Siebert, 2002) with a spatial resolution of 0.5° to explore the impact of climate change on net crop irrigation requirements for the areas across the globe that were equipped with irrigation until 1995. Recently, Fischer *et al.* (2007) projected an increase in global net irrigation requirements of 20% by the 2080s, considering positive effects of increased CO_2 on crop water use efficiency. About 65% of this increase in the net irrigation requirement was considered as a consequence of higher crop water demand and the remaining 35% was contributed by the extended crop calendar. They also reported about 40% reduction in the agricultural water requirement in the case of the climate scenario with and without mitigation for climate change. On the other hand, water stress (the ratio of irrigation withdrawal to renewable water resources) is projected to amplify in the Middle East and South-east Asia (Arnell, 2004; Fischer *et al.*, 2007). In the developing countries of Asia, water use is projected to increase by 40% in the next two decades to feed the growing population (see Sivakumar, 2006).

The influence of socio-economic developments (development paths as specified by IPCC SRES), with special reference to emission of greenhouse gases (GHGs) into the atmosphere on the irrigation water

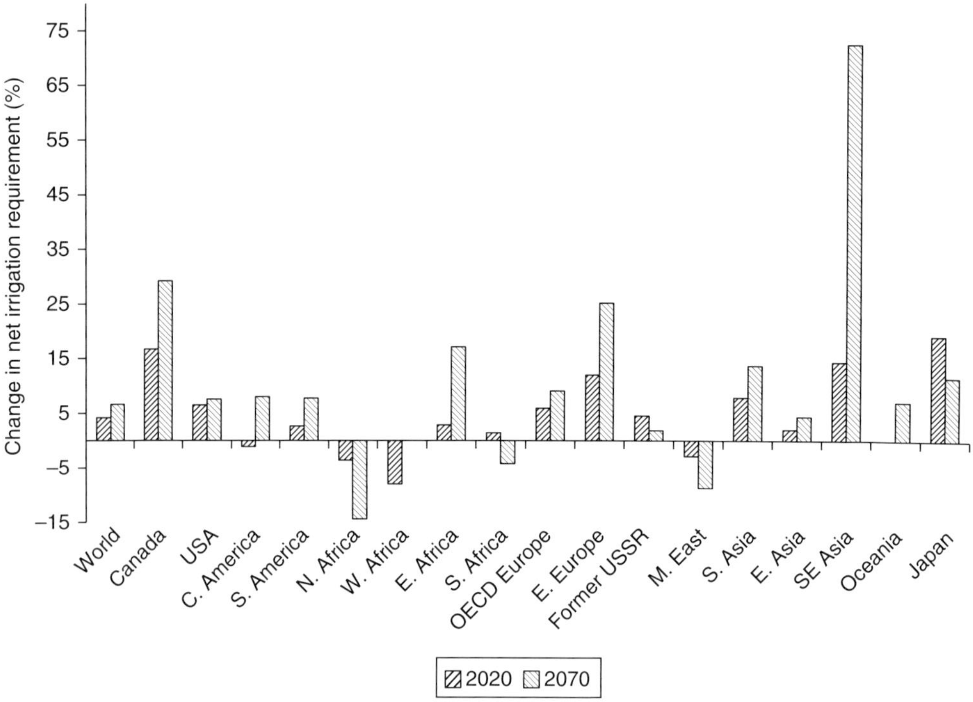

Fig. 2.3. Change in net irrigation water requirement (IRnet) for the 2020s (2020–2029) and the 2070s (2070–2079) of the world regions equipped with irrigation in 1995 (change in IRnet projected from ECHAM4 and HadCM3 climate change scenarios were averaged; data from Döll, 2002).

requirement, may vary significantly across the region and much remains to be done in predicting the irrigation demand resulting from the interaction of socio-economic and climate change scenarios. For developing countries, the increase in net irrigation water requirement from socio-economic developments (A2r scenario, i.e. SRES A2 scenario with revised population projection) is higher than the increase from climate change (HadCM3). However, the reverse is the case for developed countries (Fig. 2.4). Fischer *et al.* (2007) assumed that the BLS projected an increase in irrigation water requirement from socio-economic developments (A2r scenario), which is proportional to the estimated additional irrigated land. Hence, the large proportion of the projected additional irrigated area from the developing countries (112 million ha (Mha) out of 122 Mha for 2080) will result in a significant increase in the irrigation water demand under socio-economic development.

2.3 Adaptation Options in Agriculture and Agricultural Water Management

Adapting agriculture and agricultural practices (including agricultural water management) to climate change is a complex, multi-dimensional and multi-scale process (Bryant *et al.*, 2000). Climate change is expected to increase the variability in climate by shifting and intensifying extreme weather events and introducing higher uncertainty in the quality and quantity of water supply. Thus, adaptation strategies should incorporate both traditional and new technologies to cope with climate change and variability as well as the changes in agronomic practices. Moreover, water resources management options implemented to cope with current climate variability will also assist to better prepare for increased variability expected in the future. Furthermore, the social and technological aspects of vulnerability, such as obtainable

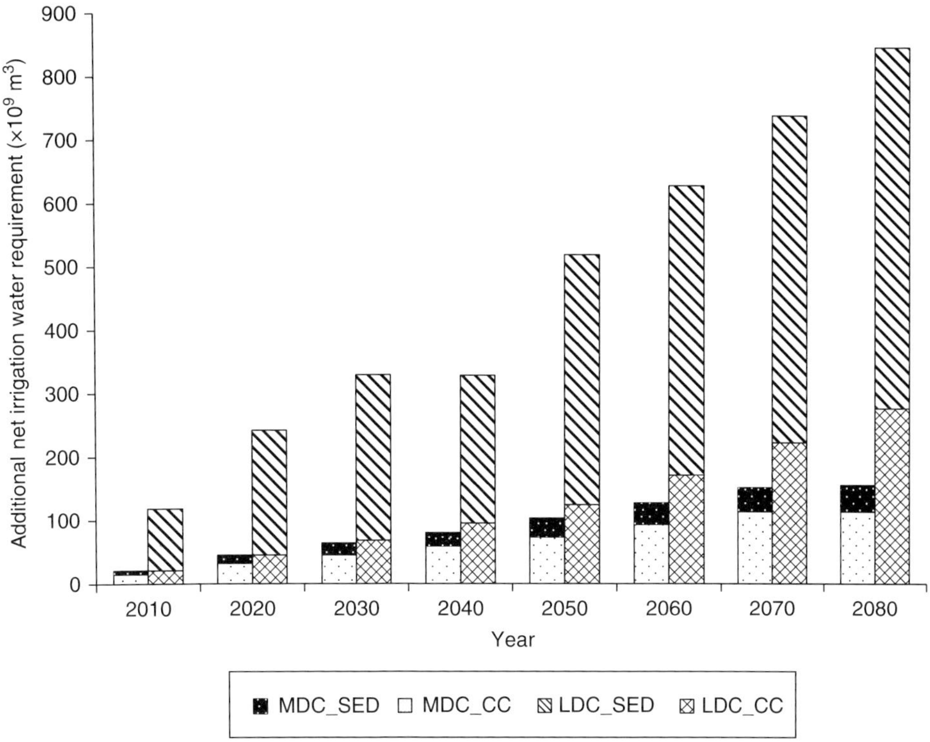

Fig. 2.4. Agro-ecological zone (AEZ) projected an additional net irrigation water requirement (with reference to irrigation water requirement in 2000) from socio-economic development (SED) and climate change (CC) (Hadley) for MDC and LDC under the A2r scenario (from Fischer *et al.*, 2007).

adaptive capacity in a region and the complexity of adaptation for specific crops, should be incorporated while developing adaptation strategies (Lobell *et al.*, 2008). More importantly, such adaptation options must be easily available to the farmers.

There exists a large array of possible adaptation options in response to diversity of agricultural practices depending on the range of climate, cultural, economic and environmental variables (Howden *et al.*, 2007). However, the response of a particular cropping system to a specific adaptation strategy can vary significantly depending on the location and climate scenario. Changing crop varieties, efficient water use, and altering the timing or location of cropping activities are some of the widely suggested adaptation strategies. The full list of the options suggested in the literature is large. For ease in discussion, we classified them into a number of groups that range from water and crop management options to technology developments and government programmes. These adaptation options together with their pros and cons are listed in Table 2.1. Although fully implemented simple adaptation options like shifting planting date and switching to existing cultivars can reduce the negative impacts (Mendelsohn and Dinar, 1999), the pronounced benefit will likely result from more costly measures such as developing new crop varieties and expanding irrigation (Rosenzweig and Parry, 1994). The fact is that, whether we like it or not, most of the widely effective measures require substantial investment from the farmers, development organizations, governments and scientists. Such strategies are also time-consuming (developing new varieties may take up to decades) and may be constrained by other sectors

Table 2.1. Adaptation options in agriculture and agricultural water management (from LEISA, 2000; Desjardins *et al.*, 2002; Kurukulasuriya and Rosenthal, 2003; Verchot *et al.*, 2007).

Adaptation options	Examples	Pros (P) and cons (C)
1. Water management		
Water access (increasing water supply and ecosystem services)	Water transfer schemes Storage reservoirs Rainwater harvesting Groundwater extraction (wells) Reuse of wastewater	P: Addresses the uncertainty associated with natural precipitation regime and assists to cope with increased climate variability. C: High implementation cost and is applicable only in regions without physical scarcity.
Water demand (decreasing water demand and increasing use efficiency)	Remove invasive non-native vegetation Use of drought-resistant crops Maintenance of irrigation infrastructure Change in irrigation techniques Crop management (change in cropping pattern and timing of farm operations)	P: Efficient use of available water resources, which is relatively cheaper than supply management. P: Increases the tolerance and suitability of plants to temperature, moisture and other relevant climatic conditions. C: May not have pronounced benefits in all conditions.
2. Information systems		
Weather and climate information systems and knowledge management	Implement systems to use daily and seasonal weather forecasts	P: Improves efficiency of agricultural management by providing information early enough to adjust the critical decision. C: The information may not be always achieved in time and scale relevant to farmers.
3. Socio-economic		
Agricultural subsidy and support	Subsidy/support programmes to influence farm-level production	P: Reduces the risk of climate-related income loss and can motivate for positive change in farm-level management. C: Limited by the government subsidy and support programme.
Insurance	Insurance schemes to address crop damage from climate-related events, e.g. drought	P: Reduces vulnerability at the farm level. C: Limited by the government subsidy and support programme.
4. Farm production practices		
Land use	Change in location of crop production Change from rainfed to irrigated agriculture Use of alternate fallow and tillage practices	P: Conserve moisture and nutrients. C: Cost of the support system and infrastructure will be high for change in location and shift to irrigated agriculture.
Land topography	Change land topography (land contouring and terracing)	P: Reduces erosion, improves the retention of moisture and nutrient and improves water uptake.
5. Diversifying production system	Agroforestry	P: Maintains production during both wetter and drier years and acts as a buffer against income risks associated with climate variability. C: Government help is required to smallholder farmers mainly during the initial years.
6. Traditional knowledge and indigenous practices	Traditional water-harvesting technologies, grass-mulching, etc.	P: More likely to be accepted by the community and feasible to be adopted without external help.

(e.g. the inter-sectoral competition for resources may constrain expansion of irrigation).

2.4 Modelling-based Assessment of Adaptation Options

A continuous assessment of the impacts, particularly concerning the impact of higher temperature, change in precipitation patterns (Watanabe and Kume, 2009) and climate variability including short-term extreme events constitute the basis for developing a sound adaptation strategy for sustainable crop production. The nature of the stimuli and allied vulnerability establish the relevancy of adaptation options (Pittock and Jones, 2000). It is imperative to recognize the climate variables to which a particular adaptation option is the most suitable and to take into account the role of non-climatic factors that influence the sensitivity of agriculture to climate change. Typically, adaptation options are evaluated using a crop growth simulation model, with or without a coupled hydrological model, forced with climate projections from one or more global climate models. The use of a hydrological model to couple with the crop growth models is not very common. Only a few studies reported the coupling with the Variable Infiltration Capacity (VIC) hydrological model. When a hydrological model is integrated with a crop model for the impact analysis on crop yield, the crop model benefits from the dynamic input of available water on the temporal scale to which the hydrological model works, typically daily. On the other hand, changes in crop characteristics may also modify the hydrological impacts of climate change such as the risk of drought and flooding (Betts, 2005) on a much smaller scale than the climate models allow. Table 2.2 summarizes the recent literature on model-based evaluation of various adaption measures indicating the study region, types of crops analysed, models used for climate projections and crop growth simulations, types of adaptation options evaluated and climatic variables considered for assessing the adaptation measures. Key results and conclusions reported in the literature are also summarized. The literature covers a wide geographic range and crop types. CERES (Ritchie *et al.*, 1998), DSSAT (Jones *et al.*, 1998) and CropSys (Stöckle *et al.*, 2003) are the most commonly used crop growth models or modelling systems in these studies. The commonly assessed adaptation options reported in these studies are the following:

- sowing dates;
- crop varieties (hybrids, slow maturing, etc.);
- level of fertilizer application;
- crop density, different crop rotations and double-cropping;
- expansion of irrigation and soil moisture conservation;
- improvement in agricultural technology; and
- land use and water allocation policies, etc.

The adaptation responses are commonly evaluated with respect to the improvement in the crop yields alone. In practice, some of these adaptation options may not be always feasible either due to the constraints from other sectors (e.g. competition for resources, socio-economic, institutional, technical constraints) or due to high underlying cost. These modelling-based methods constitute what is commonly known as the 'impact approach'. Although less commonly reported in the literature, adaptation options for agriculture are also evaluated using the 'capacity approach' in which the existing capacities and vulnerabilities of socio-economic groups are the basis for developing politically and economically feasible adaptation options given the plausible future climate projections (Vermeulen *et al.*, 2013). To bridge the gap between science and policy and planning long-term adaptation, integration of impact and capacity approaches is essential. Although the uncertainty due to climate scenarios and selection of General Circulation Models (GCMs) are normally considered in most of the climate-change adaptation studies, the uncertainty (input data, model structure and parameters) of the impact (crop and hydrological) model and

Table 2.2. Development and evaluation of adaption options.

Study region, crops and source	Climate model and scenarios	Adaptation options evaluated and climatic variables considered	Key results and conclusion
South-eastern USA Maize, wheat, soybean and groundnut Alexandrov and Hoogenboom (2000)	CMS: Geophysical Fluid Dynamics Laboratory (GFDL-R15), Canadian Centre for Climate Modelling and Analysis (CGCM1), Max-Planck Institute for Meteorology (ECHAM4), UK Hadley Center for Climate Prediction and Research (HadCM2) and Australian Commonwealth Scientific and Industrial Research Organization (CSIRO-Mk2b) (Mitchell *et al.*, 1995; Hirst *et al.*, 1996; Haywood *et al.*, 1997; Johns *et al.*, 1997; Bacher *et al.*, 1998; Flato *et al.*, 1999). CGM: CERES (Ritchie *et al.*, 1998) and CROPGROW (Boote *et al.*, 1998).	Changing sowing dates, hybrids and cultivar and fertilization. Temperature, precipitation, solar radiation and CO_2 levels.	Increased temperature projected a shorter vegetative and reproductive growing season for maize for 2020. Assuming the direct benefits of elevated CO_2 level, simulations indicated an increase in soybean and groundnut yield under all GCM climate change scenarios for 2020. Alteration of sowing dates, cultivars and fertilization could minimize the negative impact of future warming.
Northern Thailand Rice Babel *et al.* (2011)	CMS: ECMWF atmospheric general circulation model coupled with the University of Hamburg's ocean circulation model (ECHAM4) A2 (Roeckner *et al.*, 1996), providing regional climates for impact studies (PRECIS). CGM: CERES (Ritchie *et al.*, 1998).	Changing sowing dates, nitrogen application, tillage practices and cultivars. Temperature and CO_2 levels.	Under future climate, duration between anthesis and maturity was reduced resulting in reduced yield. Delayed sowing avoids high temperature during the grain-filling phase. However, the alteration of sowing date is limited by water availability. Modification of fertilizer application schedule and use of cultivars having longer maturity duration, lower photoperiod sensitivity and higher temperature tolerance has a positive impact on yield under future climatic condition.

Romania Winter wheat and rain-fed maize Cuculeanu *et al.* (1999)	CMS: Canadian Climate Centre model (CCCM) (McFarlane *et al.*, 1992) and Goddard Institute for Space Studies (GISS) (Hansen *et al.*, 1988). CGM: CERES (Godwin *et al.*, 1989; Ritchie *et al.*, 1998).	Changing crop varieties, sowing dates, crop density, and level of fertilization. Temperature, precipitation and CO_2 levels.	Winter wheat and rain-fed maize benefit from the climate change but irrigated maize shows negative response to climate change. The negative impact on maize was reduced by the use of longer-maturing hybrids, change in sowing date and plant density and increasing fertilization level. The effect of doubling CO_2 on photosynthesis and water use varies according to the plant species, which is still an important research question.
Keith, South Australia Wheat Luo *et al.* (2009)	CMS: CSIRO-conformal cubic atmospheric model (C-CAM) for 2080 (Sadourny, 1972). CGM: Agricultural Production Systems sl Mulator (APSIM)-Wheat (Keating *et al.*, 2003).	Early sowing, changing fertilizer application rate and use of different cultivars. Mean rainfall, temperature, solar radiation, wet spells, dry spells and temperature variability.	Early sowing is effective in dealing with the adverse effect of climate change. In drier conditions, the early sowing needs to be supplemented by other adaptation options such as irrigation. Changing N application rate and wheat cultivars is not adequate to fully offset the negative impact of climate change.
Asia Rice Matthews *et al.* (1997)	CMS: GFDL, GISS and United Kingdom Meteorological Office (UKMO) (Wilson and Mitchell, 1987). CGM: ORYZA1 (Kropff *et al.*, 1994) and SIMRIW (Horie, 1987).	Modification of sowing/ planting dates, use of varieties with a higher tolerance of spikelet fertility to temperature. Temperature and CO_2 levels.	The average production in the region will decline but the magnitude of the impact varies with climate scenarios, regions and crop simulation models. At high altitudes where warmer temperature allowed a longer-growing season, modification of sowing date may permit double-cropping. Similarly, in case of a longer growing season, shifting of planting date will avoid high temperature at the critical stage of development.
Chile Maize (irrigated) Meza *et al.* (2008); Meza and Silva (2009)	CMS: HadCM3 for A1F1 and B2B scenario. CGM: Decision Support System for Agrotechnology Transfer (DSSAT) (Jones *et al.*, 2003).	Changing sowing dates, nitrogen fertilizer doses, plant densities and double-cropping. Temperature and precipitation.	Showed yield reduction from 10 to 30%, depending on change scenarios and hybrid used. Early sowing and N management can minimize the adverse impact of climate change. In case of a longer growing season, double-cropping outperformed other adaptation options, such as new cultivars.

Continued

Table 2.2. Continued.

Study region, crops and source	Climate model and scenarios	Adaptation options evaluated and climatic variables considered	Key results and conclusion
India Sorghum Srivastava *et al.* (2010)	CMS: HadCM3 for A2a scenarios. CGM: InfoCrop-SORGHUM (Aggarwal *et al.*, 2006a, b).	Changing crop varieties, planting dates and a combination of both. Temperature, precipitation, CO_2 levels.	More impacts were observed on winter crops in the central and south-central zone, and on monsoonal crops in the south-western zone. Simple strategies such as shifting sowing time and changing varieties can reduce vulnerability. Although better management strategies can reduce vulnerability, low-cost adaptation options must be explored for benefit in resource-constrainted situations.
North China Plain Maize Tao and Zhang (2010)	CMS: SuperEPPS using ten climate scenarios from 5 GCMs (HadCM3, PCM, CGCM2, CSIRO2 and ECHAM4) and two emission scenarios (A1F1, B1). CGM: Model to Capture the Crop–Weather relationship over a Large Area (MCWLA) (Tao *et al.*, 2009).	Early planting, late planting, fixing crop-growing duration and use of different varieties. Temperature and precipitation.	Without adaptation the maize yield could reduce by 13–19% during the 2050s. Different adaptation options (changing planting dates and fixing growing duration) showed marginal (<5%) to significant (>30%) increase in yield. The benefits are sensitive to the crop varieties. The highest benefit was obtained from the high-temperature-tolerant variety.
Switzerland (alpine region) Maize, wheat, canola Torriani *et al.* (2007)	CMS: HIRHAM4 (Christensen *et al.*, 1998) driven by HadAM3H (SRES A2 scenario). CGM: Cropping Systems Simulation Model (CropSyst) (Stöckle *et al.*, 2003).	Slow maturing variety, shifting planting date and expansion of irrigation. Temperature, precipitation, solar radiation, relative humidity and CO_2 levels.	Shifting the sowing date resulted in positive yield on maize and negative yield on wheat and canola. Slow-maturing cultivars showed a positive impact on average yield on all three crops. Adaptation responses are crop specific and difficult to generalize.
Czech Republic Barley Trnka *et al.* (2004)	CMS: ECHAM4, HadCM2, NCAR-DOE and scenario averaged over 7 GCMs. CGM: CERES-Barley (Otter-Nacke *et al.*, 1991).	Early sowing, change of cultivars, change in N-fertilizer and soil moisture conservation. Temperature, precipitation and CO_2 levels.	Simulations showed generally positive impacts on yield, considering the effect of doubled CO_2 concentration. Early planting and use of cultivars with longer growing season will further increase the yield under doubled CO_2 concentration. Soil water conservation is important for sustainable production mainly in the low rainfall areas.

Modena and Foggia (Italy) Maize, wheat, soybean, barley, sorghum, sunflower (in rotations) Tubiello *et al.* (2000)	CMS: GISS (Hansen *et al.*, 1988) and Geophysical Fluid Dynamics Laboratory (GFDL) (Manabe and Weatherland, 1987). CGM: Cropping Systems Simulation Model CropSyst (Stöckle *et al.*, 2003).	Early planting, use of slow-maturing variety and expansion of irrigation. Temperature and precipitation.	Warmer temperature showed crop yield reduction by 10–40%. Combination of early planting for spring–summer crops and slower-maturing winter cereal cultivars is able to maintain yield. High temperature increases evaporative demand and reduces irrigation water use efficiency. For irrigated crops, 60–90% more irrigation water was required to maintain yield.
USA Wheat, potato, maize, citrus Tubiello *et al.* (2002)	CMS: CCGS and HCGS (CC: Canadian Centre Model; HC: Hadley Centre Model; GS: greenhouse gases with sulfate aerosols). CGM: DSSAT.	Changing planting dates and cultivars. Temperature, precipitation and CO_2 levels.	Yield response to climate change varied significantly in magnitude and even direction for the climate change scenarios considered, due to difference in projected precipitation. For all the crops simulated, precipitation and elevated CO_2 each contributed about half of the yield increase.
China Rice, maize, wheat Wei *et al.* (2009)	CMS: PRECIS (Jones *et al.*, 2004) based on SRES A2 and B2. CGM: CERES (Ritchie *et al.*, 1989). HM: Variable Infiltration Capacity VIC (Liang *et al.*, 1994, 1996).	Improvement in agricultural technology, land-use change policy and water allocation policy. Temperature, precipitation, solar radiation, CO_2 levels.	The absolute effects of climate change are relatively modest, but climate scenarios combined with socio-economic developments lead to a decrease in total production. Policy options related to land, water and agricultural technology can offset the negative impact and the combination of these policy options presents a better result.

CMS, Climate Model and Scenario; CGM, Crop Growth Model; HM, Hydrological Model.

socio-economic uncertainty (price fluctuation, international trade) is mostly neglected. Therefore, a more comprehensive study should prioritize the adaptation strategies giving due attention to the availability of resources, financial and social constraints, responses of stakeholders and farmers and associated uncertainties under the changed environment. Moreover, such adaptation measures should be cost effective and easily available to the farmers.

Adaptation success is closely linked with the alternatives available to the farmers. Integration of two or more feasible adaptation measures is generally expected to give higher benefits (e.g. Srivastava *et al.*, 2010). The impact on crop yield may also come from secondary factors such as higher weed and pest infestations (e.g. Hossain *et al.*, 2003), which are more likely under higher exposure to warm temperatures (Tubiello *et al.*, 2000). Such indirect consequences of climate change are normally not considered in most models used in adaptation studies. Thus, it is likely that the estimated benefits of some of these adaptation measures may not be achieved in the farmers' fields. Moreover, adaptation assessment studies should not just consider the level of crop yield but need to evaluate the trade-offs between crop production and resources availability that considerably influence farmers' decision making and profitability.

The existing simulation studies signify the progress in our understanding of how adaptation measures can be useful to curtail the likely effect of future climate on crop yields in different geographic and climatic regions. However, it is still poorly understood how variation in crop production and water availability, as a consequence of climate change, will interact with other socio-economic pressures. Moreover, the estimated impact of climate change on crop production and significance of adaptation can depend largely on the crop model used, particularly the approach used for simulating the impact of extreme events (Tao and Zhang, 2010). Thus, evaluation of adaptation responses using two or more crop models might help to minimize the uncertainty due to the crop model structure.

2.5 Challenges in Climate Change Impact and Adaptation Research

The impacts of climate change on crops (vegetation), catchment hydrology and water management systems underline the need for integrative studies. However, the issue of scale and uncertainty is a challenge for such integration (Betts, 2005). The difference in temporal as well as spatial scales between climate and crop models is one of the major difficulties of integration, which is also discussed by Osborne *et al.* (2006).The integrated climate-crop models if used appropriately can play an important role to identify potential adaptation strategies. However, their role may be limited to support agricultural climate risk management (Hansen, 2005). Although parameterization of some components of hydrological models can be uncertain due to inadequate data, the relationship between climate, human activities and water resources can be investigated with the hydrological models (Jothityangkoon *et al.*, 2001) forced with the climate model results (predictions), normally with downscaling of the climatic variables.

The climate scenarios used can alter the magnitude and even the direction of the impact on crop yield irrespective of the location and type of crops studied (e.g. see Reilly *et al.*, 2003) due to the variation of the projected change in climatic variables, especially precipitation. A large part of this ambiguity in precipitation is due to the coarse resolution of the GCM, as it does not sufficiently represent specific regional land features (such as mountains and lakes). Such regional or local features can significantly influence the local climates (Hu *et al.*, 2013a, b). A widely recognized approach to address the uncertainty related to the choice of GCMs is to employ an ensemble of a range of models, but obviously it adds complexity in modelling and analysis. Managing the present risk and building capacity to deal with unpredictable future events is key for the adaptation to climate change. Moreover, the relative importance of the uncertainty associated with climate change may vary spatially and temporally. Vermeulen *et al.* (2013) presented a framework for prioritizing adaptation approaches with particular reference to uncertainty linked to the time frames considered. They illustrated the importance of timescale in applying a suitable approach: impact approach or capacity approach or a combination of the two.

Scale and geography are also important for determining the crop yield. The balance between the generality and specificity in region and scale is yet another challenge to predict the response of crops to climate change (Challinor *et al.*, 2009). Variation in commodity prices, trade agreements, resources use rights and government subsidies and support programmes may obscure the adaptation process (Smit *et al.*, 1996). Generally, social and technical constraints in developing countries may restrict sustainable production in the long run (Parry *et al.*, 1999). In certain circumstances, such socio-economic complexity may even outweigh the climatic uncertainty in evaluating the feasible adaptation measures (Eakin, 2005; Vincent, 2007). On the other hand, the adaptation capability is low in developing countries due to limited access to market for crop inputs or outputs and lack of appropriate infrastructure (Reilly and Hohmann, 1993). In order to address these challenges, an adaptation framework needs to equitably involve farmers, agribusiness and policy makers (Howden *et al.*, 2007) and cover a range of scales and issues which should be integrated with a comprehensive and dynamic policy approach.

2.6 Conclusions and Recommendations

Sensitivity of agriculture to climate change varies across the globe. Warmer temperatures tend to reduce the crop yields in many regions, mainly due to reduction of crop duration associated with water stress during the critical stages of crop development. Developing countries, where more than 800 million people are already undernourished, will be hardest hit. Hence, adaptation in the agriculture sector is essential in order to feed the world's growing population. Even without climate change, inherent climate variability and socio-economic development mean that transformation of agricultural systems is inevitable, but the urgency of timely adaptation has been amplified due to climate change.

Crop growth and production and water resources distribution will be affected by the interaction between increasing atmospheric CO_2 concentration, higher temperature, varying patterns of precipitation, altered frequency and severity of extreme events, land-use change and regional socio-economic development. Although efforts have been made to understand better the climate–crop relationship, the interactions that are still not fully described include: (i) field response of crop to higher CO_2 concentration; (ii) response to increased extreme events under climate change; (iii) influences of local/regional socio-economic drivers on the climate crop relationship; (iv) economics of adaptation at the regional/local scale; and (v) significance of the uncertainty of the impact model.

To fully understand the resulting impacts of these interactions requires an integrated approach that incorporates the physics of climate change with the biology of crop development and socio-economic dimension of the region. Furthermore, the resulting impacts are highly dependent on regional variability of biophysical conditions (Tan and Shibasaki, 2003). Limited knowledge of this variability constrains our capacity to determine optimal responses for adaptation. Therefore, future adaptation studies should consider, among other things, the influence of extreme temperature at the critical stages of crop development, regional socio-economic development and significance of the uncertainty of the impact model. Similarly, there is a need to expand the number of field experiments to understand the influence of increased CO_2 concentration on a range of crops.

Integrated technical and policy adaptation measures including no-regret options based on both traditional and new technologies are deemed to be favourable to cope with climate and other global changes. Any adaptation options proposed must be easily available and economically acceptable to the farmers. Thus, there is a need to expand the number of studies that focus on the acceptability of adaptation options in terms of factors important to all stakeholders. Furthermore, the social and technological aspects of vulnerability, such as obtainable adaptive capacity in a region and the complexity of adaptation for specific crops, should be incorporated while developing adaptation strategies (Lobell *et al.*, 2008).

Most of the adaptation studies reported in the literature have focused on the modification of existing practices (such as shifting planting dates, using existing cultivars, application of irrigation, etc.) to improve crop yield. Such studies should be extended to evaluate and prioritize a range of other possible management and policy options, taking into account social, technical, financial, institutional and resource constraints in the modelling framework. Trade-offs between crop production and resource availability, which influence the farmers' decision making and profitability, have not received substantial attention so far. To adequately address the impacts of climate and other changes on agriculture, evaluation of diverse adaptation options is needed, covering a range of scales from farm to policy level and with consideration of the availability of resources, constraints and associated uncertainties. Interaction with the farmers and stakeholders is also essential to evaluate the employability of any adaptation options and to understand the dynamics of traditional practices to cope with the changing environment.

Acknowledgements

The review presented in this chapter was carried out as part of the AGloCAP (Adaptation to Global Change in Agricultural Practices) project, which is a joint initiative of UNESCO-IHE Institute for Water Education, the Netherlands, Asian Institute of Technology, Thailand and Department of Irrigation, Nepal. The project is funded by the Netherlands Ministry of Development Cooperation (DGIS) through the UNESCO-IHE Partnership Research Fund. The chapter has not been subjected to peer and/or policy review by DGIS or the project partner institutions and, therefore, does not necessarily reflect the views of these institutions.

References

Aggarwal, P.K., Banerjee, B., Daryaei, M.G., Bhatia, A., Bala, A. *et al.* (2006a) InfoCrop: a dynamic simulation model for the assessment of crop yields, losses due to pests, and environmental impact of agro-ecosystems in tropical environments. II. Performance of the model. *Agricultural Systems* 89, 47–67.

Aggarwal, P.K., Kalra, N., Chander, S. and Pathak, H. (2006b) InfoCrop: a dynamic simulation model for the assessment of crop yields, losses due to pests, and environmental impact of agro-ecosystems in tropical environments. I. Model description. *Agricultural Systems,* 89, 1–25 and *Agronomy* 18 (3–4), 267–288.

Ainsworth, E.A. and Long, S.P. (2005) What have we learned from fifteen years of free-air CO_2 enrichment (FACE)? A meta-analytic review of the responses of photosynthesis, canopy properties and plant production to rising CO_2. *New Phytologist* 165, 351–372.

Alexandrov, V.A. and Hoogenboom, G. (2000) Vulnerability and adaptation assessment of agricultural crop under climate change in Southeastern USA. *Theoretical and Applied Climatology* 67, 45–63.

Amthor, J.S. (2001) Effects of atmospheric CO_2 concentration on wheat yield: review of results from experiments using various approaches to control CO_2 concentration. *Field Crops Research* 73, 1–34.

Arnell, N.W. (2004) Climate change and global water resources: SRES emissions and socio-economic scenarios. *Global Environmental Change* 14, 31–52.

Arnell, N.W., Cannell, M.G.R., Hulme, M., Kovats, R.S., Mitchell, J.F.B., Nicholls, R.J., Parry, M.L., Livermore, M.T.J. and White, A. (2002) The consequences of CO_2 stabilization for the impacts of climate change. *Climatic Change* 53, 413–446.

Babel, M.S., Agarwal, A., Swain, D.K. and Herath, S. (2011) Evaluation of climate change impacts and adaptation measures for rice cultivation in Northeast Thailand. *Climate Research* 46, 137–146.

Bacher, A., Oberhuber, J.M. and Roeckner, E. (1998) ENSO dynamics and seasonal cycle in the tropical Pacific as simulated by the ECHAM4/OPYC3 coupled general circulation model. *Climate Dynamics* 14, 431–450.

Betts, R.A. (2005) Integrated approach to climate-crop modelling: needs and challenges. *Philosophical Transactions of the Royal Society B* 360, 2049–2065.

Betts, R., Falloon, P., Challinor, A. and Hemming, D. (2006) *Assessing uncertainties in key climate change impacts indicators. Defra Milestone 04/07 13.06.05, Physical, chemical, biological effects of climate change; 2006. Report to Defra.* Met Office Hadley Centre, Exeter, UK.

Boote, K., Jones, J. and Pickering, N. (1998) The CROPGRO model for grain legumes. In: Tsuji, G., Hoogenboom, G. and Thornton, P. (eds) *Understanding Options for Agricultural Production.* Kluwer Academic Publishers, Dordrecht, the Netherlands, pp. 99–128.

Boxall, A.B.A., Hardy, A., Beulke, S., Boucard, T., Burgin, L. *et al.* (2009) Impacts of climate change on the health risks of pathogens and chemicals from agriculture. *Environmental Health Perspectives* 117(4), 508–514.

Brown, M.E. and Funk, C.C. (2008) Food security under climate change. *Science* 319, 580–581.

Brown, R.A. and Rosenberg, N.J. (1999) Climate change impacts on the potential productivity of corn and winter wheat in their primary United States growing regions. *Climate Change* 41, 73–107.

Bryant, R.C., Smit, B., Brklacich, M., Johnston, R.T., Smithers, J. *et al.* (2000) Adaptation in Canadian agriculture to climatic variability and change. *Climate Change* 45, 181–201.

Caldwell, C.R., Britz, S.J. and Mirecki, R.M. (2005) Effect of temperature, elevated carbon dioxide, and drought during seed development on the isoflavone content of dwarf soybean [*Glycine max* (L.)Merrill] grown in controlled environments. *Journal of Agricultural Food Chemistry* 53, 1125–1129.

Challinor, A.J., Ewert, F., Arnold, S., Simelton, E. and Fraser, E. (2009) Crops and climate change: Progress, trends, and challenges in simulating impacts and informing adaptation. *Journal of Experimental Botany* 60(10), 2775–2789.

Christensen, O.B., Christensen, J.H., Machenauer, B. and Botzet, M. (1998) Very high-resolution regional climate simulations over Scandinavia – present climate. *Journal of Climate* 11, 3204–3229.

Cuculeanu, V., Marica, A. and Simota, C. (1999) Climate change impact on agricultural crops and adaptation options in Romania. *Climate Research* 12, 153–160.

Desjardins, R.L., Smith, W., Grant, B., Campbell, C., Janzen, H. *et al.* (2002) Management strategies to sequester carbon in agricultural soil and to mitigate greenhouse gas emissions. Paper presented at the *International Workshop on Reducing Vulnerability of Agriculture and Forestry to Climate Variability and Climate Change, 2002,* Ljubljanu, Slovenia.

Döll, P. (2002) Impact of climate change and variability on irrigation requirements: a global perspective. *Climate Change* 54, 269–293.

Döll, P. and Siebert, S. (2002) Global modeling of irrigation water requirements. *Water Resources Research* 38(4), 8.1–8.10.

Eakin, H. (2005) Institutional change, climate risk, and rural vulnerability: cases from Central Mexico. *World Development* 33(11), 1923–1938.

Easterling, W.E., Aggarwal, P.K., Batima, P., Brander, K.M., Erda, L., Howden, S.M., Kirilenko, A., Morton, J., Soussana, J.-F., Schmidhuber, J. and Tubiello, F.N. (2007) Food, fibre and forest products. In: Parry, M.L., Canziani, O.F., Palutikof, J.P., Linden, P.J. and Hanson, C.E. (eds) *Climate Change 2007: Impacts, Adaptation and Vulnerability. Contribution of Working Group II to the Fourth Assessment Report of the Intergovernmental Panel on Climate Change.* Cambridge University Press, Cambridge, UK, pp. 273–313.

Erda, L., Wei, X., Hui, J., Yinlong, X., Yue, L. *et al.* (2005) Climate change impacts on crop yield and quality with CO_2 fertilization in China. *Philosophical Transactions of the Royal Society B* 360, 2149–2154.

Ewert, F., Van Oijen, M. and Porter, J.R. (1999) Simulation of growth and development processes of spring wheat in response to CO_2 and ozone for different sites and year in Europe using mechanistic crop simulation models. *European Journal of Agronomy* 10, 231–247.

Fallon, P. and Betts, R. (2010) Climate impacts on European agriculture and water management in the context of adaptation and mitigation – the importance of integrated approach. *Science of the Total Environment* 408, 5667–5687.

Fischer, G., Shah, M. and van Velthuizen, H. (2002) *Climate change and agricultural vulnerability. Special report to the UN world summit on sustainable development Johannesburg.* International Institute for Applied Systems Analysis, Austria.

Fischer, G., Tubiello, F.N., van Velthuizen, H. and Wiberg, D. (2007) Climate change impacts on irrigation water requirements: effects of mitigation, 1990–2080. *Technological Forecasting and Social Change* 74, 1083–1107.

Flato, G.M., Boer, G.J., Lee, W.G., McFarlane, N.A., Ramsden, D. *et al.* (1999) The Canadian Centre for Climate Modelling and Analysis global coupled model and its climate. *Climate Dynamics* 16(6), 451–467.

Gifford, R.M. (2004) The CO2 fertilising effect – does it occur in the real world? *New Phytologist* 163(2), 221–225.

Godwin, D.C., Ritchie, J.T., Singh, U. and Hunt, L. (1989) *A User's Guide to CERES-Wheat: v2.10.* International Fertilizer Development Center, Muscle Shoals, Alabama.

Hansen, J., Fung, I., Rind, D., Lebedeff, S., Ruedy, R. and Russel, G. (1988) Global climate changes as forecast by the Goddard Institute for Space Studies three-dimensional model. *Journal of Geophysical Research* 93(D8), 9341–9364.

Hansen, J.W. (2005) Integrating seasonal climate prediction and agricultural models for insight into agricultural practice. *Philosophical Transactions of the Royal Society B* 360, 2037–2047.

Hansen, J.W., Challinor, A., Ines, A., Wheeler, T. and Moron, V. (2006) Translating climate forecasts into agricultural terms: advances and challenges. *Climate Research* 33, 27–41.

Haywood, J.M., Stouffer, R.J., Wetherald, R.T., Manabe, S. and Ramaswamy, V. (1997) Transient response of a coupled model to estimated changes in greenhouse gas and sulphate concentrations. *Geophysical Research Letters* 24, 1335–1338.

Hirst, A.C., Gordon, H.B. and O'Farrell, S.P. (1996) Global warming in a coupled climate model including oceanic eddy induced advection. *Geophysical Research Letters* 23, 3361–3364.

Horie, T. (1987) A model for evaluating climatic productivity and water balance of irrigated rice and its application to Southeast Asia. *Southeast Asian Studies* 25, 62–74.

Hossain, I., Epplin, M. and Krenzer, E.G. Jr (2003) Planting date influence on dual-purpose winter wheat forage yield, grain yield, and test weight. *Agronomy Journal* 95, 1179–1188.

Howden, S.M., Soussana, J.F., Tubiello, F.N., Chhetri, N., Dunlop, M. *et al.* (2007) Adapting agriculture to climate change. *PNAS* 104(50), 19691–19696.

Hsiao, T.C. and Jackson, R.B. (1999) Interactive effects of water stress and elevated CO_2 on growth, photosynthesis, and water use efficiency. In: Luo, Y. and Mooney, H.A. (eds) *Carbon Dioxide and Environmental Stress.* Academic Press, San Diego, California, pp. 3–31.

Hu, Y., Maskey, S. and Uhlenbrook, S. (2013a) Downscaling daily precipitation over the Yellow River source region in China: a comparison of three statistical downscaling methods. *Theoretical and Applied Climatology* 112, 447–460.

Hu, Y., Maskey, S. and Uhlenbrook, S. (2013b) Expected changes in future temperature extremes and their elevation dependency over the Yellow River source region. *Hydrology Earth Systems Sciences* 17, 2501–2514.

IPCC (Intergovernmental Panel on Climate Change) (2000) *IPCC Special Report on Emission Scenarios – Summary for Policymakers.* Cambridge University Press, Cambridge, UK, 599 pp.

IPCC (Intergovernmental Panel on Climate Change) (2007) *Climate change 2007: Impacts, adaptation and vulnerability. Contribution of Working Group II to the fourth assessment report of the intergovernmental panel on climate change.* Cambridge University Press, Cambridge, UK, 976 pp.

Johns, T.C., Carnell, R.E., Crossley, J.F., Gregory, J.M., Mitchell, J.F.B. *et al.* (1997) The second Hadley Centre coupled ocean-atmosphere GCM: model description, spin up and validation. *Climate Dynamics* 13, 103–134.

Jones, J.W., Tsuji, G.Y., Hoogenboom, G., Hunt, L.A., Thornton, P.K. *et al.* (1998) Decision support system for agrotechnological transfer DSSAT v3. In: Tsuji, G.Y., Hoogenboom, G. and Thornton, P.K. (eds) *Understanding Options for Agricultural Production.* Kluwer Academic Publishers, Dordrecht, the Netherlands, pp. 157–177.

Jones, J.W., Hoogenboom, G., Porter, C.H., Boote, K.J., Batchelor, W.D. *et al.* (2003) The DSSAT cropping system model. *European Journal of Agronomy* 18, 235–265.

Jones, R.G., Noguer, M., Hassell, D.C., Hudson, D., Wilson, S.S. *et al.* (2004) *Generating High Resolution Climate Change Scenarios Using PRECIS.* Met Office Hadley Centre, Exeter, UK, 35p.

Jothityangkoon, C., Sivapalan, M. and Farmer, D.L. (2001) Process controls of water balance variability in a large semi-arid catchment: downward approach to hydrological model development. *Journal of Hydrology* 254, 174–198.

Keating, B.A., Carberry, P.S., Hammer, G.L., Probert, M.E., Robertson, M.J. *et al.* (2003) An overview of APSIM, a model designed for farming systems simulation. *European Journal of Agronomy* 18(3–4), 267–288.

Kimball, B.A. (1983) Carbon dioxide and agricultural yield. An assemblage of analysis of 430 prior observations. *Agronomy Journal* 75, 779–788.

Kimball, B.A., Kobayashi, K. and Bindi, M. (2002) Responses of agricultural crops to free-air CO_2 enrichment. *Advances in Agronomy* 77, 293–368.

Kropff, M.J., van Laar, H.H. and Matthews, R.B. (eds) (1994) ORYZAI: an ecophysiological model for irrigated rice production. *SARP Research Proceedings*, AB-DLO/TPE-WAU/IRRI, 110 pp.

Kurukulasuriya, P. and Rosenthal, S. (2003) *Climate Change and Agriculture. A review of impacts and adaptations.* Climate Change Series. Paper No. 91. Environment Department and Agriculture and Rural Development Department, The World Bank, Washington, DC.

Kutcher, H.R., Warland, J.S. and Brandt, S.A. (2010) Temperature and precipitation effects on canola yields in Saskatchewan, Canada. *Agricultural Forest Meteorology* 150, 161–165.

LEISA (Low External Input and Sustainable Agriculture) (2000) Grassroots innovations. *ILEIA Newsletter* 16(2), 40.

Liang, X., Lettenmaier, D.P., Wood, E.F. and Burges, S.J. (1994) A simple hydrologically based model of land surface water and energy fluxes for general circulation models. *Journal of Geophysical Research* 99, 14415–14428.

Liang, X., Lettenmaier, D.P. and Wood, E.F. (1996) Surface soil moisture parameterization of the VIC-2L model: evaluation and modification. *Global and Planetary Change* 13(3–4), 195–206.

Lobell, D.B., Burke, M.B., Tebaldi, C., Mastrandrea, M.D., Falcon, W.P. and Naylor, R.L. (2008) Prioritizing climate change adaptation needs for food security in 2030. *Science* 319, 607.

Long, S.P., Ainsworth, E.A., Rogers, A. and Ort, D.R. (2004) Rising atmospheric carbon dioxide plant FACE the future. *Annual Review of Plant Biology* 55, 591–628.

Long, S.P., Ainsworth, E.A., Leakey, A.D.B. and Morgan, P.B. (2005) Global food insecurity. Treatment of major food crops with elevated carbon dioxide or ozone under large-scale fully open-air conditions suggests recent models may have overestimated future yields. *Philosophical Transactions of the Royal Society B* 360, 2011–2020.

Luo, Q., Bellotti, W., Williams, M. and Wang, E. (2009) Adaptation to climate change of wheat growing in South Australia: Analysis of management and breeding strategies. *Agricultural Ecosystems Environment* 129, 261–267.

Manabe, S. and Weatherland, R.B. (1987) The effects of doubling the CO_2 concentration on the climate of a general circulation model. *Journal on the Atmospheric Sciences* 44, 1211–1235.

Matthews, R.B., Kropff, M.J., Horie, T. and Bachelet, D. (1997) Simulating the impact of climate change on rice production in Asia and evaluating options for adaptation. *Agricultural Systems* 54(3), 399–425.

McFarlane, N.A., Boer, G.J., Blanchet, J.P. and Lazare, M. (1992) The Canadian Climate Centre second-generation general circulation model and its equilibrium climate. *Journal of Climate* 5, 1013–1044.

Mendelsohn, R. and Dinar, A. (1999) Climate change, agriculture, and developing countries: Does adaptation matter? *World Bank Research Observer* 14(2), 277–293.

Meza, F.J. and Silva, D. (2009) Dynamic adaptation of maize and wheat production to climate change. *Climate Change* 98, 143–156.

Meza, F.J., Silva, D. and Vigil, H. (2008) Climate change impacts on irrigated maize in Mediterranean climates: Evaluation of double cropping as an emerging adaptation alternative. *Agricultural Systems* 98, 21–30.

Mitchell, J.F.B., Johns, T.C., Gregory, J.M. and Tett, S. (1995) Climate response to increasing levels of greenhouse gases and sulphate aerosols. *Nature* 376, 501–504.

Nelson, G.C., Rosegrant, M.W., Koo, J., Robertson, R., Sulser, T. *et al.* (2009) *Climate Change: Impact on agriculture and costs of adaptation.* International Food Policy Research Institute, Washington, DC.

Osborne, T.M., Lawrence, D.M., Challinor, A.J., Slingo, J.M. and Wheeler, T.R. (2006) Development and assessment of a coupled crop-climate model. *Global Change Biology* 12, 1–15.

Otter-Nacke, S., Rirchie, J.T., Godwin, D.C. and Singh, U. (1991) *A User's Guide to CERES Barley – V2.10.* International Fertilizer Development Center, Muscle Shoals, Alabama.

Parry, M., Rosenzweig, C., Iglesias, A., Fischer, G. and Livermore, M. (1999) Climate change and world food security: a new assessment. *Global Environmental Change* 9, S51–S67.

Parry, M.L., Arnell, N.W., McMichael, A.J., Nicholls, R.J., Martens, P. *et al.* (2001) Millions at risk: defining critical climate change threats and targets. *Global Environmental Change* 11, 181–183.

Parry, M.L., Rosenzweig, C., Iglesias, A., Livermore, M. and Fischer, G. (2004) Effects of climate change on global food production under SRES emissions and socio-economic scenarios. *Global Environmental Change* 14, 53–67.

Pittock, B. and Jones, R.N. (2000) Adaptation to what and why? *Environmental Monitoring Assessment* 61, 9–35.

Porter, H. (1992) Interspecific variation in the growth response of plants to an elevated ambient CO_2 concentration. *Vegetation* 104–105, 77–97.

Porter, J.R. and Semenov, M.A. (2005) Crop responses to climatic variation. *Philosophical Transactions of the Royal Society B* 360(1463), 2021–2038.

Reilly, J. and Hohmann, N. (1993) Climate change and agriculture: The role of international trade. *American Economic Association Papers and Proceedings* 83, 306–312.

Reilly, J., Tubiello, F., McCarl, B., Abler, D., Darwin, R. *et al.* (2003) US agriculture and climate change: new results. *Climate Change* 57, 43–69.

Riahi, K., Grübler, A. and Nakicenovic, N. (2006) *IIASA Greenhouse Gas Initiatives. Long-Term Emissions and Climate Stabilization Scenarios. Interim Report.* International Institute for Applied Systems Analysis, Austria.

Ritchie, J., Singh, U., Godwin, D. and Bowen, W. (1998) Cereal Growth, Development and Yield. In: Tsuji, G., Hoogenboom, G. and Thornton, P. (eds) *Understanding Options for Agricultural Production.* Kluwer Academic Publishers, Dordrecht, the Netherlands, pp. 79–98.

Ritchie, J.T., Singh, U., Godwin, D. and Hunt, L. (1989) *A User's Guide to CERES-Maize v. 2.10.* International Fertilizer Development Center, Muscle Shoals, Alabama.

Roeckner, E., Arpe, K., Bengtsson, L., Christoph, M., Claussen, M. *et al.* (1996) *The Atmospheric General Circulation Model ECHAM-4: Model Description and Simulation of Present-Day Climate. Report No.218.* Max-Planck Institute for Meteorology, Hamburg, Germany, 90 pp.

Rosenzweig, C. and Parry, M.L. (1994) Potential impact of climate change on world food supply. *Nature* 367, 133–138.

Sadourny, R. (1972) Conservative finite-difference approximations of the primitive equations on quassi-uniform spherical grids. *Monthly Weather Review* 100, 136–144.

Sivakumar, M.V.K. (2006) Climate prediction and agriculture: current status and future challenges. *Climate Research* 33, 3–17.

Slingo, J.M., Challinor, A.J., Hoskins, B.J. and Wheeler, T.R. (2005) Introduction: food crops in changing climate. *Philosophical Transactions of the Royal Society B* 360, 1983–1989.

Smit, B., McNabb, D. and Smithers, J. (1996) Agricultural adaptation to climate variation. *Climate Change* 33, 7–29.

Srivastava, A., Naresh Kumar, S. and Aggarwal, P.K. (2010) Assessment on vulnerability of sorghum to climate change in India. *Agriculture Ecosystems & Environment* 138(3–4), 160–169.

Stöckle, C.O., Donatelli, M. and Nelson, R. (2003) CropSyst, a cropping systems simulation model. *European Journal of Agronomy* 18, 289–307.

Tan, G.X. and Shibasaki, R. (2003) Global estimation of crop productivity and the impacts of global warming by GIS and EPIC integration. *Ecological Modelling* 168, 357–370.

Tao, F. and Zhang, Z. (2010) Adaptation of maize production to climate change in North China Plain: Quantify the relative contributions of adaptation options. *European Journal of Agronomy* 33, 103–116.

Tao, F., Yokozawa, M., Hayashi, Y. and Lin, E. (2003) Future climate change, the agricultural water cycle and agricultural production in China. *Agriculture Ecosystems & Environment* 95, 203–215.

Tao, F., Yokozawa, M. and Zhang, Z. (2009) Modelling the impacts of weather and climate variability on crop productivity over a large area: a new process-based model development, optimization, and uncertainties analysis. *Agricultural Forest Meteorology* 149, 831–850.

Torriani, D.S., Calanca, P., Schmid, S., Beniston, M. and Fuhrer, J. (2007) Potential effects of changes in mean climate and climate variability on the yield of winter and spring crops in Switzerland. *Climate Research* 34, 59–69.

Trnka, M., Dubrovský, M. and Žalud, Z. (2004) Climate change impacts and adaptation strategies in spring barley production in the Czech Republic. *Climate Change* 64, 227–255.

Tubiello, F.N., and Ewert, F. (2002) Simulating the effects of elevated CO2 on crops: approaches and applications for climate change. *European Journal of Agronomy* 18, 57–74.

Tubiello, F.N. and Fischer, G. (2007) Reducing climate change impacts on agriculture: Global and regional effects of mitigation, 2000–2080. *Technological Forecasting and Social Change* 74, 1030–1056.

Tubiello, F.N., Donatelli, M., Rosenzweig, C. and Stockle, C.O. (2000) Effects of climate change and elevated CO_2 on cropping systems: model predictions at two Italian locations. *European Journal of Agronomy* 13, 179–189.

Tubiello, F.N., Rosenzweig, C., Goldberg, R.A., Jagtap, S. and Jones, J.W. (2002) Effects of climate change on US crop production: simulation results using two different GCM scenarios. Part i: wheat, potato, maize and citrus. *Climate Research* 20, 259–270.

Tubiello, F.N., Amthor, J.S., Boote, K.J., Donatelli, M., Easterling, W. *et al.* (2007) Crop response to elevated CO_2 and world food supply – a comment on 'food for thought . . .' by Long *et al.* *Science* 321, 1918–1921.

Uhlenbrook, S. (2009) Climate and man-made changes and their impact on catchments. In: Kovar, P., Maca, P. and Redinova, J. (eds) Water Policy 2009 – Water as a Vulnerable and Exhaustible Resource. *Proceedings of the Joint Conference of APLU and ICA*, 23–26 June 2009, Czech University of Life Sciences, Prague, Czech Republic, pp. 81–87.

UN Millennium Project (2005) *Investing in Development: A Practical Plan to Achieve the Millennium Development Goals, Overview.* Communications Development Inc., Washington, DC.

Verchot, L.V., Noordwijk, M.V., Kandji, S., Tomich, T., Ong, C. *et al.* (2007) Climate change: linking adaptation and mitigation through agroforestry. *Mitigation and Adaptation Strategies for Global Change* 12, 901–918.

Vermeulen, S.J., Challinor, A.J., Thorton, P.K., Campbell, B.M., Eriyagama, N. *et al.* (2013) Addressing uncertainty in adaptation planning for agriculture. *PNAS* 110(21), 8357–8362.

Vincent, K. (2007) Uncertainty in adaptive capacity and the importance of scale. *Global Environmental Change* 17(1), 12–24.

Watanabe, T. and Kume, T. (2009) A general adaptation strategy for climate change impacts on paddy cultivation: Special reference to the Japanese context. *Paddy Water Environment* 7, 313–320.

Wei, X., Declan, C., Erda, L., Yinlong, X., Hui, J. *et al.* (2009) Future cereal production in China: The interaction of climate change, water availability and socio-economic scenarios. *Global Environment Change* 19, 34–44.

Wilson, C.A. and Mitchell, J.F.B. (1987) A doubled CO_2 climate sensitivity experiment with a GCM including a simple ocean. *Journal of Geophysical Research* 92, 13315–13343.

Xiao, G., Liu, W., Xu, Q., Sun, Z. and Wang, J. (2005) Effect of temperature increase and elevated CO_2 concentration, with supplement irrigation, on the yield of rain-fed spring wheat in semiarid region of China. *Agricultural Water Management* 74, 243–255.

3 Global Water Requirements of Future Agriculture: Using WATERSIM

Aditya Sood*
International Water Management Institute, Colombo, Sri Lanka

Abstract

Currently, agriculture is the biggest consumer of water globally. About 60% of global consumptive water demand is from the agriculture sector. With the world population increasing to about 9 billion, and the per capita GDP rising, the demand for domestic and industrial water will increase in future and compete with the agriculture water demand. Changing climate will exert additional pressure on the agriculture sector. For this study, WATERSIM, a water accounting and global food trade model, was run for three socio-economic scenarios and two climate change scenarios to analyse the water demand of the agriculture sector till 2050. The changes in the agriculture sector's consumptive water demand due to changing population and GDP is examined in relation to other sectors (domestic, industrial and livestock). The increase in water requirement of the agriculture sector under different climate change scenarios is also analysed at regional and global scales.

3.1 Back on the Food Agenda

During the last half of the 20th century, global agriculture grew at roughly 2.1–2.3% annually in terms of value (Lundqvist, 2010). Global yields also showed upward trends, on average, although there is a gap in the actual yield between developed and developing countries. Cultivated land (food-crop and non-food-crop land) has also increased by about 12% from 1961 to 2000 (INRA and CIRAD, 2009). The gains in food productivity, which was higher than the growth rate of the human population, led to a fall in prices of most food commodities. With growing per capita income of the population and falling food prices, the food became more accessible, the average per capita food availability in developing countries rose from 2110 kcal per person day^{-1} to about 2650 kcal per person day^{-1} from 1976 to 2006 (FAO, 2006). The increase in food production can be mainly attributed to improvement in technology (for example: Green Revolution) and investment in agricultural infrastructure. As policy makers became more complacent, their focus shifted to other issues. The waning interest in agriculture led to falling investment in the sector in general. The World Bank figures on investment in irrigation shows the drastic reduction since the early 1990s (Fig. 3.1).

Encouraged by the positive trends in falling world hunger, the global community at the 1996 World Summit on Food Security agreed to set a goal to halve the number of people who suffer from hunger by 2015 (from 1990). In 2000, at the Millennium Summit of the United Nations, the member states agreed to halve the proportion of people who suffer from hunger by 2015. One difference in the goals is that while the World Summit talked about halving the absolute number of people, the Millennium

* Corresponding author, e-mail: a.sood@cgiar.org

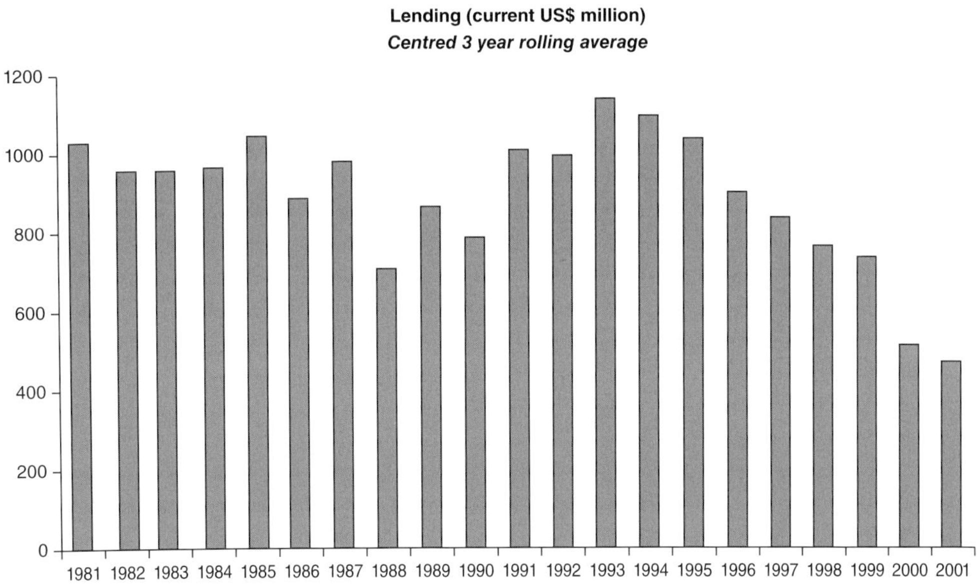

Fig. 3.1. World Bank lending for irrigation (from FAO Corporate Document Repository: The Irrigation Challenge, Issue Paper 4, 2003).

Summit talked about halving the proportion of people suffering from hunger. For the first few years the trends were heading in the right direction: the under-nourished population decreased from 20% in 1990–1992 to 16% in 2005–2007. Although food reached more people than before, the actual number of people who were undernourished went up from 817 million in 1990–1992 to 830 million in 2005–2007 (UN, 2010). The numbers of under-nourished were going down till 2000–2002 and then the progress stopped, and culminated in the 2007–2008 food crises. In a span of just a few months the prices of cereals almost doubled. There were riots in as many as 30 countries. Along with the reduced interest and investment in agriculture, the other reasons that led to this crisis were drastic increases in global fuel prices, higher demand for biofuels and the precautionary trade restrictions put in place by some countries (Nelson *et al.*, 2010). Cereal prices did fall after the crisis but started to rise again in 2009. The 2007–2008 crisis was a wakeup call to the policy makers from around the globe not to lose focus on agriculture. Food security issues are back on the agenda of the global community.

Over the last decade there has been impetus in looking at the future trends in agriculture production, food commodities prices and the impact of food availability on human wellbeing. This chapter reviews the recent studies carried out on projecting future food demand. It then introduces a model developed by the International Water Management Institute (IWMI), the Water, Agriculture, Technology, Environmental and Resource Simulation Model (WATERSIM), to show how modelling can be used to analyse complex issues of food security and water security in the future. The model links river basin hydrology with food trade between the economic regions of the world. It is used to analyse three socio-economic scenarios, developed based on population growth and GDP growth, and focuses on water demands till 2050. The analysis looks at the competition for water between agriculture, industry and domestic demand. It looks at the consumptive use of water in the three sectors for all the three scenarios. Finally, it calculates the change in consumptive water demand for the agriculture sector due to climate change.

3.2 Looking Into the Future

There have been multiple studies that have looked at food security issues. Due to the complexity of the issue, these studies have focused on scenario analysis. Scenario analysis is a process in which different possible outcomes are analysed by changing possible variables that can impact the future world. It is especially helpful where the future outcome is dependent on many complex systems. Due to the uncertainties within such systems and complexity of their interactions with each other, it is difficult to predict the future direction. In such cases, scenario analysis helps in developing stories that define the boundary conditions for a likely future state of affairs, based on some rational assumptions. By considering the most pessimistic and the most optimistic scenarios, we can obtain a range of possible outcomes for the future. Conditions under different scenarios are forecast using models that rely on historic data for calibration and validation and then run for future timelines. Recent high profile examples of such scenario developments are: (i) Intergovernmental Panel on Climate Change (IPCC) Special Report on Emission Scenarios (SRES) that created four storylines – A1, A2, B1 and B2 – by combining economic growth, environmental values, increased globalization and increased regionalization; and (ii) global scenarios created for the Millennium Ecosystem Assessment – Global Orchestration (GO), Order from Strength, Techno garden and Adapting Mosaic – by combining globalization, regionalization, reactivity and proactivity to ecosystem problems. The goal of such scenario development is to provide basic guidelines for interdisciplinary research. It also provides a uniform framework so that it is easy to interpret and compare outputs from different research work. Making a storyline helps policy makers to understand the situation better.

For agriculture, Agrimonde developed scenarios to analyse the future for feeding the world by 2050 (INRA and CIRAD, 2009). The panel for the Agrimonde study created a scenario called Agrimonde GO (AGO), which is a modified version of the MA GO scenario (depicting the pathway of liberalized global trade with major technological development leading to large reductions in poverty and malnutrition but reactive approach to the increasing risk to the ecosystem). They then developed another scenario, called Agrimonde 1 (AG1), which shows the future pathway of a world system that uses technology to advance agriculture production especially in those countries that lack the capital required to invest in such production systems, while also being proactive in terms of protecting the ecosystem. According to the analysis, the food consumption by 2050 will reach between 3000 kcal per capita day^{-1} (AG1 scenario) and 3590 kcal per capita day^{-1} (AGO scenario). The increase in the food demand is met by increasing the cultivated land (39%, AG1 scenario to 23%, AGO scenario) area and improvement in the crop yields.

The International Food Policy Research Institute (IFPRI) has also developed scenarios and conducted analysis to look at the possible future resource limitations and policy options. In a study conducted in 2002 (Rosegrant *et al.*, 2002), IFPRI in collaboration with IWMI looked at the linkage between food security and water security till 2025. Three scenarios were explored in this analysis: (i) the Business-As-Usual scenario (BAU); (ii) the Water-Crisis scenario (CRI); and (iii) the Sustainable Water Use scenario (SUS). BAU assumes similar agriculture policies and investments in future as prevalent in 2002, with falling interest in the agriculture sector, lack of high technological innovations, and limited institutional and management reforms, and a greater pressure from the environment to meet its demand. For CRI, it is presumed that due to a failing economy, the budget cuts will hit investments in water and the agriculture sector and the irrigation systems would be turned over to the farmers without proper capacity building. Due to an increase in deforestation and lack of proper watershed management plans, the land quality will deteriorate further. Lack of proper research would lead to very little increase in food productivity. For SUS, it is assumed that there is greater protection for the environment and greater social equity. It considers more investment in R&D, technology development and adequate water pricing to improve agricultural productivity and to conserve water. These scenarios were analysed using a global modelling framework, which was made up of a combination of two models: the International Model for Policy Analysis of Agriculture Commodities and Trade (IMPACT) and Water Simulation Model (WSM). Their analysis showed that the competition between agriculture and other sectors will increase in future due to a rapid increase in industrial and domestic water demand, which would affect developing countries more. For the BAU scenario, the future food production will increase due to increase in crop yields and improvement in water productivity. The higher demand for water will also put pressure on the water needed for the ecosystem services. For the CRI scenario, water scarcity would increase, thus not only endangering ecosystem services but also constraining food production. This will lead to a large jump in food commodity prices, thus negatively impacting the drive against malnutrition.

In a later study conducted by IFPRI (Nelson *et al.*, 2010), the focus was once more on food security but it also looked at climate change till 2050. In this global study, the scenarios created were based on changes in population, gross domestic product (GDP) and climate change. For the population and GDP, three scenarios were considered: (i) pessimistic, i.e. low GDP growth rate with high population increase; (ii) baseline, i.e. average GDP growth rate (based on World Bank EACC study) with medium population growth; and (iii) optimistic, i.e. high GDP growth rate with low population growth rates. For the climate change, the study considered four scenarios, i.e. CSIRO[1] A1B and CSIRO B1 (depicting a dry and relatively cool future), MIROC[2] A1B and MIROC B1 (depicting a wet and relatively warm future). Their findings suggest imbalance in supply and demand of food commodities till 2050 leading to an increase in food prices. The buying capacity of people will increase, in general, but due to the negative impact of climate change and changing diets of the increasing population, the demand will outpace supply, leading to an increase in prices (calculated as 31.2% for rice to 100.7% for maize). In this scenario, this imbalance can be reduced by proper R&D and improved global trade. There will be little scope to increase the agriculture area in future (without impacting the environment) but there is still scope to improve crop yields (for e.g. 2% per year for maize, wheat and cassava). Even in the best case scenario, prices still increase by 18.4% for rice to 34.1% for maize. Overall the malnutrition among children under 5 years of age will decrease till 2050 but the percentage of reduction varies from 45% in the optimistic scenario to only 2% in the pessimistic scenario. Climate change will cause an increase of 8.5–10.3% of malnutrition among the same group compared to their 'perfect mitigation' case. The study suggests a more severe impact of climate change after 2050, when the population will stabilize but the impact of climate change can be substantial.

Another major study by the Food and Agriculture Organization of the United Nations (FAO) in 2003 looked at world agriculture at 2015/2030 (FAO, 2003), and has been updated twice, first in 2006 to extend the time period to 2030/2050 (FAO, 2006) and second, in 2012, for the same time period but with more recent data (FAO, 2012). In this study, only one scenario was considered, where, at global scale, the growth in future agriculture will slow down because of the stabilization in future food demand as the population stabilizes and the diets converge towards 3000 plus kcal/capita day^{-1}. The growth in the agriculture sector will be about 60% from 2005/7 to 2050 for its baseline scenario. But the global numbers hide regional variations. Although the global population stabilizes, there will be growth in developing countries, which are already undernourished, that will be counterbalanced with decline in the population of developed nations. This will turn many developing countries from net exporter to net importer. Similar to the IFPRI study, their analysis shows that most of the future increase in food production will come from improved crop yields. There would be some addition to agriculture land, which would come at the expense of the pasture land, and is expected to increase by about 70 million ha by 2050. The analysis does not see any persistent threat to agriculture due to climate change at least till 2030 (FAO, 2003). There may be local/regional variation that can be mitigated by agricultural management. Overall, in the short term, the agriculture sector might have a positive impact on climate change due to carbon sequestration.

A similar global study was conducted by IWMI in 2007 to answer the question of whether there is enough water to produce food to meet the demands of the growing population till 2050 (IWMI, 2007). In short, the outcome of the work suggested that there is enough fresh water to meet the demand of growing agriculture, but only with proper water management and improvement in yield, especially in rainfed agriculture. More than 75% of the increase in food demand can possibly be met by improvement in yields rather than increase in agricultural area, which will also help safeguard ecosystems from further damage. Currently 70% of the global freshwater withdrawal is for irrigation, but the proportion of water withdrawal for domestic and industrial uses will increase in future. If water productivity is not considered, the amount of water consumed (evapotranspiration) by agriculture will increase by 70–90% by 2050 (i.e. from 7130 km^3 to 12,000–13,500 km^3). The IWMI analysis considered four scenarios: (i) rainfed scenario, by investing to increase rainfed agriculture production by increasing productivity, improving land management and increasing rainfed agriculture area; (ii) irrigation scenario, by investing in irrigation to provide more irrigated water and improving irrigation efficiency; (iii) trade scenario, by having more liberal trade policies between countries; and (iv) comprehensive assessment scenario, by incorporating features from all the above scenarios to suit each region. In the rainfed scenario, a 1% increase in agriculture yield per year can help in meeting the future food demand with only a 7% increase in agriculture area. A 40% increase in irrigation withdrawal in the irrigation scenario could help irrigated agriculture cater to 55% of total food demand in 2050. On the other hand, trade could help bridge gaps in the regions where supplies cannot meet the demands of a growing population. A combination of all these scenarios can lead to the most sustainable path to meet future food demand with the least damage to the ecosystem. But even in such a scenario, the water withdrawals for irrigation would increase by 13% and cropped area would increase by 9%.

As seen in the 2007/2008 crises, biofuels can play a critical role in the future agriculture scene. Analysis done by the International Institute for Applied System Analysis (IIASSA) shows that based on the current biofuel targets, biofuels will provide 12% of transportation fuel in developed countries and 8% in developing countries by 2030. Due to the increase in demand for biofuels, food prices would rise by 30% by 2020, creating a risk of under-nourishment of 140–150 million additional people (OFID, 2009). The increased demand for biofuels will also put pressure on agriculture land. To meet the future biofuel demand till 2030, an additional 37 million ha of land would be required. As discussed before, global agriculture trade and improvement in water productivity, crop yields and water efficiency can help reduce pressure on natural resources while also meeting demands in the future.

Except for the IWMI study, most of the above studies have focused on food but not so much on the resources behind it per se. Water is one of the most critical resources for food production. Water security is closely linked to food security. How would an increase in future food demand impact the water sector and vice versa? A recent study done by the Organization for Economic Cooperation and Development (OECD) to look at environmental issues till 2050 predicts that if no action is taken, then by 2050, about 40% of the world population will be living in the river basins that are under severe water stress (OECD, 2012). The water demand in 2050 will increase by 55%, most of which will be coming from the industrial and domestic sector. The higher temperature due to climate change could exacerbate the water scarcity problem even more.

3.3 Developing Scenarios

To develop scenarios, the variables that impact the situation need to be identified. There are multiple variables or drivers that impact food security. They are also common to water security. The next section looks at these drivers of change.

3.4 Defining Drivers of Change

Drivers of change are the entities that can impact and modify the path of the future course of action, thus causing a change in the future outcomes. Such drivers may fall into one of the following categories: social or cultural, political, technological, natural and economic. Hazell and Wood (2008) classify drivers of change in agriculture at three scales, global, country and local. The global scale drivers include trade, energy prices and agriculture policies. Country scale drivers include per capita income, urbanization and changing market chains. The local scale drivers are poverty, population pressure, health, technology, property rights, infrastructure and market access, and non-farm opportunities. Some of the examples of major drivers of change in the agriculture sector are population, GDP and climate change, which are also considered here. The global population is expected to reach 9 billion by 2050 and then stabilize. Although there will not be a net increase in the global population, there will be spatial variation in future population growth. Most of the future population growth will take place in the developing countries, which will be countered by the negative growth in the developed countries. Most of the future urbanization will also take place in developing countries. These changing dynamics of the global population will have consequences on the future global players in the food trade, which will also have an impact on natural resources in the developing world. In these regions the water demand for domestic consumption and food demand will increase substantially. The increasing GDP could lead to more buying power for the people, thus changing their dietary habits. Existing data show that in general, diets shift from cereal based to more meat based as the economic conditions of people improve. This could also have serious implications on water resources, as a meat-based diet requires more water than a cereal-based diet. Finally, climate change would have a direct impact on agriculture. The changing rainfall pattern and the increasing temperature will impact the water requirement and crop yields of the crops.

Three socio-economic scenarios were considered. The socio-economic scenarios were created from the GDP and population data provided by IFPRI and used in their study (Nelson *et al.*, 2010). The BAU scenario considers regular growth in GDP and population (i.e. 2008 UN population forecast, medium variant). The optimistic scenario (OPT) considers higher GDP growth (i.e. the highest of the GDP scenarios in the Millennium Ecosystem Assessment GDP scenarios including BAU scenario) and lower population growth (i.e. 2008 UN population forecast, low variant), whereas the pessimistic scenario (PES) considers lower GDP growth (i.e. the lowest of the GDP scenarios in the Millennium Ecosystem Assessment GDP scenarios including the BAU scenario) with higher population growth (i.e. 2008 UN population forecast, high variant) (Nelson *et al.*, 2010). Globally, for the BAU scenario, the GDP is predicted to grow by 9.14% annually and population by 1.2% by 2050, which would lead to annual per capita GDP growth of 6.3%. For OPT, by the year 2050 the global GDP will increase by 11.1% and population by 0.77% annually, leading to an annual per capita GDP growth of 9.93%. For PES, the annual growth is projected to be 4.11%, 1.78% and 1.86%, respectively.

Two climate change scenarios, SRESA2 and SRESB1, were also considered. SRESA2 describes a situation in which the world is made up of heterogeneous regional economies that are growing at a slower pace but the global population is increasing. SRESB1 describes a more globalized world, where global population will peak in the mid-century and then decline, and the global economy will move towards less resource-intensive but clean technologies. For each climate change scenario, the hydrological model was run using climate data from four global circulation models (GCMs). These are MPI-ECHAM5, MIROC3.2, CSIRO Mk 3.0 and CNRM-CM3.[3] The output from the GCM runs were averaged to obtain a single output for each of the two climate change scenarios.

In the WATERSIM model, global trade, through prices, controls the food demand and to some extent the supply (as there is positive shift in yields and increase in area due to rising prices). For the current project, the year 2000 was considered the base year and the model was run till 2050. The hydrological data of the model were acquired from the Lund-Potsdam-Jena managed Land (LPJmL) global hydrological model developed by Potsdam Institute for Climate Impact Research (PIK), Germany. The baseline hydrological data were considered as 30-year average monthly values. The average values from 1970 to 2000 were used in this case.

3.5 Water-accounting and Economic Model – WATERSIM

WATERSIM is an integrated water accounting and food trade balance model, developed by IWMI to answer questions related to future availability of water and food security issues at regional and global scales. Thus the model consists of two modules: (i) water supply and demand module, which is based on IWMI's water accounting framework (Molden, 1997); and (ii) food demand and supply module, which is adapted from IFPRI's IMPACT model (Fig. 3.2).

For the 'water supply and demand' module, the world is divided into 125 prominent river basins. Monthly 'water supply' information is supplied to the model as an input. The water supply consists of precipitation, surface runoff and groundwater recharge. These inputs are acquired from a

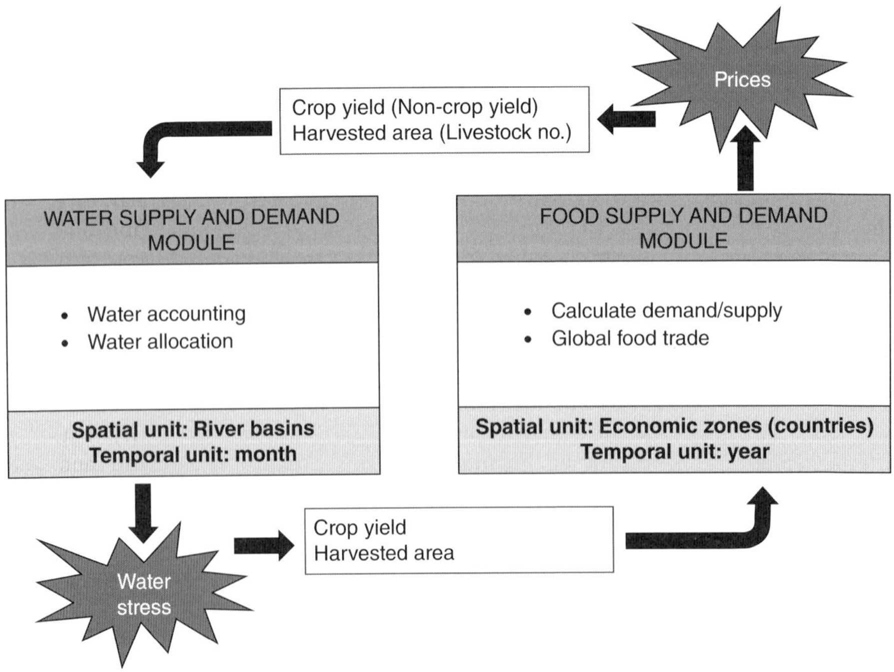

Fig. 3.2. WATERSIM framework.

third-party global hydrological model. The precipitation information is used to calculate the effective precipitation (i.e. the precipitation that is converted to crop evapotranspiration; ET) for the crops. Effective precipitation is calculated by the SCS method (USDA-SCS, 1967). Each basin is defined with a surface storage capacity and groundwater pumping capacity, i.e. the maximum amount of water that can be stored in surface storage structures and pumped out of aquifers, respectively. The surface runoff is used to see how much water is available at monthly time steps to fill the available surface storage capacity. The model conducts surface water balance, wherein the surface runoff is used to fill up available storage capacity and the remaining flows out of the basin as an 'overflow' and is unavailable to meet the water demands for that basin. The 'groundwater recharge potential' (i.e. proportion of water withdrawal that is met by groundwater) for each region is used to calculate the potential water available in the form of groundwater. In the model, the water actually consumed in each sector is controlled by the depletion factor. Depletion factor is defined as a ratio between water that is consumed to water that is supplied. During the optimization process, the depletion factor is allowed to vary between 0.4 and 0.75. Water not depleted is returned to the flow going out of the basin. Some of the irrigation water helps in recharging the groundwater and is accounted for in the model. The 'water demand' in the module is divided into four sectors: domestic, industrial, livestock and irrigation water demand. Domestic and industrial water demands are a function of population and per capita income. The relationships between domestic and industrial water demands and population and per capita income were developed by IFPRI by regression analysis of historic data. Base-year data are used to calculate the intercepts. The livestock water demand is dependent upon the number of livestock multiplied by the water demand per livestock. The agriculture water demand is irrigation water demand, which is calculated as the difference between the crop potential evapotranspiration and the effective rainfall in the irrigated harvested area. Some of the constraints such as environmental flow requirements and flow committed due to treaties are also considered.

Once the supply and consumptive demand of water is calculated at a river basin for each month, an optimization routine is run to maximize the ratio of depletive supply (i.e. the available water that is used to meet the consumptive demand) over consumptive water demand. The range of this ratio is from 0 to 1; 0 implies that no demand is met and 1 implies that all the demand is met. The routine is based on a reservoir operation model, in which the goal is to maximize the reservoir yield while also trying to maximize the storage. The optimization routine also attempts to maximize surface storage while meeting the environmental flow and any committed flow constraints. If for some month the water available is less than the total water demand, the depletive supply is partitioned between different sectors either proportional to each sector's demand or on priority basis (i.e. highest priority to domestic, then industrial, then livestock and finally to agriculture). Within the agriculture sector, the scarce water is further allocated to crops based on the profitability, sensitivity to water stress and net irrigation demand of the crop. The reduction in available water for crops leads to reduction in crop area and also reduction in yields.

The modified crop area and yields from the water supply and demand module provides the total supply of crop commodities for a year. The supply of commodities is effectively harvested area of crops multiplied by the yields of the crop. In the case of livestock, it is the number of livestock multiplied by the yields per livestock. The area and yields of each commodity that is calculated in the 'water demand and supply' module is fed into the 'food demand and supply' module of WATERSIM.

For the 'food demand and supply', the globe is divided into 115 economic regions. The model is based on the concept of partial equilibrium and connects the regions through trade. The equilibrium in trade is reached at annual basis. Demand is a function of price, income and population and the supply is influenced by prices and income. The difference between supply and demand generates excess supply or demand, which is aggregated at global level. This helps determine the world market clearing price, i.e. the equilibrium world price at which the total amount of imports is equal to the total amount of exports. This module also determines the new yields and harvested areas as influenced by price and income, which feed back into the water supply and demand module.

Multiple iterations are done between the 'water supply and demand' module and the 'food supply and demand' module for each year till equilibrium is developed. After the model has reached the equilibrium, the final supply and demand of commodities is considered. The working unit of WATERSIM is called a Food Producing Unit (FPU), which is the intersection of the 125 river basins and 115 economic zones. There are 281 FPUs in WATERSIM.

3.6 What the Future Holds

3.6.1 Water resource implications

WATERSIM calculates the consumptive water demand for each of the sectors. Consumptive water demand is the demand of actual water required for consumption, i.e. water that is lost by evaporation and/or transpiration or degraded in quality as to not be useful without substantial treatment. In the agricultural sector, the consumptive water can be classified in two groups, blue water and green water. Blue water is the water that is extracted from surface or groundwater storage whereas green water is the water that is retained in the soil (as soil moisture) due to precipitation. Thus blue water refers to water supplied by irrigation. Figure 3.3a shows consumptive water demand for the three scenarios, for different regions of the world, considering consistent climatic conditions. Figure 3.3b shows the same demand but divided by the sectors. As per the model, the consumptive water demand at global scale will increase from about 2400 km^3 in 2010 to about 5250 km^3 in 2050 for the BAU scenario. For the OPT socio-economic scenario, the total consumptive water demand increases to 7230 km^3 by 2050 whereas for the PES scenario, it is as little as 3820 km^3. In the year 2010, the percentage of consumptive water demand for agriculture is about 72% of the total demand. Due to increase in population and GDP, this demands goes down to 37% in BAU, 27% in OPT socio-economic condition and 50% in the case of PES socio-economic condition. Thus the greatest increase in demand for consumptive water comes from the domestic and industrial sectors. Even among the two sectors, the demand for the industrial sector outpaces the demand from the other sectors. IWMI's Comprehensive Assessment Report (IWMI, 2007) had predicted that the non-agriculture withdrawal of water is expected to increase by a factor of 2.2 from 2000 to 2050. At this stage, the model does not consider increase in water use efficiency in the domestic and industrial sectors. In reality, as the economics of the countries improve, there would be a much greater improvement in their water use efficiency.

In this model run, the increase in agriculture area was not based on historic trends. Instead, the change in the agriculture area is because of the changes in the prices of crops in the 'food supply and demand' module. In this case the model assumes perfect trade conditions and does not consider any trade barriers. The harvested area of the crops changes from 1050 million ha to 1220 million ha, which is an increase of about 16%. It is to be noted that this is an increase in harvested area and not an increase in agriculture area (which should be less than this). Consequently, the green water demand also goes up from 2010 to 2050. The green water usage (i.e. effective rainfall used in rainfed and irrigated crops) for 2010 is about 4580 km^3 and it increases to 5450 km^3 by 2050. This increase in green water usage is due to the increase in the harvested area for the time period.

Figure 3.4 shows the sectorial water demand for the BAU scenario. There is a big variation in regions. While in OECD countries the agriculture water demand drops from less than 40% to less than 20%, in MNA countries it still constitutes close to 90% of total water demand till 2050. In South Asia, the agriculture water demand reduces from 90% to about 70% by the year 2050. For all these regions, the trend is towards higher water demand for the industrial and domestic sectors, which would lead to higher competition with agriculture demand.

The environmental demand for the regions is calculated based on the

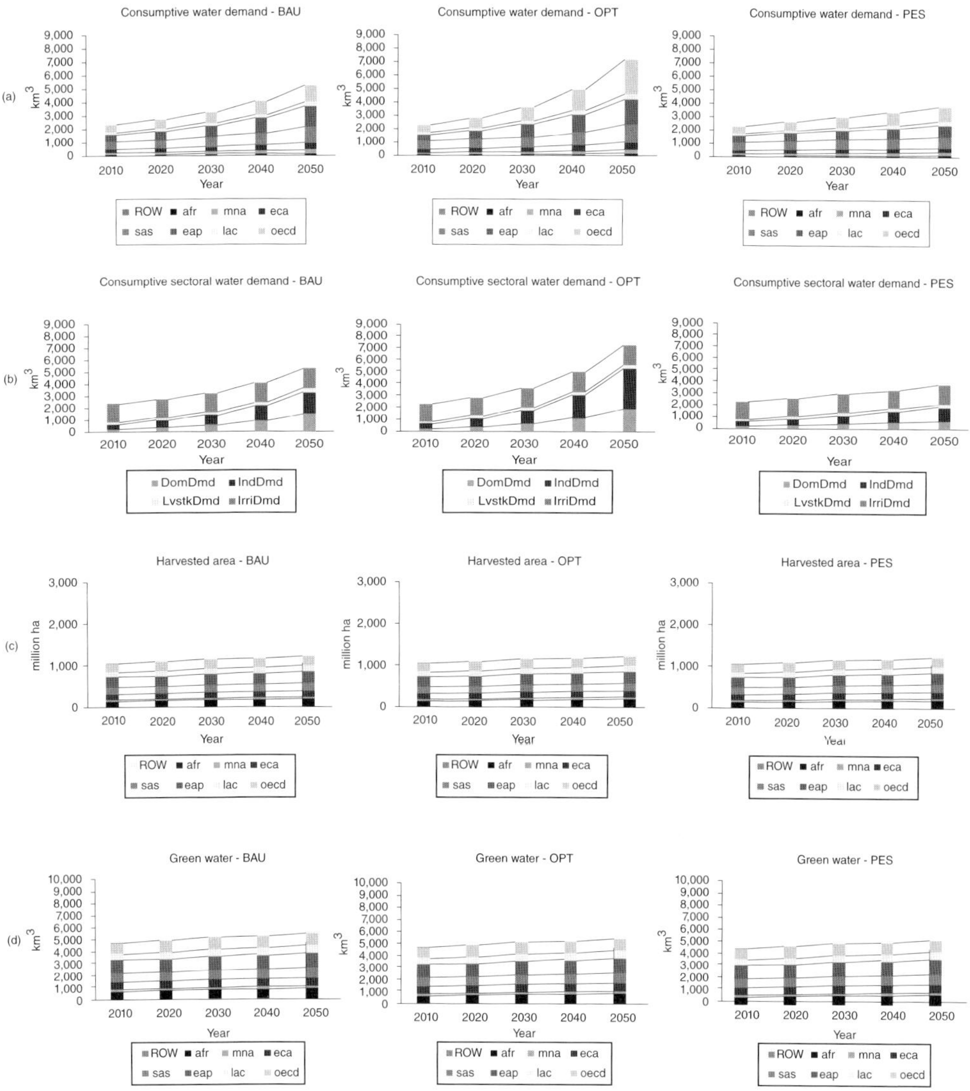

Fig. 3.3. WATERSIM outputs for baseline climate scenarios for three socio-economic conditions: business as usual (BAU), optimistic (OPT) and pessimistic (PES) scenarios. (a) Consumptive water demand by regions; (b) consumptive water demand by sectors; (c) harvested area; and (d) green water. ■ ROW, rest of world; ■ afr, sub-Saharan Africa; ■ mna, Middle East and North Africa; ■ eca, Eastern Europe and Central Asia; ■ sas, South Asia; ■ eap, East Asia and Pacific; ■ lac, Latin America and Caribbean; ■ OECD, Organisation for Economic Co-operation and Development; ■ DomDmd, domestic demand; ■ IndDMD, industry demand; ■ LvstkDmd, livestock demand; ■ IrriDmd, irrigation demand.

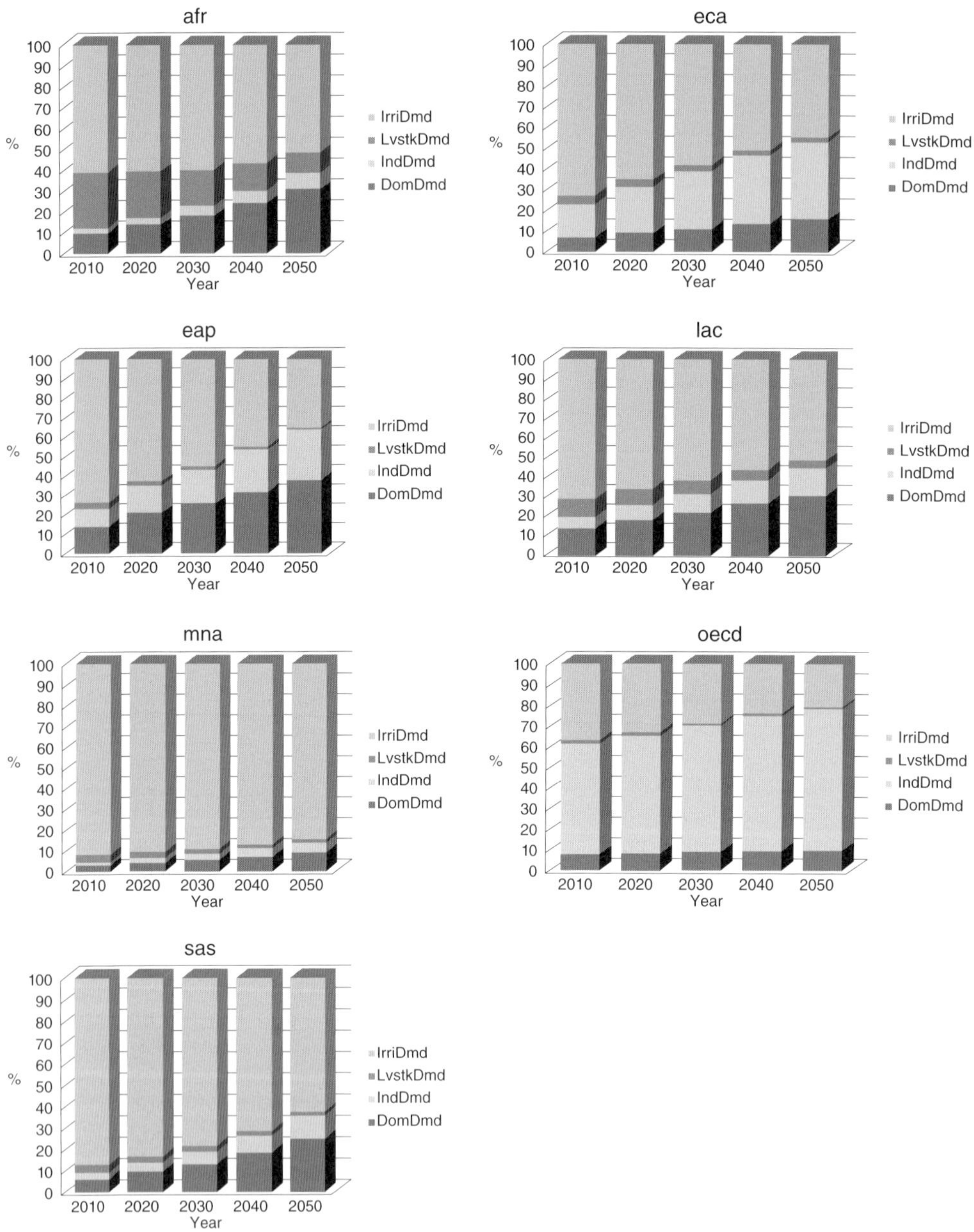

Fig. 3.4. Sectorial breakdown of consumptive water demand for the BAU scenario till 2050 (for abbreviations, see Fig. 3.3).

environmental flow requirement as per Smakhtin *et al.* (2004). Figure 3.5 shows environmental demand for different regions in the world. It varies from about 20% for the 'Rest of the World' to about 33% for ECA region. The highest environment demand of about 3500 km^3 is for the LAC region and the least (about 40 km^3) for the MNA region.

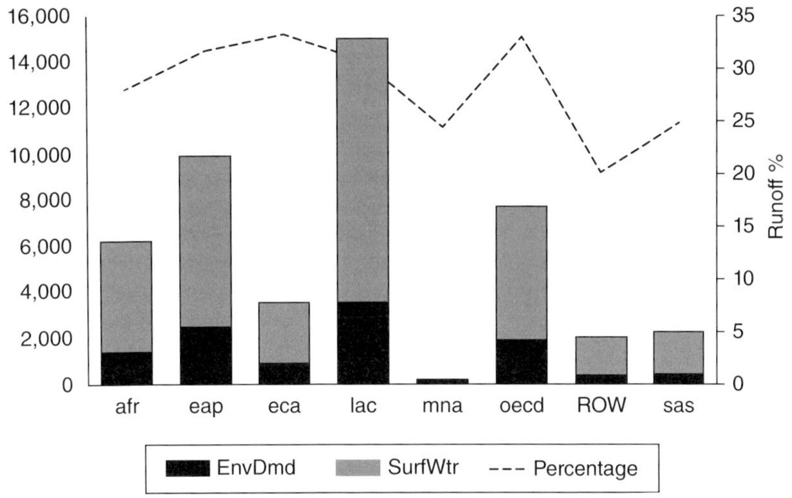

Fig. 3.5. Environmental demand at regional level (in absolute values and in percentage of the total surface runoff (EnvDmd, environmental demand; SurfWtr, surface water; for other abbreviations, see Fig. 3.3).

3.6.2 Climate change scenarios

WATERSIM was then run for SRESA2 (A2) and SRESB1 (B1) climate change scenarios. Figure 3.6 shows total rainfall, potential evapotranspiration (PET) and effective precipitation (i.e. rainfall that is converted to ET by the crops), and difference between PET and effective precipitation at the global scale for A2 and B1 scenarios (using average of four GCMs as discussed above) as compared to the baseline.

The analysis shows that at the global scale, there is no clear trend in the total rainfall. On the other hand, PET, which is dependent upon the temperature, increases, with a much sharper increase after 2040. This leads to higher effective precipitation (i.e. increased water consumption by crops, which depends both on increased precipitation and higher temperature). But if the difference between PET and effective precipitation (which represents the shortfall in water requirement for agriculture, both rainfed and irrigated) is considered, it is much higher during the A2 and B1 climate scenarios as compared to the baseline. Also A2 climate change scenario has a greater shortfall than the B1 climate change scenario. At the regional level (shown in Fig. 3.7), Africa, East Asia and Pacific, Eastern Europe and Central Asia show lower rainfall till 2020 and then increase till the mid-2040s before decreasing again. The Middle East and North Africa receive lower rainfall whereas there are no discernible trends for OECD and South Asia. The difference between PET and effective precipitation is above the baseline in most of the regions, which indicates a greater demand for irrigation water in future scenarios.

Based on the analysis done using WATERSIM, for the year 2050, for the irrigated area, the gap between PET and effective rainfall will be about 19% higher than the baseline for the A2 climate change scenario, whereas it will be about 16% higher for B1 climate change scenario. This will put extra stress on demand for irrigation water. The changes in the cropping pattern or changes in the length of cropping season have not been considered in this study as that may impact the actual requirement for irrigated water. At regional level, the gap between PET and effective rainfall due to climate change (on average, from 2010 to 2050) is shown in Table 3.1.

The climate change will have a larger impact on water demand for agriculture in Africa and the OECD regions and lesser impact in Eastern Europe and Central Asia.

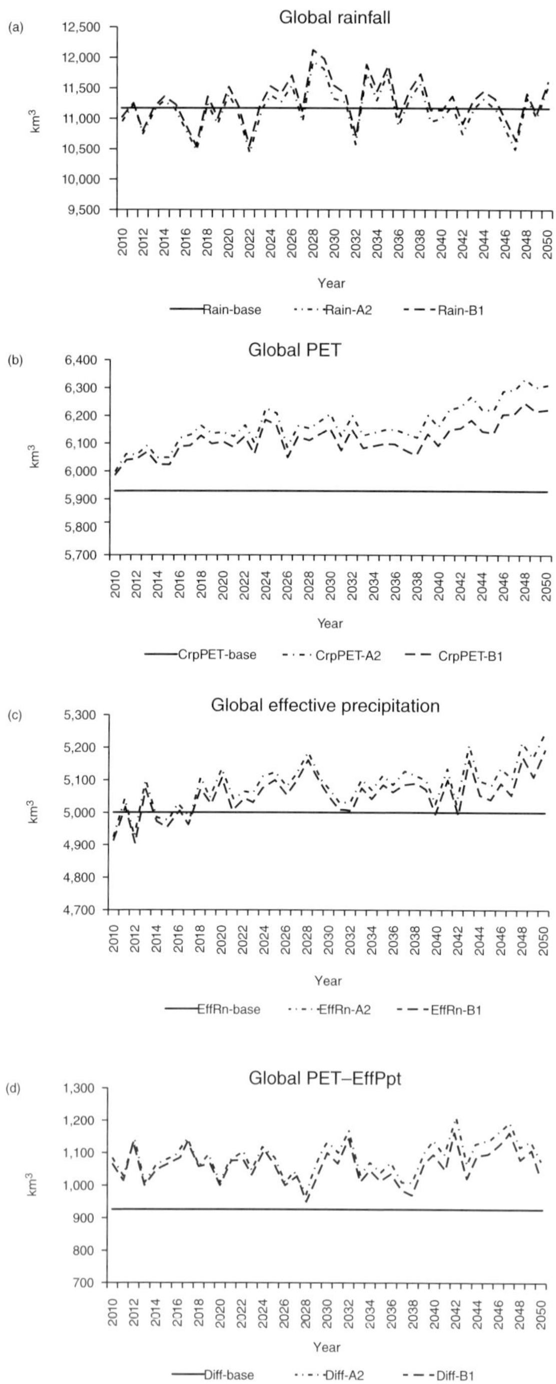

Fig. 3.6. Changes in (a) rainfall, (b) PET, (c) effective precipitation, and (d) difference between PET and effective precipitation at global scale.

3.7 Conclusions

After the first green revolution in the 1960s, the world became complacent with the progress in the agriculture sector. They were encouraged by the falling food commodity prices and improving food consumption of the global population. However, since the beginning of the 21st century, the trends in food prices have reversed. Since then, many studies have been conducted by renowned organizations from around the world to look

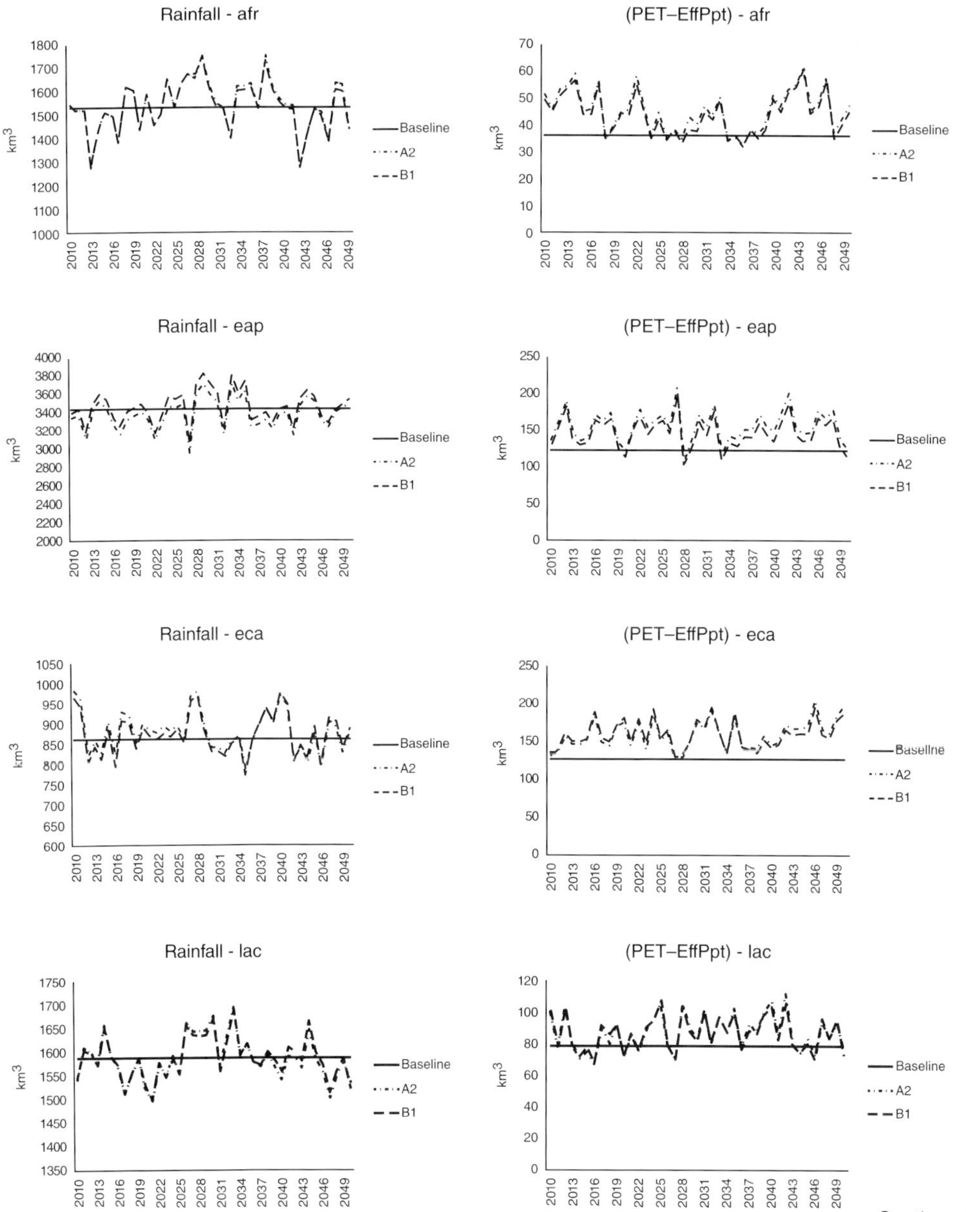

Continued

Fig. 3.7. Changes in rainfall and difference between PET and effective precipitation at regional level (for abbreviations, see Fig. 3.3).

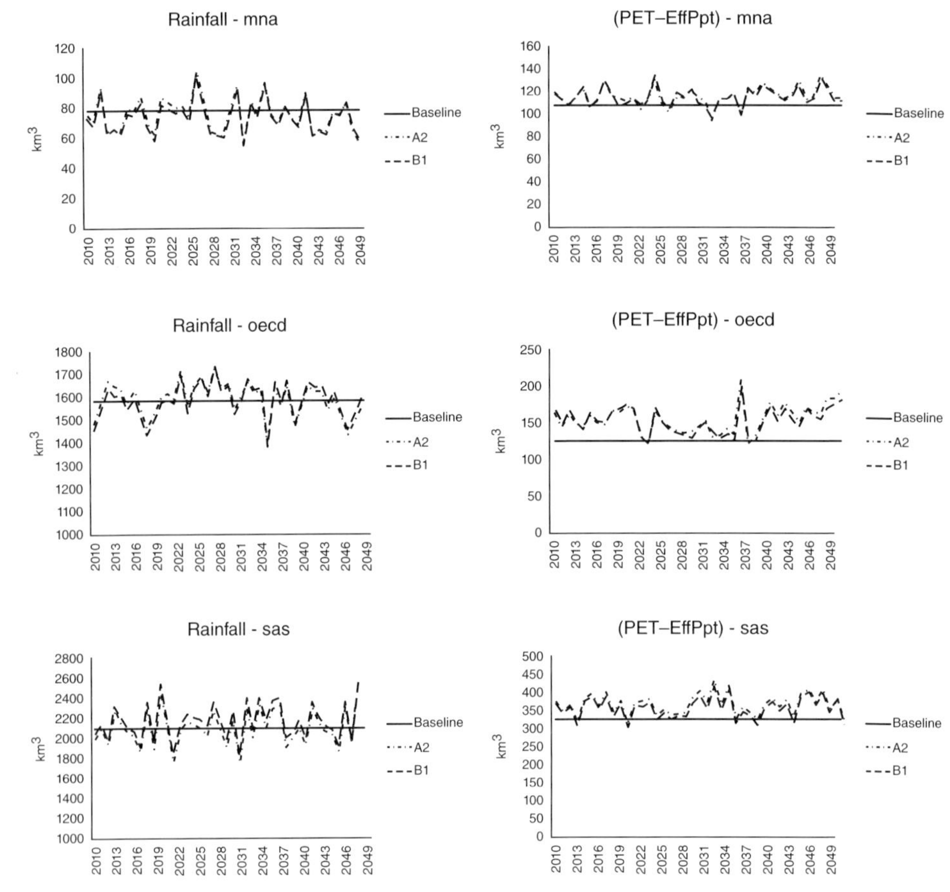

Fig. 3.7. Continued.

Table 3.1. Percentage change (from baseline) in the gap between PET and effective rainfall in 2050.

Regions	A2 (% increase)	B1 (% increase)
Sub-Saharan Africa	25.07	21.64
East Asia and Pacific	26.58	19.34
Eastern Europe and Central Asia	26.03	25.72
Latin America and Caribbean	11.04	10.70
Middle East and North Africa	7.52	6.89
OECD	25.30	23.33
South Asia	10.58	7.72

at future food and natural resources scenarios. The common message that comes out of all these studies is that with proper management of food cycles, technical innovation and the wise use of natural resources, the future demand for the population can be met, which includes reducing malnourishment. There is a large scope for improving yields in rainfed agriculture and in reducing waste within the food cycle. The globalization of food trade will also help to mitigate some of the shortfalls in regions with increasing populations.

Water is a critical resource that will be impacted by the increasing population, GDP and climate change. The water demand for the industrial and domestic sector will increase (without considering the efficiency) due to increase in population and per capita income. This would lead to higher competition with the agriculture sector from the other sectors. The results emphasize the growing demand that will come from industrial and domestic users, particularly under an 'optimistic' scenario of limited population growth but significantly increasing GDP per capita. Climate change scenarios suggest that if we are to maintain optimum yields, between 16% and 19% more irrigation water will be required on average due to higher evaporative demand. This estimate is more conservative as the growing seasons of the crops may shorten, which has not been considered in this research. The impact of climate change may become much more dominant after 2050, although most of the current studies have only focused till 2050.

Acknowledgements

The hydrological data at the global scale for the analysis was provided by the Potsdam Institute for Climate Impact Research (PIK). The population and the GDP data were provided by the International Food Policy Research Institute (IFPRI).

Notes

1 Commonwealth Scientific and Industrial Research Organization.

2 Medium Resolution General Circulation Model.

3 MPI-ECHAM5: Max Planck Institute for Meteorology European Centre for Medium-Range Weather Forecasts (ECMWF) Hamburg; MIROC3.2: Center for Climate System Research (The University of Tokyo), National Institute for Environmental Studies, and Frontier Research Center for Global Change; CSIRO Mk 3.0: The Commonwealth Scientific and Industrial Research Organisation; CNRM-CM3: Center National Weather Research.

References

FAO (2003) *World Agriculture: Towards 2015/2030.* Food and Agriculture Organization of the United Nations, Rome.

FAO (2006) *World Agriculture: Towards 2030/2050.* Interim Report. Global Perspective Studies Unit, Food and Agriculture Organization of the United Nations, Rome.

FAO (2012) *World Agriculture: Towards 2030/2050: The 2012 Revision.* Food and Agriculture Organization of the United Nations, Rome.

Hazell, P. and Wood, S. (2008) Drivers of change in global agriculture. *Philosophical Transactions of the Royal Society B* 363, 495–515.

INRA and CIRAD (2009) Paillard, S., Treyer, S. and Dorin, D. (eds) *Agrimonde – Scenarios and Challenges for Feeding the World in 2050.* INRA/CIRAD, Paris.

IWMI (2007) *Water for Food Water for Life: A Comprehensive Assessment of Water Management in Agriculture.* Earthscan, London, UK and International Water Management Institute (IWMI), Colombo, Sri Lanka.

Lundqvist, J. (2010) Producing more or Wasting Less. Bracing the food security challenge of unpredictable rainfall. In: Martínez-Cortina, L., Garrido, G. and López-Gunn, L. (eds) *Rethinking Water and Food Security: Fourth Marcelino Botín Foundation Water Workshop.* Taylor & Francis, London.

Molden, D. (1997) *Accounting for Water Use and Productivity.* SWIM Paper 1. International Irrigation Management Institute (IIMI), Colombo, Sri Lanka.

Nelson, G.C., Rosegrant, M.W., Palazzo, A., Gray, I., Ingersoll, C., Robertson, R., Tokgoz, S., Zhu, T., Sulser, T.B., Ringler, C., Msangi, S. and You, L. (2010) *Food Security, Farming, and Climate Change to 2050: Scenarios, Results, Policy Options.* International Food Policy Research Institute (IFPRI), Washington, DC.

OECD (2012) *OECD Environmental Outlook to 2050: The Consequences of Inaction.* OECD Publishing, Paris.

OFID (2009) *Biofuels and Food Security. Final Report to Sponsor: The OPEC Fund for International Development (OFID), Vienna, Austria (March 2009).* International Institute for Applied Systems Analysis (IIASA), Laxenburg, Austria. DOI: 10.1111/j.1467-9353.2008.00425.x.

Rosegrant, M.W., Cai, X. and Cline, S.A. (2002) *World Water and Food to 2025: Dealing with Scarcity.* International Food Policy Research Institute (IFPRI), Washington, DC.

Smakhtin, V., Revenga, C. and Döll, P. (2004) A pilot global assessment of environmental water requirements and scarcity. *Water International* 29(3), 307–317.

UN (2010) *The Millennium Development Goals Report 2010.* Department of Economic and Social Affairs of the United Nations Secretariat, New York.

USDA-SCS (United States Department of Agriculture, Soil Conservation Service) (1967) *Irrigation Water Requirement.* Technical Release No. 21. USDA, Washington, DC.

4 Impacts of Climate Change on Crop Water Requirements in Huang-Huai-Hai Plain, China

Qin Liu,[1,2] Changrong Yan,[1,2*] Jianying Yang,[1,2] Xu-rong Mei,[1,2] Weiping Hao[1,2] and Hui Ju[1,3]

[1]*Institute of Environment and Sustainable Development in Agriculture, Chinese Academy of Agricultural Sciences, Beijing;* [2]*Key Laboratory of Dryland Agriculture, Ministry of Agriculture, Beijing;* [3]*Key Laboratory of Agricultural Environment, Ministry of Agriculture, Beijing, China*

Abstract

Climate change will have important implications in the agriculture of water-short regions, such as Huang-Huai-Hai Plain (3H Plain), where expected warmer and drier conditions might augment crop water requirement (ET_c). To evaluate the effect of climate change, a data set consisting of observed daily values of air temperature, relative humidity, sunshine duration and wind speed from five selected weather stations in the 3H Plain and covering the period 1981–2009 was used for estimating reference evapotranspiration (ET_0). ET_0 was calculated using FAO-56 Penman–Monteith equation; then sensitivity coefficient of ET_c of major climatic variables and regional responses of precipitation deficit to climate change were conducted in the 3H Plain. The results showed that a clear drop in solar radiation (SR) was detected and temperatures increased, especially the minimum temperature. Wind speed (WS) decreased significantly in most of the stations, especially from seeding to jointing stages for wheat. No significant change was detected for relative humidity (RH) in 1981–2009. Temperature was the most sensitive variable in general for the plain, followed by SR, WS and RH. The decrease of sensitivity coefficient of solar radiation (S_{SR}) mainly occurred in seeding to jointing stages and heading to maturity stages of winter wheat. Sensitivity coefficient of temperatures (S_T) increased in Beijing, Xinxiang, Xuzhou and Yanzhou stations, which means that an increase in temperatures may lead to a larger increase of ET_c in these four stations. However, S_T decreased in the Shijiazhuang station. With the decrease of WS, ET_c will decrease due to the positive coefficient in Beijing and Xinxiang stations. Trends of S_{RH} showed no significant changes in the time series analysis. A positive relationship was detected between precipitation deficit and relative humidity, and the latter was considered the most correlative factor for precipitation deficit.

4.1 Introduction

Climate change with the characteristic of global warming has become a hotspot of research in the field of water resources, agriculture, ecology and other disciplines. According to the IPCC report, in the recent 100 year period (1906–2005) the global temperature has risen by 0.74°C (IPCC, 2007). Climate model projections summarized in the 2007 IPCC report indicate that the global surface temperature is likely to rise further by 1.1–6.4°C during the 21st century.

Agriculture is directly affected by climatic conditions and changes. Changes in

* Corresponding author, e-mail: yanchangrong@caas.cn

climatic elements such as temperature, precipitation, radiation, humidity and WS could have profound implications on hydrologic processes in agriculture (McKenney and Rosenberg, 1993). Therefore it is essential to understand the impact of climate change on agricultural water resources for sustainable agriculture and to study methods to minimize the negative effects caused by such changes. It is also necessary to evaluate effects of climate changes in terms of temperature rise, amounts of SR, WS and RH changes. Xiong *et al.* (2010) reported that there would be insufficient water for agriculture in China in the coming decades, due primarily to increases in water demand for non-agricultural uses, which would have significant implications for adaptation strategies and policies for agricultural production and water management.

Crop water requirement is decided from reference evapotranspiration (ET_0) estimation methods suggested, usually, by Food and Agriculture Organization of the United Nations (FAO). In 1977, FAO presented guidelines for predicting crop water requirements. The guidelines suggested methods to derive crop water requirements and discussed the application of data on crop water requirements in irrigation project planning, design and operation. ET_0 refers to the crop evapotranspiration in the open short grassland where the soil moisture is adequate, the ground is completely covered, and grass grows normally at a similar height (grass height is about 8–15 cm). ET_0 is an integrated climate parameter that gives a measure of the evaporation demand of the air and is essentially dependent on four meteorological variables: air temperature, SR, RH and WS (Allen *et al.*, 1998). One or more of these four meteorological variables can be taken into account, depending on the ET_0 calculation method selected. The main advantage of the Penman–Monteith approach is that it takes into account the most significant variables, so that the influence of each of them to ET_c can also be analysed (Blaney and Criddle, 1952; Hargreaves and Samani, 1982) on physically based equations requiring daily data for temperature and RH of the air, SR and WS (Allen *et al.*, 1998). Several studies have carried out sensitivity analyses of ET_0 to determine meteorological data in different climates (Rana and Katerji, 1998; Goyal, 2004; Irmak *et al.*, 2006), but they were restricted to a single station. Furthermore, what has been reported to be the most effective variable detected is WS (Todisco and Vergni, 2008), SR (Gao *et al.*, 2006; Wang *et al.*, 2007) and RH (Gong *et al.*, 2006) in other papers; however, these studies were almost restricted to monthly, seasonal or annual ET_0. Liu, Y. *et al.* (2010) reported that annual ET_0 and its constituents (ET_{rad} and ET_{aero}) had significantly declined, while the spring ET_{aero} value was the highest across the North China Plain (NCP). Song *et al.* (2009) also reported that for the whole NCP, annual ET_0 showed a statistically significant decrease of 11.92 mm per decade over the 46 years of data collection and that the decreasing net radiation and WS had a bigger impact on ET_0 rates than the increases observed by the maximum and minimum temperatures. However, studies about sensitivity analyses of ET_0 then ET_c during the typical crop growing season and its trend in variation are rarely seen in the 3H Plain.

The objectives of this chapter are: (i) to investigate the trends for ET_c for winter wheat in the 3H Plain in the past; (ii) to evaluate the major factors related to the changes in ET_c; and (iii) to compare the temporal variations of climatology sensitivity coefficients in different stations, in an attempt to understand the relative roles of the main climatic variables.

4.2 Materials and Methods

4.2.1 Study area and climate data

The 3H Plain, one of the largest plains in China, is located in the north of the country and extends from 31°14′ to 40°25′ N and from 112°33′ to 120°17′ E, stretching over an area of about 350,000 km^2. The plain is mainly formed by the alluvial deposits of the Yellow, Huai and Hai rivers. Almost the entire plain is found at an altitude below 50 m above sea level and the slope gradient is less than 3°. The climate is temperate, subhumid, and continental monsoon with a cumulative temperature (>0°C) of 4200–5500°C, a frost-free period of about 170–220 days and average annual precipitation ranging between 500 and 800 mm (Ren *et al.*, 2008). The annual rainfall concentrates in the summer period, from July to September. On the other hand, it is characterized by a lack of water for agricultural production in winter. Although precipitation is insufficient for cultivation in the 3H Plain, it is the largest agricultural production area in China, accounting for around 50% of the wheat (Wang *et al.*, 2009). Li *et al.* (2010) reported that the 3H Plain provided 42.3% of the total national winter wheat and summer maize production with intensive management characterized by the application of sufficient irrigation water and fertilizers.

Data from five weather stations provided by China Meteorological Administration (CMA) were used in this chapter (Fig. 4.1). Daily data observed from 1981 to 2009 on maximum temperature (T_{max}) and minimum air temperature (T_{min}), precipitation (P), wind speed (WS) measured at 10 m height, with average relative humidity (RH) and daily sunshine duration (representing SR) were available. The weather stations were selected by the two following criteria: (i) the spatial distribution had to guarantee such a coverage that could be representative of irrigated lands in the 3H Plain; and (ii) the time series had to be long enough to obtain statistically significant results in trend analyses.

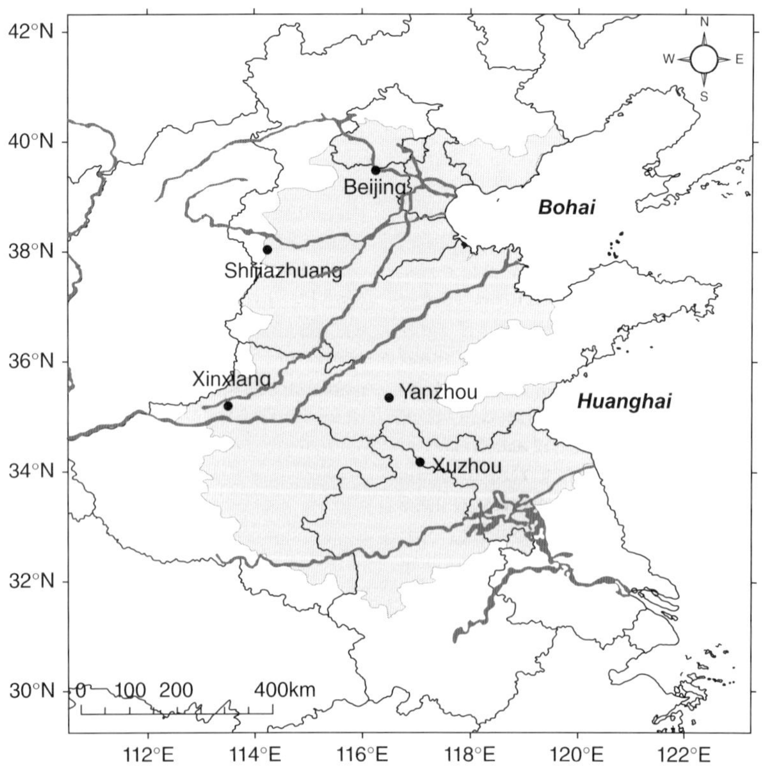

Fig. 4.1. Location of the weather stations selected in this chapter.

Table 4.1. Annual variation tendency and statistics of the five phenological phases for winter wheat at the five selected stations.

Name of station	Phenological phase	Maximum value	Average value	Minimum value	Slope[a]
Beijing	Sowing date	287	275	262	0.57**
	Seeding date	297	268	285	0.64**
	Jointing date	126	104	112	−0.33**
	Heading date	140	123	131	−0.24**
	Maturity date	174	159	168	−0.20**
Shijiazhuang	Sowing date	288	278	271	0.09
	Seeding date	300	286	277	0.17
	Jointing date	108	99	87	−0.24**
	Heading date	129	120	112	−0.26**
	Maturity date	164	160	156	0.05
Yanzhou	Sowing date	299	281	272	0.12
	Seeding date	308	288	280	0.14
	Jointing date	100	90	78	−0.42**
	Heading date	124	114	101	−0.31**
	Maturity date	165	156	150	−0.08
Xinxiang	Sowing date	293	284	276	0.04
	Seeding date	303	292	283	0.10
	Jointing date	102	87	75	−0.55**
	Heading date	123	113	102	−0.33**
	Maturity date	158	152	147	−0.12
Xuzhou	Sowing date	297	285	276	0.17
	Seeding date	308	293	281	0.19
	Jointing date	99	82	66	−0.64**
	Heading date	121	113	102	−0.34**
	Maturity date	162	155	149	−0.21**

[a]Linear coefficients significant at *$p<0.05$, **$p<0.01$ and ***$p<0.001$.

4.2.2 Phenological data

Data on the date of sowing, seeding, jointing, heading and maturity of winter wheat (representing Julian Day) were provided by CMA for the period 1981–2009. As described in Table 4.1, information on the latest period for sowing, seeding, jointing, heading and maturity dates of winter wheat are found in Yanzhou and Beijing stations, respectively. Similarly, information on the earliest period for these five dates is found in Beijing, Shijiazhuang, Xuzhou, Yanzhou and Xinxiang stations. Sowing and seeding dates in the five stations are observed to be delayed with the biggest amplitude of 0.57 and 0.64 days year^{-1} in the Beijing Station. However, jointing, heading and maturity dates are nearly observed to advance with the biggest amplitude of 0.64, 0.34 and 0.21 days year^{-1} in the Xuzhou Station.

4.2.3 Crop coefficient approach

Crop coefficients (K_c) vary for different crops and the growing stage of the crop. In this chapter, coefficient of winter wheat was developed by FAO (Table 4.2), and the coefficient of daily winter wheat can be obtained in relation to climatic conditions for phenological phases for winter wheat (Allen *et al.*, 1998).

Table 4.2. Crop coefficients of winter wheat in different phases.

Phenological phase	K_c
K_{cini}	0.70
K_{cmid}	1.15
K_{cend}	0.40

4.2.4 Calculation of precipitation deficit by the FAO Penman–Monteith method

In this chapter, the precipitation deficit is defined as the difference between effective precipitation and the crop water requirement in the duration of the whole crop and during the four different stages i of winter wheat.

$$PD = P_e - ET_c \tag{4.1}$$

where PD is the precipitation deficit (mm), P_e is the effective precipitation (mm) and ET_c is the crop water requirement (mm).

Effective precipitation is defined as the total precipitation minus deep percolation, runoff and evaporation. The method of effective precipitation analysed in this chapter is from the US Department of Agriculture Soil Conservation Service. It is one of the most popular methods to calculate effective precipitation and has proven to be effective in many research studies (Smith, 1992; Döll, 2002).

$$\begin{aligned} &P_e = P \times (4.17 - 0.2 \times P) / 4.17 \\ &\quad P < 8.3\ \text{mm day}^{-1} \\ &P_e = 4.17 + 0.1 \times P \quad P \geq 8.3\ \text{mm day}^{-1} \end{aligned} \tag{4.2}$$

where P is precipitation detected in weather station.

ET_0 is the evapotranspiration from disease-free, well-fertilized crops grown in large fields under optimum soil water conditions and achieving full production under the given climatic conditions. The crop water requirement can be calculated from climatic data and by directly combining the crop resistance, albedo and air resistance factors under standard conditions (Allen *et al.*, 1998).

$$ET_c = k_c \times ET_0 \tag{4.3}$$

where ET_c is the crop water requirement (mm), k_c is the crop coefficient and ET_0 is the reference crop evapotranspiration (mm).

The FAO Penman–Monteith method to estimate ET_0 can be derived from the original Penman–Monteith equation and the equations of the aerodynamic and surface resistance (Allen *et al.*, 1998).

$$ET_0 = \frac{0.408\Delta(R_n - G) + \gamma \frac{900}{T + 273} U_2 (e_s - e_a)}{\Delta + \gamma(1 + 0.34U_2)} \tag{4.4}$$

where, ET_0 is reference evapotranspiration (mm day^{-1}), Δ is the slope of the saturation vapour pressure curve (kPa °C^{-1}), R_n is the net radiation at the surface (MJ m^{-2} day^{-1}), G is the soil heat flux (MJ m^{-2} day^{-1}), $R_n - G$ is the available energy (MJ m^{-2} day^{-1}), γ the psychrometric constant (kPa °C^{-1}), T is the mean temperature at 2 m height (°C), U_2 is the mean daily wind speed at 2 m height (m s^{-1}), e_s is the saturation vapour pressure (kPa), e_a is the actual vapour pressure (kPa) and $e_s - e_a$ is the vapour pressure deficit (kPa). The computation of all data required for calculating daily reference evapotranspiration followed the procedures in the FAO irrigation paper 56 (Allen *et al.*, 1998). To obtain the total monthly evapotranspiration, Eqn 4.4 was multiplied by the number of days in a given month.

All the above variables can be calculated from daily meteorological observation data. For the calculation of ET_0, wind speed measured at 2 m above the ground surface is required. It can be converted from the normal measurement at 10 m WS based on the equation given by the FAO Penman–Monteith method (Allen *et al.*, 1998) as follows:

$$U_2 = \frac{U_z \times 4.87}{\log_e(67.8 \times 10 - 5.42)} \tag{4.5}$$

where, U_2 is the wind speed at 2 m above ground surface (m s^{-1}), U_z is the measured wind speed at 10 m above ground surface (m s^{-1}).

4.2.5 Sensitivity analysis of crop water requirement to meteorological variables

In order to evaluate the effect of meteorological parameters on ET_c, a sensitivity analysis is performed to find more sensitive parameters.

For a general definition of sensitivity, the variable V is considered, which is a function of the input variables $x_1, x_2, x_3 \ldots x_n$:

$$V = f(x_1, \ldots, x_n) \tag{4.6}$$

If the variables $x_1, x_2, x_3 \ldots x_n$ are independent of V, it may be written:

$$V+\Delta V=f(x_1+\Delta x_1,\ldots,x_n+\Delta x_n) \quad (4.7)$$

Neglecting higher-order terms, from a Taylor series expansion we have:

$$\Delta V+\frac{\partial V}{\partial X_1}\Delta X_1+\ldots+\frac{\partial V}{\partial X_n}\Delta X_n \quad (4.8)$$

By definition, the partial differentials, $\frac{\partial V}{\partial X_i}$, are the sensitivities and SX_i is the dependent variable V of the independent input variable X_i (McCuen, 1974; Saxton, 1975; Beven, 1979; McCuen and Beighley, 2003). They denote the change in V per unit change in X_i.

From Eqn 4.8 we have:

$$SX_i=\frac{\partial V}{\partial X_i}=\frac{\Delta V}{\Delta X_i} \quad (4.9)$$

which shows that SX_i may be obtained by directly calculating the value of the partial differential, or by applying a step change in X_i, while leaving the variables other than X_i constant. Here, SX_i may be sensitive to the relative magnitude of V and X_i. Therefore, SX_i may be divided by the ratio V/X_i, which leads to the relative sensitivity or sensitivity coefficient RSX_i:

$$RSX_i=\frac{\partial VX_i}{\partial X_i V} \quad (4.10)$$

Now, the relative change in V can be expressed as Eqn (4.11) (Saxton, 1975), which shows that the relative sensitivity coefficient denotes the part of the relative change in X_i that is transferred to the relative change in V. If, for example, $RSX_i=25\%$, a 10% change in X_i will result in a 2.5% change in V.

$$\frac{\Delta V}{V}=RSX_1\frac{\Delta X_1}{X_1}+\ldots+RSX_n\frac{\Delta X_n}{X_n} \quad (4.11)$$

4.2.6 Time series analysis method

In order to understand the temporal variation of the climate data, the linear trend and the associated periods were analysed by a linear fitted model.

The least-squares linear model is the most common method used for statistical diagnosis in modern climatic analysis studies (Zeng and Heilman, 1997; Liu, Y. *et al.*, 2010), and is a fundamental technology to forecast in modern climate analysis studies. The linear trend was chosen because of being the simplest model for an unknown trend. The level of adequacy of the model fitted was measured by the percentage of variance explained by it. Linear trends for the series of annual total precipitation were calculated by the least squares regression. The estimated slopes were tested against the hypothesis of null slope by means of a two-tailed T-test at a confidence level of 95% (Serrano *et al.*, 1999).

A series $y_1, y_2 \ldots y_i \ldots y_n$, can be expressed by the polynomial:

$$\hat{y}_n(t)=a_0+a_1t+\ldots+a_mt^n \quad (4.12)$$

where t is year. Generally, the linear trend of a time series can be estimated by the least square method and can be expressed by the linear regression equation as:

$$\hat{y}_n(t)=a_0+a_1t \quad (4.13)$$

where, the slope a_1 is the estimated trend.

4.3 Results

4.3.1 Trends and persistence of crop water requirement in different phenological phases

In order to appreciate the temporal variability of ET_c changes and their magnitudes, ET_c series of phenological phases at each station were tested. As described in Table 4.3, the mean value of ET_c in the whole growth stage is from 391.9 to 462.2 mm, with the minimum value detected in the Yanzhou station and maximum value in the Beijing station. The ET_c in the four stations showed a negative trend, with ET_c decreasing significantly only in the Yanzhou station. However, there is also a significant increasing trend found in the analysis, such as ET_c in the Xinxiang station, in which station was detected a significantly increasing trend in the whole growth stage of winter wheat, with a slope of 1.57 mm year^{-1}.

ET_c in seeding to jointing stages was detected with a significantly decreasing trend in the stations selected, and the minimum value, that is 1.81 mm for the slope, was found in the Beijing station. However, in the heading to maturity stages of winter wheat, the ET_c increased in the time series. The maximum value for slope, which is 1.61 mm, was detected in the Xinxiang station. In the jointing to heading stages of winter wheat, no significant trend was detected in this time series.

4.3.2 Inter-annual variation of relative moisture index

Relative moisture index is one of the common drought indexes, such as Z (Ju *et al.*, 1998), BMDI (Bhalme and Mooley, 1980) and PDSI (Szinell *et al.*, 1998), to evaluate dry condition in the 3H Plain (Li *et al.*, 2012). It can be concluded that there were varying degrees of drought in sowing to seeding, seeding to jointing, jointing to heading, heading to maturity and sowing to maturity for winter wheat from the characteristics and trends of relative moisture index of the 3H Plain (Table 4.4). According to the average value of five stations, each relative moisture index was less than –0.4, which means that sowing to seeding and heading to maturity are in light drought condition, and that sowing to maturity, seeding to jointing, and jointing to heading are in moderate drought condition. As for each station, a wet condition is detected in sowing to seeding stages in Xinxiang, Xuzhou and Yanzhou stations, and a serious drought condition is detected during seeding to jointing stages in Beijing and during jointing to heading stages in the Shijiazhuang station. As described, a wet trend was observed during jointing to heading

Table 4.3. Long-term trends over the entire studied period of crop water requirement (ET_c, mm) in different phenological phases for the selected weather stations.

	Sowing to maturity		Sowing to seeding		Seeding to jointing		Jointing to heading		Heading to maturity	
Station	Mean	Slope	Mean	Slope	Mean	Slope	Mean	Slope	Mean	Slope
Beijing	462.24	–0.67	8.42	–0.04	278.29	–1.81**	79.78	0.50	95.74	0.69*
Shijiazhuang	404.69	–0.44	6.01	0.05	194.68	–1.04*	89.07	–0.42	114.92	0.96*
Xinxiang	391.93	1.57**	6.09	0.09**	166.09	–0.26**	92.17	0.13	127.58	1.61**
Xuzhou	391.99	–0.59	6.85	0.01	152.13	–1.59**	99.56	0.05	133.44	0.93*
Yanzhou	396.00	–1.11*	6.10	–0.01	171.12	–1.31**	86.43	–0.35	132.34	0.57

* and ** represent linear coefficients significant at $p<0.05$ and $p<0.01$, respectively.

Table 4.4. The characteristics and trends of relative moisture index (M) in the selected weather stations.

	Sowing to maturity		Sowing to seeding		Seeding to jointing		Jointing to heading		Heading to maturity	
Station	Mean	Slope (/10 yr)	Mean	Slope (/10 yr)	Mean	Slope (/10 yr)	Mean	Slope (/10 yr)	Mean	Slope (/10 yr)
Beijing	–0.797	0.014	–0.766	0.000	–0.854	0.000	–0.743	0.071	–0.713	0.01
Shijiazhuang	–0.746	0.012	–0.610	0.185	–0.774	–0.003	–0.807	0.033	–0.674	0.000
Xinxiang	–0.697	–0.037	–0.311	–0.329	–0.738	–0.018	–0.791	0.085*	–0.598	–0.111
Xuzhou	–0.498	–0.011	–0.137	–0.833*	–0.436	0.050	–0.653	0.023	–0.467	–0.040
Yanzhou	–0.644	0.025	–0.264	0.208	–0.664	–0.003	–0.712	0.181**	–0.599	–0.025

* and ** represent linear coefficients significant at $p<0.05$ and $p<0.01$, respectively.
Slope is calculated for each 10-year period.

stages in Xinxiang and Yanzhou stations. However, the relative moisture index of winter decreased significantly from sowing to seeding stages in Xuzhou.

4.3.3 Trends and persistence of meteorological variables

Investigation of trends and persistence in historical meteorological dates is helpful in understanding the effect of climate change on crop water requirement in the stations. An analysis of the average climate factors was carried out in different phenological phases for winter wheat (Table 4.5). As described in Table 4.5, the highest P_e is found in Xuzhou station, followed by Yanzhou, Xinxiang, Shijiazhuang and Beijing stations, but no significantly increasing or decreasing trend is detected for every phenological phase in these five stations. Likewise, the highest SR is found in the Beijing station, followed by Yanzhou, Shijiazhuang, Xuzhou and Xinxiang stations, and a decreasing trend was detected in annual SR for the analysed locations in sowing to maturity stages. Furthermore, the average trend was −145.04 MJ m^{-2} 10 year^{-1} with a value range of between −87.64 MJ m^{-2} 10 year^{-1} for the Shijiazhuang station and −225.36 MJ m^{-2} 10 year^{-1} for the Beijing station. The decreased reduction in SR was higher and statistically significant in the regional scale from seeding to jointing stages compared with other stages. However, no significant change was detected in RH except in the Xinxiang station and a significantly decreasing trend was found ($p < 0.01$) for RH in the whole growth stage of winter wheat, and a slope of −1.56. T_{min} in the four stations (except Beijing station) was found to have a significant trend such that it increases. The increase in T_{min} was higher and more statistically significant from seeding to jointing stages, with 0.57°C 10 year^{-1}. Compared to T_{min}, only the Yanzhou station has been detected to have a significantly increasing trend in T_{max}. Generally, the presence of increasing tendencies is higher in T_{min} than in T_{max}. WS in the Shijiazhuang and Xuzhou stations decreased significantly ($p < 0.01$), while it increased significantly in the Xinxiang station. No significant changes were detected in the Beijing and Yanzhou stations for WS.

4.3.4 Sensitivity coefficients of crop water requirement for climatic variables

A large number of research studies showed that a negative correlation exists between RH and ET_0 (Zeng and Heilman, 1997; Gong *et al.*, 2006) and a positive correlation exists between WS, SR, temperature and ET_c. The most effective meteorological factor that impacts ET_c is diverse in different stations and phenological stages for winter wheat. It cannot be denied that sensitivity coefficients exhibit large fluctuations during the growth stage of winter wheat. The primary meteorological factor and its influence on ET_c in the five stations were determined by a sensitivity correlation analysis. ET_c has a significant correlation with T_{min}, T_{max}, RH, WS and SR. According to the long-term trends analysis over the entire studied period of meteorological variables and sensitivities in the selected weather stations, RH has a negative correlation with ET_c, namely, that a decrease of RH has led to an increase of ET_c. SR, T_{min} and T_{max}, and WS showed a positive correlation with ET_c, that is, a decrease of WS, T_{min} and T_{max} has led to a decrease of ET_c. Temperature was the most sensitive variable to the ET_c in the whole growth stage of winter wheat, followed by SR, WS and RH. However, ET_c was observed to be more sensitive to T_{max} than to T_{min}. ET_c is more sensitive to temperatures in the sowing to seeding stages. As for SR, ET_c is more sensitive from sowing to seeding stages, jointing to heading stages and heading to maturity stages. ET_c in seeding to jointing stages was more sensitive to WS than in other phenological phases. A negative correlation exists between RH and ET_c, especially from seeding to jointing stages.

Table 4.6 shows the results of sensitive changes in ET_c in five stations selected in the 3H Plain. Trends of sensitivity coefficient of

Table 4.5. Long-term trends over the entire studied period of the main meteorological variables in the selected weather stations.

Station	Sowing to maturity		Sowing to seeding		Seeding to jointing		Jointing to heading		Heading to maturity	
	Mean	Slope	Mean	Slope	Mean	Slope	Mean	Slope	Mean	Slope
P_e (mm 10 year^{-1})										
Beijing	67.09	0.55	3.09	−0.40	31.48	−2.19	9.27	1.84	23.25	1.29
Shijiazhuang	70.52	0.58	2.41	0.20	35.46	−3.26	7.78	0.55	24.88	3.09
Xinxiang	74.80	−4.40	5.43	−2.08	36.81	−3.08	9.13	2.72	23.43	−1.96
Xuzhou	117.80	−4.11	4.75	−2.62	64.24	−1.31	15.99	0.76	32.83	−0.94
Yanzhou	86.63	0.36	3.81	0.78	45.61	−3.40	9.50	1.86	27.71	1.12
SR (MJ m^{-2} 10 year^{-1})										
Beijing	3572.5	−225.4**	134.9	−7.4	2222.1	−201.5**	402.4	0.2	813.2	−16.6
Shijiazhuang	3239.8	−87.6*	101.1	3.6	1890.1	−122.9**	402.5	−21.7	846.2	53.5**
Xinxiang	2992.4	−95.7**	99.9	4.9	1645.9	−139.9**	457.1	13.3	789.4	25.9
Xuzhou	3117.6	−123.9**	116.5	4.0	1615.9	−157.3**	538.1	28.3	847.1	1.0
Yanzhou	3282.6	−192.5**	102.3	−7.1	1834.3	−179.1**	447.2	−9.4	898.9	3.1
RH (per cent 10 year^{-1})										
Beijing	48.51	−0.33	61.88	0.23	47.31	−0.17	45.86	−0.45	52.78	−2.06
Shijiazhuang	56.33	−1.99	65.63	−0.78	56.54	−2.17	50.84	−0.47	56.51	−2.70
Xinxiang	63.48	−1.56*	71.86	−1.82	63.55	−1.36	59.04	−2.39	64.33	−1.64
Xuzhou	65.43	0.79	69.23	−1.30	66.65	1.42	60.41	−0.61	63.93	0.43
Yanzhou	66.20	1.01	72.30	1.22	66.66	1.07	60.14	1.25	66.69	0.98
T_{min} (°C 10 year^{-1})										
Beijing	2.99	0.19	11.68	−0.18	−0.79	0.11	11.46	−0.65*	16.21	−0.44*
Shijiazhuang	3.89	1.01**	12.31	1.08*	0.22	1.02**	9.83	−0.16	15.56	0.20
Xinxiang	3.99	0.68**	11.63	0.83	0.49	0.65**	8.29	−0.35	14.27	−0.13
Xuzhou	4.92	0.55**	12.36	0.61	1.35	0.61**	7.66	−0.56	14.76	−0.30
Yanzhou	2.91	0.41**	11.30	0.97	−0.55	0.46**	6.69	−0.59*	13.08	−0.48**
T_{max} (°C 10 year^{-1})										
Beijing	13.66	−0.13	22.78	−1.35*	9.52	−0.25	23.37	−0.87**	28.02	−0.51
Shijiazhuang	14.45	0.19	23.83	−0.88	10.33	0.17	21.68	−0.82**	27.28	−0.53*
Xinxiang	14.64	0.21	22.33	0.56	10.83	0.22	19.88	−1.09*	25.67	−0.72**
Xuzhou	14.70	0.19	23.33	0.61	10.67	0.24	18.29	−0.90	25.50	−0.88**
Yanzhou	14.57	0.29**	23.78	0.18	10.72	0.38*	19.58	−0.80*	25.56	−0.69**
WS (m s^{-1} 10 year^{-1})										
Beijing	2.48	−0.03	2.01	0.00	2.44	−0.05	2.94	−0.08	2.59	0.08
Shijiazhuang	1.75	−0.23**	1.38	−0.17*	1.65	−0.24**	2.24	−0.34**	2.00	−0.19**
Xinxiang	2.15	0.19**	1.66	0.10	2.02	0.17**	2.64	0.19*	2.41	0.28**
Xuzhou	2.24	−0.17**	1.71	−0.25*	2.06	−0.19**	2.66	−0.13*	2.67	−0.19**
Yanzhou	2.48	−0.15	1.76	−0.06	2.36	−0.13	2.96	−0.20	2.81	−0.22*

P_e, effective precipitation; SR, solar radiation; T_{max}, maximum temperature; T_{min}, minimum temperature; RH, relative humidity; WS, wind speed.
* and ** represent linear coefficients significant at $p<0.05$ and $p<0.01$, respectively.

solar radiation (S_{SR}) are negative in Beijing and Xinxiang stations in the whole growth stage of winter wheat, with a $p < 0.01$ level of significance. The decrease of S_{SR} mainly occurred from seeding to jointing stages and heading to maturity stages of winter wheat, which means that the fluctuation of SR controlled less the changes of ET_c from seeding to jointing stages and from heading to maturity stages of winter wheat in Beijing and

Table 4.6. Long-term trends over the entire studied stage of meteorological variables and sensitivities in the selected weather stations.

	Sowing to maturity		Sowing to seeding		Seeding to jointing		Jointing to heading		Heading to maturity	
Station	Mean	Slope	Mean	Slope	Mean	Slope	Mean	Slope	Mean	Slope
S_{SR}										
Beijing	0.19	−0.02**	0.41	−0.05**	0.09	−0.02**	0.42	−0.02	0.52	−0.03**
Shijiazhuang	0.26	−0.01	0.47	−0.01	0.15	−0.02**	0.46	0.00	0.57	−0.01
Xinxiang	0.31	−0.03**	0.48	−0.02	0.21	−0.03**	0.46	−0.05**	0.58	−0.04**
Xuzhou	0.33	0.00	0.50	0.01	0.24	0.00	0.46	−0.02**	0.56	0.00
Yanzhou	0.31	0.00	0.51	−0.01	0.22	−0.01	0.46	−0.01	0.57	0.00
S_{RH}										
Beijing	−0.01	0.00	0.00	0.00	−0.01	0.00	0.00	0.00	0.00	0.00
Shijiazhuang	−0.01	0.00	0.00	0.00	−0.01	0.00	0.00	0.00	0.00	0.00
Xinxiang	−0.01	0.00	0.00	0.00	−0.01	0.00	0.00	0.00	0.00	0.00
Xuzhou	−0.01	0.00	0.00	0.00	−0.01	0.00	0.00	0.00	0.00	0.00
Yanzhou	−0.01	0.00	0.00	0.00	−0.02	0.00	0.00	0.00	0.00	0.00
S_{Tmin}										
Beijing	0.34	0.02	0.53	0.07**	0.30	0.01	0.42	0.01	0.45	0.02*
Shijiazhuang	0.35	−0.02**	0.44	0.01	0.33	−0.03*	0.38	−0.02**	0.39	−0.02**
Xinxiang	0.45	0.04**	0.56	0.04	0.43	0.04**	0.45	0.03	0.47	0.05**
Xuzhou	0.53	0.01	0.55	−0.08	0.53	0.02	0.50	−0.01	0.54	−0.01
Yanzhou	0.53	0.02	0.54	0.04	0.54	0.03	0.47	0.00	0.52	0.00
S_{Tmax}										
Beijing	0.75	0.02	1.10	0.06	0.67	0.01	0.92	0.01	0.94	0.03
Shijiazhuang	0.76	−0.08**	0.95	−0.10**	0.73	−0.09**	0.84	−0.08**	0.83	−0.07**
Xinxiang	0.94	0.05*	1.11	0.06	0.93	0.05*	0.97	0.03	0.97	0.07**
Xuzhou	1.04	0.00	1.08	−0.14	1.04	0.02	1.02	−0.03	1.08	−0.05*
Yanzhou	1.20	0.04	1.19	0.02	1.23	0.06	1.11	0.00	1.14	−0.02
S_{WS}										
Beijing	0.34	0.02**	0.22	0.02*	0.38	0.02**	0.23	0.01	0.19	0.02*
Shijiazhuang	0.27	0.01	0.18	0.00	0.30	0.02	0.20	0.00	0.15	0.01
Xinxiang	0.21	0.02**	0.15	0.02	0.25	0.03**	0.17	0.03**	0.12	0.02*
Xuzhou	0.19	0.00	0.15	0.01	0.22	0.00	0.15	0.01	0.12	0.00
Yanzhou	0.20	0.00	0.14	0.00	0.24	0.00	0.15	0.01	0.10	−0.01

S_{SR}, sensitivity coefficient for solar radiation; S_{RH}, sensitivity coefficient for relative humidity; S_{Tmin}, sensitivity coefficient for minimum temperature; S_{Tmax}, sensitivity coefficient for maximum temperature; S_{WS}, sensitivity coefficient for wind speed. These sensitivity coefficients are indices without units.
* and ** represent linear coefficients significant at $p<0.05$ and $p<0.01$, respectively.

Xinxiang stations. Sensitivity coefficients of temperatures (S_T) increased in Beijing, Xinxiang, Xuzhou and Yanzhou stations, while it decreased in the Shijiazhuang station. The increase of S_T mainly occurred from seeding to jointing stages and heading to maturity stages in the Xinxiang station, with a statistically significant increase at the $p < 0.01$ level of significance. Decreases in sensitivity coefficients to temperatures were detected in the Shijiazhuang station and it mainly occurs from seeding to jointing stages and jointing to heading stages of winter wheat. It can be concluded that the fluctuation of temperature had a more positive effect on ET_c in Beijing, Xinxiang, Xuzhou and Yanzhou stations, which means that the increase in temperatures may lead to a greater increase of ET_c in these four stations. Sensitivity coefficients of wind speed (S_{WS}) increased significantly in

Beijing and Xinxiang stations, with statistically significant increases at the $p < 0.01$ level of significance. With the decrease of WS, ET_c will decrease due to the positive coefficient. Trends of S_{RH} showed no significant changes in the time series analysis.

4.3.5 Variation of precipitation deficit and regional response to climate change

The precipitation deficit series of phenological phases at each station were detected with the purpose of appreciating the temporal variability of precipitation deficit changes and their magnitudes. As described in Table 4.7, the mean value of precipitation deficit in the whole growth stage is from −395.2 mm to −274.2 mm, with the minimum value detected at the Beijing station and the maximum value at the Xuzhou station. In the whole growth stage of winter wheat it was only at the Xinxiang station that a significantly decreasing trend was observed, with a slope of −2.01 mm year^{-1}; in other words, where the problem of water shortage becomes more severe. The precipitation deficit accounts for around 44.7% in the total precipitation deficit in seeding to jointing stage for five stations, followed by heading to maturity stages and jointing to heading stages. Moreover, the precipitation deficit increases significantly with the magnitude of 1.60 mm year^{-1} from seeding to jointing stages in the Beijing station. On the other hand, the precipitation deficit decreases significantly with the magnitude of −1.80 mm year^{-1} from the heading to maturity stages in the Xinxiang station.

As described in Table 4.8, the analysis of responses of precipitation deficit to climate change in different phenological phases was carried out in the selected weather stations. A negative relationship was found between precipitation deficit and SR and WS, which means that the increase in SR and WS had occurred with aggravation of precipitation deficit. Temperature in most stations showed a negative correlation to precipitation deficit, except minimum temperature in the Xuzhou and Yanzhou stations. A positive relationship was detected between precipitation deficit and RH, and RH was considered the most correlative factor to precipitation deficit, followed by SR, temperature and WS. A higher correlation was especially found in the heading to maturity stages of winter wheat between RH and precipitation deficit. As for SR, precipitation deficit in seeding to jointing stages, jointing to heading stages and heading to maturity stages of winter wheat showed a higher correlation with SR. A negative correlation was detected between temperature and precipitation in most stations, except minimum temperature in the Xuzhou and Yanzhou stations, in which stations a positive correlation was found, especially in sowing to seeding stages and seeding to jointing stages of winter wheat. The fact that the highest value for correlation coefficient of WS was detected as described in Table 4.8 indicates that WS is responsible for precipitation deficit of winter wheat.

Table 4.7. Long-term trends over the entire studied period of precipitation deficit (mm) in different phenological phases for the selected weather stations.

	Sowing to maturity		Sowing to seeding		Seeding to jointing		Jointing to heading		Heading to maturity	
Station	Mean	Slope	Mean	Slope	Mean	Slope	Mean	Slope	Mean	Slope
Beijing	−395.15	0.72	−5.33	0.00	−246.81	1.60*	−70.51	−0.31	−72.50	−0.56
Shijiazhuang	−334.16	0.50	−3.60	−0.03	−159.23	0.71	−81.30	0.47	−90.04	−0.65
Xinxiang	−317.14	−2.01*	−0.66	−0.29	−129.28	−0.05	−83.04	0.14	−104.15	−1.80**
Xuzhou	−274.19	0.18	−2.10	−0.27	−87.89	1.45	−83.57	0.02	−100.62	−1.03
Yanzhou	−309.37	1.14	−2.29	0.09	−125.51	0.97	−76.94	0.54	−104.64	−0.45

* and ** represent linear coefficients significant at $p < 0.05$ and $p < 0.01$, respectively.

4.4 Discussion and Conclusions

Changes in the meteorological variables are important factors in the assessment of crop water requirement with climate change. The simultaneous consideration of the effects of meteorological variables such as SR, temperature, RH and WS on the crop water requirement with climate change is necessary. The present chapter has quantified changes in a set of meteorological variables and in ET_c under the weather conditions of the 3H Plain during the last 29 years.

A clear drop in SR was detected, which is usually called global dimming, probably caused by man-made aerosols and other air pollutants that have changed the optical properties of the atmosphere, in particular those of clouds (Mozny *et al.*, 2009; Espadafor *et al.*, 2011). The value of RH in several stations in the 3H Plain decreased due to the decline of precipitation and these results are in agreement with those from previous studies, but the tendency is not serious. The maximum temperature and minimum temperature were observed to increase asymmetrically and the minimum temperature was more responsible for average temperature than maximum temperature increase. This agrees with the asymmetry of the increase in maximum and minimum temperatures pointed out by several researchers

Table 4.8. The response of precipitation deficit (PD) to climate change in different phenological phases.

Station	Sowing to maturity	Sowing to seeding	Seeding to jointing	Jointing to heading	Heading to maturity
Solar radiation (MJ m^{-2} year^{-1})					
Beijing	–0.39	0.12	–0.70	–0.83	–0.39
Shijiazhuang	–0.50	–0.16	–0.46	–0.68	–0.67
Xinxiang	–0.19	–0.15	–0.58	–0.53	–0.81
Xuzhou	–0.10	–0.53	–0.57	–0.16	–0.54
Yanzhou	–0.31	–0.63	–0.55	–0.73	–0.46
Relative humidity (%)					
Beijing	0.84	0.37	0.77	0.67	0.79
Shijiazhuang	0.82	0.46	0.79	0.73	0.83
Xinxiang	0.78	0.35	0.69	0.48	0.83
Xuzhou	0.81	0.56	0.86	0.84	0.80
Yanzhou	0.78	0.61	0.79	0.40	0.82
Minimum temperature (°C)					
Beijing	–0.17	0.13	–0.13	0.30	–0.05
Shijiazhuang	–0.03	–0.13	0.10	–0.10	–0.12
Xinxiang	–0.23	0.13	0.17	–0.06	0.36
Xuzhou	0.46	0.40	0.60	–0.16	0.11
Yanzhou	0.46	0.55	0.52	0.34	0.43
Maximum temperature (°C)					
Beijing	–0.43	–0.07	–0.39	–0.02	–0.51
Shijiazhuang	–0.59	–0.38	–0.46	–0.56	–0.36
Xinxiang	–0.45	–0.24	–0.25	–0.25	–0.17
Xuzhou	–0.09	–0.02	0.01	–0.37	–0.39
Yanzhou	–0.20	–0.23	–0.12	0.13	–0.28
Wind speed (m s^{-1})					
Beijing	–0.56	–0.03	–0.58	–0.34	–0.47
Shijiazhuang	–0.47	–0.43	–0.53	–0.48	–0.51
Xinxiang	–0.54	0.27	–0.32	–0.18	–0.57
Xuzhou	–0.08	0.31	–0.12	–0.29	0.06
Yanzhou	–0.25	–0.01	0.04	–0.40	–0.21

(Karl *et al.*, 1993; Easterling *et al.*, 1997; IPCC, 2007). The increase in the minimum temperature was higher and statistically more significant from seeding to jointing stages, with 0.57°C 10 year^{-1}. This is partially in agreement with the findings of Brunet *et al.* (2007), which indicate that the greatest contribution to the higher annual warming was winter and summer over the period 1901–2005. Due to the decreases in SR and RH, and the increase in the temperatures, a decrease in ET_c for the selected weather stations was detected in the whole growth stage of winter wheat. ET_c in seeding to jointing stages was observed with a significantly decreasing trend in the stations selected, while in the heading to maturity stages of winter wheat, the ET_c increased in the time series.

Sensitivities of ET_c to five major climatic variables were studied in the 3H Plain using a 29-year data set. Long-term average sensitivities were analysed in different phenological phases of winter wheat. This chapter showed that temperature was the most sensitive variable in general for the plain, followed by SR, WS and RH. However, this contrasts with the results of Gao *et al.* (2006) and Wang *et al.* (2007). They pointed to SR reduction along with WS as the main contributing variables in the year. From the results obtained in this chapter, changes in meteorological variables and their sensitivities may lead to different results on ET_c. The variability of the sensitivity coefficients indicated that the response of ET_c of winter wheat in the 3H Plain to climate change will differ with location and phenological phases. The decrease of S_{SR} mainly occurred from seeding to jointing stages and from heading to maturity stages of winter wheat. S_T increased in Beijing, Xinxiang, Xuzhou and Yanzhou stations, while it decreased in the Shijiazhuang Station. Fluctuation of temperature had a more positive effect on ET_c in Beijing, Xinxiang, Xuzhou and Yanzhou stations, which means that increase in temperature may lead to a greater increase of ET_c in these four stations. With the decrease of WS, ET_c will decrease due to the positive coefficient in Beijing and Xinxiang stations. Trends of S_{RH} showed no significant changes in the time series analysis.

The precipitation deficit series of phenological phases at each station were observed for the purpose of appreciating the temporal variability of precipitation deficit changes and their magnitudes. The precipitation deficit increases significantly with the magnitude of 1.60 mm year^{-1}, probably due to the decline of SR by 201.5 MJ m^{-2} 10 year^{-1} from seeding to jointing stages in the Beijing station. In contrast, the precipitation deficit decreases significantly with a magnitude of −1.80 mm year^{-1} due to the decline of RH by −1.64% 10 year^{-1} from heading to maturity stages in the Xinxiang station. A positive relationship was detected between precipitation deficit and RH, and RH was considered the most correlative factor to precipitation deficit.

A local/regional adjustment would be required for other areas, mainly with climate conditions differing substantially from those of the present chapter. From these ET_c estimates and temporal knowledge of precipitation deficit, irrigation schedules can be defined with the climate change. With the results provided in this chapter, agronomic effects due to changes in ET_c could be inferred for irrigated agriculture in the 3H Plain. In addition, the crop cycle is expected to be modified due to changes of weather and crop water requirement. For example, in Spain, Döll (2002) has estimated a decrease in irrigation requirements for 2020 due to the possibility of sowing earlier in time when the temperature regime is more favourable (higher temperatures). Additionally, results of this chapter can be used as a theoretical basis for partial derivative in future research on the response of crop water requirement to climatic change. Under the condition of the future climate scene data considered, further analyses on the agronomic consequences of climate change in semi-arid environments such as that found in the 3H Plain are in process, with the studying of interactions of the different components in determining crop irrigation requirements and yield. Lv *et al.* (2013) conducted the responses of winter wheat phenology, spatial variation of potential and rainfed production, actual evapotranspiration and irrigation water use efficiency to climate change with the WheatGrow model in China's main wheat production regions and further described impacting factors for improvement of wheat yield under climate change projections of A2, A1 and B1 in the 3H Plain. Furthermore, in the research of Liu, S. *et al.* (2010), the effects of climate change on grain production of a winter wheat–summer maize cropping system were investigated, corresponding to the temperature rising 2°C and 5°C, precipitation increasing and decreasing by 15% and 30%, and atmospheric CO_2 enriching to 500 ppmv and 700 ppmv (parts per million by volume) in the 3H Plain.

Acknowledgements

This chapter was supported by the 12th five-year plan of National Key Technologies R&D Program (No. 2012BAD09B01), the National Basic Research Program of China (973 Program) (No. 2012CB955904), and Project of Food Security and Climate Change in the Asia-Pacific Region: Evaluating Mismatch between Crop Development and Water Availability and Project of National Non-profit Institute Fund (No. BSRF201206).

References

Allen, R.G., Pereira, L.S., Raes, D. and Smith, M. (1998) *Crop Evapotranspiration – Guidelines for Computing Crop Water Requirements.* FAO Irrigation and Drainage Paper 56. FAO, Rome.

Beven, K. (1979) A sensitivity analysis of the Penman-Monteith actual evapotranspiration estimates. *Journal of Hydrology* 44(3–4), 169–190.

Bhalme, H.N. and Mooley, D.A. (1980) Large-scale droughts/floods and monsoon circulation. *Monthly Weather Review* 108(8), 1197–1211.

Blaney, H.F. and Criddle, W.D. (1952) *Determining Water Requirements in Irrigated Areas from Climatological and Irrigation Data.* US Department of Agriculture, Washington, DC.

Brunet, M., Sigró, J., Jones, P.D., Saladié, O., Aguilar, E., Moberg, A., Lister, D. and Walther, A. (2007) Long-term changes in extreme temperatures and precipitation in Spain. *Contributions to Science* 3, 331–342.

Döll, P. (2002) Impact of climate change and variability on irrigation requirements: a global perspective. *Climatic Change* 54(3), 269–293.

Easterling, D.R., Horton, B., Jones, P.D., Peterson, T.C., Karl, T.R., Parker, D.E., Salinger, M.J., Razuvayev, V., Plummer, N., Jamason, P. and Folland, C.K. (1997) Maximum and minimum temperature trends for the globe. *Science* 277(5324), 364–367.

Espadafor, M., Lorite, I.J., Gavilán, P. and Berengena, J. (2011) An analysis of the tendency of reference evapotranspiration estimates and other climate variables during the last 45 years in Southern Spain. *Agricultural Water Management* 98(6), 1045–1061.

Gao, G., Chen, D., Ren, G., Chen, Y. and Liao, Y. (2006) Spatial and temporal variations and controlling factors of potential evapotranspiration in China: 1956–2000. *Journal of Geographical Sciences* 16(1), 3–12.

Gong, L., Xu, C., Chen, D., Halldin, S. and Chen, Y.D. (2006) Sensitivity of the Penman–Monteith reference evapotranspiration to key climatic variables in the Changjiang (Yangtze River) basin. *Journal of Hydrology* 329(3), 620–629.

Goyal, R.K. (2004) Sensitivity of evapotranspiration to global warming: a case study of arid zone of Rajasthan (India). *Agricultural Water Management* 69, 1–11.

Hargreaves, G.H. and Samani, Z.A. (1982) Estimating potential evapotranspiration. *Journal of the Irrigation and Drainage Division* 108(3), 225–230.

Intergovernmental Panel on Climate Change (IPCC) (2007) *Summary of Policymakers of Synthesis Report of the IPCC Fourth Assessment Report.* Cambridge University Press, Cambridge, UK.

Irmak, S., Payero, J.O., Martin, D.L., Irmak, A. and Howell, T.A. (2006) Sensitivity analyses and sensitivity coefficients of standardized daily ASCE-Penman-Monteith equation. *Journal of Irrigation and Drainage Engineering* 132, 564–578.

Ju, X.S., Zou, X.H. and Zhang, Q. (1998) The method of the climatic drought-flood index and its analysis. *Journal of Natural Disasters* 7(3), 51–57 (in Chinese).

Karl, T.R., Jones, P.D., Knight, R.W., Kukla, G., Plummer, N., Razuvayev, V. and Peterson, T.C. (1993) Asymmetric trends of daily maximum and minimum temperature. *Papers in Natural Resources* 74, 1007–1023.

Li, H., Qiu, J., Wang, L., Tang, H., Li, C. and Van Ranst, E. (2010) Modelling impacts of alternative farming management practices on greenhouse gas emissions from a winter wheat–maize rotation system in China. *Agriculture, Ecosystems & Environment* 135(1), 24–33.

Li, L., Yan, D., Qin, T., Hao, C. and Zheng, X. (2012) Drought variation in Haihe river basin from 1961 to 2010. *Journal of Arid Land Resources and Environment* 26(11), 61–67 [in Chinese].

Liu, S., Mo, X., Lin, Z., Xu, Y., Ji, J., Wen, G. and Richey, J. (2010) Crop yield responses to climate change in the Huang-Huai-Hai Plain of China. *Agricultural Water Management* 97(8), 1195–1209.

Liu, Y., Wang, Y. and Yang, X.G. (2010) Trends in reference crop evapotranspiration and possible climatic factors in the North China Plain. *Acta Ecological Sinica* 30, 923–932 (in Chinese).

Lv, Z., Liu, X., Cao, W. and Zhu, Y. (2013) Climate change impacts on regional winter wheat production in main wheat production regions of China. *Agricultural and Forest Meteorology* 171–172, 234–248.

McCuen, R.H. (1974) A sensitivity and error analysis of procedures used for estimating evaporation. *Journal of the American Water Resources Association* 10(3), 486–497.

McCuen, R.H. and Beighley, R.E. (2003) Seasonal flow frequency analysis. *Journal of Hydrology* 279(1), 43–56.

McKenney, M.S. and Rosenberg, N.J. (1993) Sensitivity of some potential evapotranspiration estimation methods to climate change. *Agricultural and Forest Meteorology* 64(1), 81–110.

Mozny, M., Tolasz, R., Nekovar, J., Sparks, T., Trnka, M. and Zalud, Z. (2009) The impact of climate change on the yield and quality of Saaz hops in the Czech Republic. *Agricultural and Forest Meteorology* 149(6), 913–919.

Rana, G. and Katerji, N. (1998) A measurement based sensitivity analysis of the Penman-Monteith actual evapotranspiration model for crops of different height and in contrasting water status. *Theoretical and Applied Climatology* 60, 141–149.

Ren, J., Chen, Z., Zhou, Q. and Tang, H. (2008) Regional yield estimation for winter wheat with MODIS-NDVI data in Shandong, China. *International Journal of Applied Earth Observation and Geoinformation* 10(4), 403–413.

Saxton, K.E. (1975) Sensitivity analyses of the combination evapotranspiration equation. *Agricultural Meteorology* 15(3), 343–353.

Serrano, A., Mateos, V.L. and Garcia, J.A. (1999) Trend analysis of monthly precipitation over the Iberian Peninsula for the period 1921–1995. *Physics and Chemistry of the Earth, Part B: Hydrology, Oceans and Atmosphere* 24(1), 85–90.

Smith, M. (1992) *CROPWAT – A Computer Program for Irrigation Planning and Management.* Irrigation and Drainage Paper 46. Food and Agriculture Organization of the United Nations, Rome.

Song, Z.W., Zhang, H.L., Snyder, R.L., Anderson, F.E. and Chen, F. (2009) Distribution and trends in reference evapotranspiration in the North China Plain. *Journal of Irrigation and Drainage Engineering* 136, 240–247.

Szinell, C.S., Bussay, A. and Szentimrey, T. (1998) Drought tendencies in Hungary. *International Journal of Climatology* 18(13), 1479–1491.

Todisco, F. and Vergni, L. (2008) Climatic changes in Central Italy and their potential effects on corn water consumption. *Agricultural and Forest Meteorology* 148, 1–11.

Wang, E., Chen, C. and Yu, Q. (2009) *Modeling the Response of Wheat and Maize Productivity to Climate Variability and Irrigation in the North China Plain.* 18th World IM4CSIMODSIM Congress, Cairns, Australia, pp. 2742–2748.

Wang, Y., Jiang, T., Bothe, O. and Fraedrich, K. (2007) Changes of pan evaporation and reference evapotranspiration in the Yangtze River basin. *Theoretical and Applied Climatology* 90(1), 13–23.

Xiong, W., Holman, I., Lin, E., Conway, D., Jiang, J., Xu, Y. and Li, Y. (2010) Climate change, water availability and future cereal production in China. *Agriculture, Ecosystems and Environment* 135, 58–69.

Zeng, W. and Heilman, J.L. (1997) Sensitivity of evapotranspiration of cotton and sorghum in west Texas to changes in climate and CO_2. *Theoretical and Applied Climatology* 57(3), 245–254.

5 Impacts of Climate Change and Adaptation in Agricultural Water Management in North China

Jun Xia,[1,2]* Xingguo Mo,[2] Jinxia Wang[3] and Xinping Luo[2]

[1]State Key Laboratory of Water Resources and Hydropower Engineering Science, Wuhan University, China; [2]Key Laboratory of Water Cycle and Related Land Surface Processes, Chinese Academy of Sciences, Beijing; [3]Center for Chinese Agricultural Policy, Institute of Geographical Sciences and Natural Resources Research, Chinese Academy of Sciences, Beijing, China

Abstract

Water scarcity is one of the key problems in northern China. Efficient use and management of agricultural water resources is an important challenge in China's agricultural food production under a background of climate change. This chapter addresses issues of impacts of climate change and adaptation in agricultural water management in North China. The goal is to understand better the impacts of climate change on agricultural water resources and what measures should be taken to deal with the adverse effects in the North China Plain (NCP). First, the status of agricultural water resources in NCP was analysed. Second, considering that climate change is likely to exacerbate water stress in this area, and exploring the regional crop response to climate change, this study analysed the spatial variability and evolution of crop yield, evapotranspiration (ET) and water use efficiency (WUE) with a process-based crop model in the NCP and identified the contribution of climate change to their enhancement. Third, the impacts of future climate changes under A2 and B1 scenarios (described later in this chapter) on the wheat–maize double-cropping system are assessed. The results show that under IPCC SRES A2 and B1 scenarios, production of winter wheat will increase with slightly intensified ET; in contrast, summer maize production will slightly decline with a significant increase of ET. Also, with agricultural management, maize is more productive than wheat, in that wheat relies more on irrigation than maize, yield level of maize is higher than that of wheat, the water consumption of maize is lower, and the response of maize yield is larger than that of wheat yield to agricultural management. However, the simulation also suggests that wheat is more resilient to climate change than maize. Therefore to say if wheat or maize is more favourable in the NCP depends on the conditions in the future. Finally, in order to mitigate the impacts of climate change on agricultural water use and realize its sustainable utilization, the key adaptive water management strategies in the agriculture sector and how to improve efficiency of agricultural water use through reforming agricultural water management and policies were examined. The following measures can be implemented to reform agricultural water management and policies: improving the performance of participatory irrigation management reform, establishing a water rights system, reforming agricultural water price, and promoting the adoption of agricultural water-saving technology.

* Corresponding author, e-mail: xiaj@igsnrr.ac.cn

5.1 Introduction

The impact of climate change on the security of water resources is a widespread global concern. It is a great strategic issue in the sustainable development of China. As a big developing country, China is facing a huge challenge in managing water resources to support its economic boom. Its immense territory of 9.6 million km^2 is relatively abundant in water resources, ranking sixth in the world after Brazil, the Russian Federation, Canada, the USA and Indonesia, in that order, in terms of absolute amount of annual runoff. However, given its large population of over 1.3 billion, China's per capita water resources are very low (about one-quarter of the world average), making it one of the most water-scarce countries in the world.[1]

According to the 2nd National Comprehensive Water Resources Assessment in China that started after 2000, the annual mean precipitation in China's mainland was 6178.60 billion m^3, with an estimated annual runoff of 2774.10 billion m^3 for 1956–2000. The total available water resource in China in 2000 was 814.0 billion m^3 (Bm^3). However, the total actual water use in 2000 reached 563.2 Bm^3, where the usable water resources per capita (UWRP) in China and North China were about 628 m^3 and only 359 m^3, respectively. In 2030, China's population will reach 1.6 billion, the total actual water use will increase to 710.1 Bm^3 from 563.2 Bm^3 in the year 2000, and UWRP will go down to 508 m^3 from the 628.0 m^3 of 2000. Thus, water stress in China is significant and has become a bottleneck that limits China's sustainable development.

Water resources constitute one of the most important constraints to food security in China. It has been shown that the total agricultural water use for the whole country was 348.40 km^3 in 2000. The ratio of consumption to withdrawals varies with climatic factors, the crops grown and irrigation efficiency, and typically ranges from 50 to 80%. Basically, about 64% of national agricultural water withdrawals were consumed, resulting in a total of 221.90 km^3 of consumption for the year 2000 (Cao *et al.*, 2012). Thus, just as in most other countries of the world, agriculture is the major water consumer in China. Major agricultural productive areas in China are North China, which is also called the 3H region (Huang (Yellow) River, Huai River and Hai River), North-East China (Song-Liao River) and South China in the monsoonal area. Under the climate change condition, East China, lying in the monsoonal area, is subject to the dual perils of drought and flood. During the past 30 years, drought has become more severe in the already dry north and the water ecology has deteriorated, and extreme flood disasters have increased in the south. All these occurrences have severely restricted the sustainable development of both the economy and society. Future climate change may be expected to have a significant and possibly intensifying effect on the existing pattern of 'north drought and south floods' as well as on the distribution of water resources. This will exert an unexpected influence on the functioning of major engineering projects in China, including those launched to increase food production in the north and north-east, on inter-basin water transfers and on flood control in southern rivers.

Thus a 5-year National Basic Research Program of China (2010–14) on 'The Impact of Climate Change on Terrestrial Water Cycle, Regional Water Resources Security and the Adaptation Strategy for East Monsoon Area of China', supported by the Ministry of Science and Technology (MOST) and led by Professor Jun Xia, was launched in 2009. The project focuses on the major river basins in the eastern monsoonal region of China to investigate the mechanisms of the impact of climate change on water resources and relevant adaptation strategies. The study aims to meet the major strategic requirement of enhancing water resource security for China, focusing on the impact of climate change, vulnerability and adaptation as key issues.

This chapter will address the issue on impacts of climate change and adaptation in agricultural water management in China. The area of North China, i.e. the 3H regions with more water stress and challenges related to drought, environmental issues and social economic development, is especially selected to explain how China will face the big challenges on agricultural water issues, and how to carry out adaptive water management to cope with the impact of climate change. Finally, sustainable utilization of agricultural water resources and adaptive strategies to climate change are presented.

5.2 Climate Change during the Last 60 Years in North China

The NCP is one of the country's granaries, extending from latitude 32°00′ N to 40°24′ N and longitude 112°48′ E to 122°45′ E (Fig. 5.1). It is located in the eastern part of China with an area of 33×10^4 km^2, and is an alluvial plain developed by the intermittent flooding of the Huang (Yellow), Huai and Hai rivers. Administratively, the plain covers seven provinces (mega-cities), including Hebei, Shandong, Henan, Anhui, Jiangsu, Beijing and Tianjin. The 3H region is an important agricultural production base and occupies a decisive position in guaranteeing the nation's grain requirements. It accounts for 26% of the total grain production, although it has only 21% of the planting areas as well as 6.65% of the water resources of China (Shi, 2008). Besides soybean/millet/sorghum, the double-cropping system of

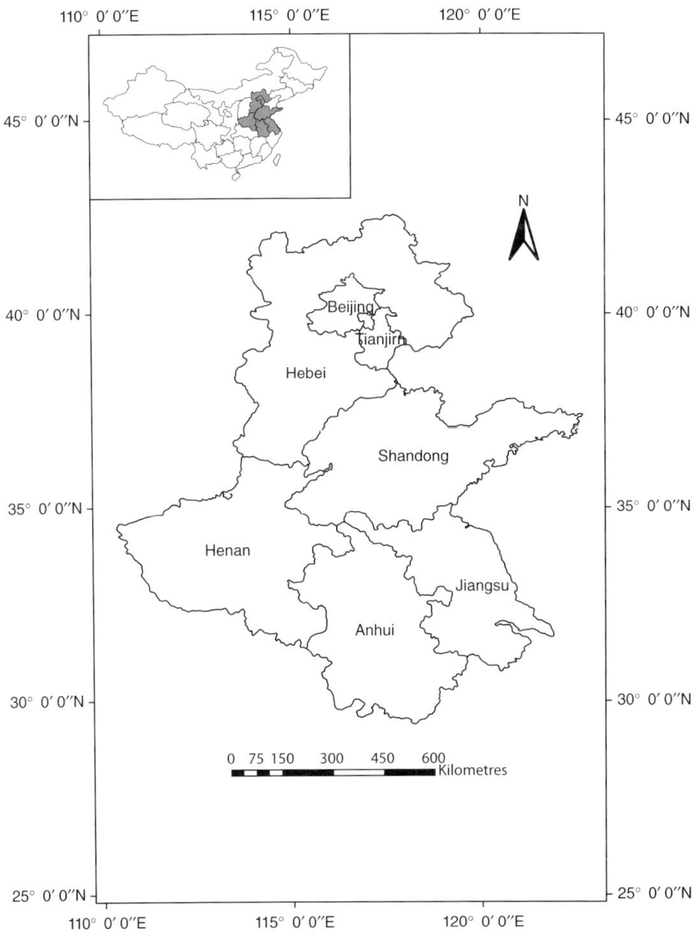

Fig. 5.1. Location map showing the NCP, China.

winter wheat–summer maize prevails in the plain, in which maize is the most common autumn harvest crop. Due to insufficient precipitation in the growing season, the spring crops (such as wheat) usually need supplemental irrigation to obtain favourable production.

The 3H region is located in the East Asian monsoonal zone. The warm temperate climate varies gradually from sub-humid in the southern to semi-arid in the northern part. The annual precipitation is about 500–1000 mm. More than 70% of precipitation falls in summer. The decrease in the regional precipitation year by year since the 1950s is due to impacts of global climate change. Regarding seasonal changes, precipitation in the spring was low in the 1970s, but increased slightly in the 1980s and the 1990s, which showed a decreasing trend. This led to more frequent spring droughts, which limited the growth of crops during the key period of water supply to them. Precipitation in summer witnessed an obvious decline, with drought, temperature and acidification becoming enhanced, raised and obvious, respectively.

Annually, both maximum and minimum air temperatures increased remarkably over the NCP from the 1950s to the 2000s with a rapid increase in the 1990s. The mean anomalies in 1997–2006 were 0.9°C and 1.28°C for the maximum and minimum temperatures, respectively, and the increments in winter are slightly higher than in summer (not shown). The mean daily sunshine duration and wind speed show a declining tendency. The decrease of sunshine is possibly related to more cloudy days and heavy aerosol conditions. The anomalies of sunshine duration and wind speed are about −0.4 h and −0.2 m s^{-1}, respectively, during the last 10 years (Fig. 5.2). It is also worth noting that there are no significant trends in either annual precipitation or water vapour pressure, but the potential evaporation is declining during this period, which mainly results from the attenuated global radiation and air movement (Mo *et al.*, 2011).

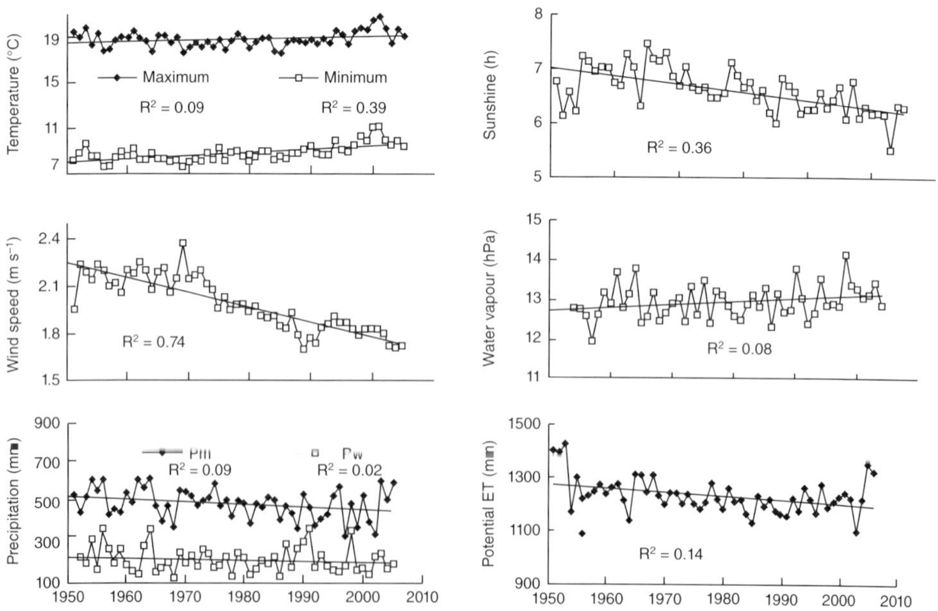

Fig. 5.2. Annual variation of the climatic variables over the NCP during 1951–2006 (potential evapotranspiration is calculated using the Penman–Monteith formula. Pw and Pm are precipitations during the growth season for wheat and maize).

5.3 Water Resources and Their Changes during the Last 60 Years in the North China Plain

Surface water and groundwater resources for a normal year in the 3H region are 128 Bm^3 (after deduction of 20 Bm^3 for silt flushing in its lower reaches). Surface water (including the amount of water entering the region) accounts for 67.5%, with groundwater accounting for 32.5%. Incoming water is quite abundant, but is declining year by year. Per capita water is 600 m^3 and average water per hectare of farmland is 7218 m^3 (Table 5.1). Per capita water resources in 2005 in all provinces of the 3H region were below 1750 m^3, the well-known international water crisis threshold (Zhang *et al.*, 2011). It was less than 500 m^3 in Beijing, Tianjin, Hebei and Shandong, which is below the threshold of severe water shortage. Per capita water resources in 2005 were 518 m^3 and the ratio of surface water to groundwater was 74.5:24.6 in the 3H region. It is forecast that by mid-21st century, per capita water resources in China will be reduced to 1750 m^3.

Water replenishment dropped remarkably due to reduced precipitation, relatively high temperature, increased actual evaporation and the impact of human activities. The primary analysis from the River Commission shows that the surface runoff generated in the whole Hai Luan River Basin during 1980–1989 was only 15.4 Bm^3, 46.5% or 13.4 Bm^3 less than 28.8 Bm^3, the multi-year average during 1956–1979. The arid zone has expanded from NCP south-westward to the upper and middle reaches of the Yellow River (Shaanxi, Gansu, Ningxia), to the Hanjiang Basin, and to the upper reaches of the Huai River since the 1990s. Annual mean water above Huayuankou of the Yellow River during the same period was 46.0 Bm^3, 17.9% or 10.0 Bm^3 less than 56.0 Bm^3, the multi-year average annual runoff. Due to the heavy drought in 1997, the runoff in the upper reaches of Huayuankou was reduced to 31.5 Bm^3, and the actual measured runoff was only 14.3 Bm^3 after deduction of water consumption by the upper/middle reaches. In the 1990s, the drought in Hailuan River was mitigated with increased precipitation close to the multi-year average. But surface runoff had little increase and was still 33.3% or 7.6 Bm^3 less than the multi-year average during 1956–1979 due to dry soil in the pro-phase, persistent high temperature and enhanced evaporation from the topsoil.

Agriculture needs a large amount of water and is facing more shortage than other sectors in the 3H region. Annual average agricultural water consumption was 9.77×10^{10} m^3 in 1997–2011 and the maximum value was 1.13×10^{11} m^3 in 1997. Agricultural water use has shown a downward trend in general, but the proportion of agricultural water use has been above 67% (Fig. 5.3).

The total water resources of this region are less than 10% of the national water resources, but water diversion for agriculture accounts for about 25% of the total amount of agricultural water use in 2000. It

Table 5.1. Water resources statistics in the 3H region in 2005 (p = 50%).

	Local water resources (100 Mm^3)		Volume entering the region (100 Mm^3)	Total water resources (100 Mm^3)	Per capita (m^3/head)	Farmland ($m^3 ha^{-1}$)
	Surface water	Groundwater				
Shanqian plain	107.4	132.2	148.6	388.2	531	7,706
Hai River low plain	44.3	81.7	98.6	224.6	554	4,949
Binhai low plain	35.1	35.9	59.1	130.1	992	11,651
Huanghuai plain	220.8	164.6	146.7	532.1	620	7,634
3H region	407.6	414.4	453	1,275	600	7,218

For downstream silt flushing of the Yellow River's waterway 20 Bm^3 have been deducted; the calculation is based on 317 counties (cities, districts) of two autonomous municipalities and five provinces with a land area of 346,000 km^2, the farmland area of 1995.

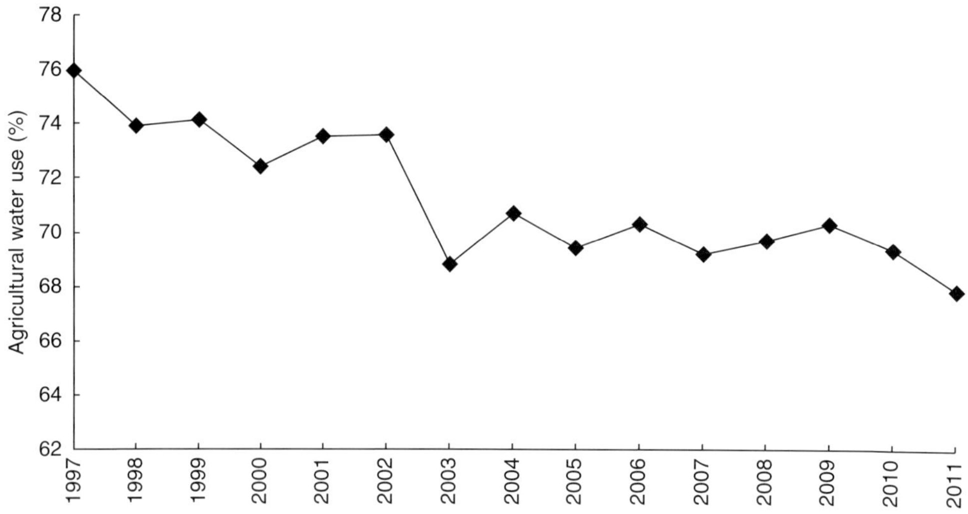

Fig. 5.3. Water diverted for agriculture in the Huang-Huai-Hai region, in 1997–2011.

is predicted that total water consumption of major crops (wheat, maize, cotton) would be $9.74\times10^{12}\,m^3$ in 2015 and the agricultural water shortage crisis will further aggravate the situation (Wang *et al.*, 2009). Compared with developed countries, water resources for agricultural use approach nearly 60–70% and irrigation efficiency is very low. The irrigation WUE in the Huaihe and Haihe river basins is 0.5 and 0.56, respectively, while it is above 0.7 in developed countries. Therefore how to improve the irrigation WUE and crop water productivity in the 3H region is an issue worth studying.

5.4 Water Use by Vegetation over the 3H Region

In general, ET showed a higher value in the south than in the north over the NCP and the ET is higher in well-irrigated locations. An example is the south of Henan, Anhui and the north of Jiangsu, which have abundant precipitation and well-covered vegetation. Annual ET of these areas is about 700–900 mm, and 850 mm of this is contributed by the paddy fields and water. Even other areas of the Hebei Plain, except the riverine, urban and mountain areas in the central regions, have an average ET of above 600–750 mm. In contrast, in Cangzhou, Hebei, there are many buildings in the urban area and floodlands along the Huanghe. Because of high salt content in groundwater, lack of irrigation water and low vegetation coverage, there is a lower ET, especially in the growth stages of winter wheat. The annual ET is about 350–500 mm, and the average transpiration of the whole area accounts for 60–70% of the ET.

For the summer and autumn harvest crops, the spatial distribution of ET was different in the crop-growing period and fitted with the seasonal distribution of precipitation. Total ET of the major crops (wheat and maize) in the crop growth period is shown in Fig. 5.4. In the wheat growth period, obviously the ET decreases from south to north. The ET values are between 350 and 450 mm in the southern areas and high values are mainly distributed in the northern part of Jiangsu and Anhui provinces. The ET is about 250 mm in the hilly areas of Shandong Province and the eastern rain-fed agricultural areas in Hebei Province, which is roughly equivalent to the synchronized rainfall. As the land lacks vegetation cover and is mostly sandy, the ET is less than 200 mm along the tidal flats of the Bohai Sea. In better irrigation locations such as the piedmont

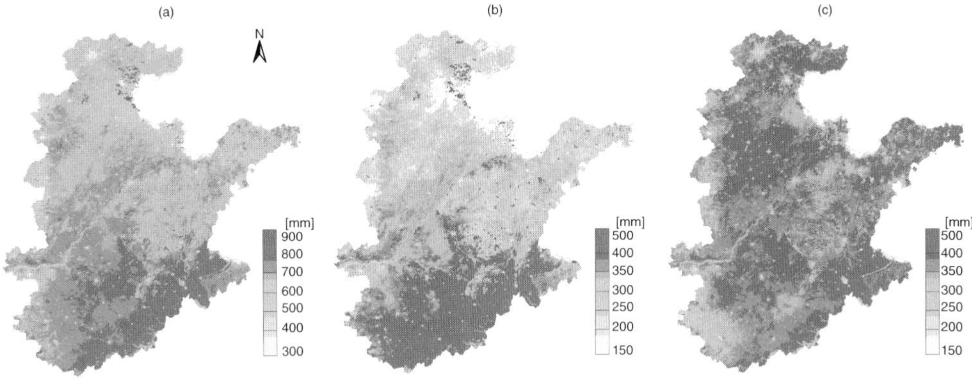

Fig. 5.4. Decadal averages of evapotranspiration (a) on annual total, (b) in winter wheat-growing period and (c) in summer maize-growing period, during 2000–2009.

plain of Taihang Mountain and the eastern part of Beijing, the ET is between 300 and 350 mm.

During the maize-growing season in summer, which is the rainy season in North China, precipitation can meet the crop growing needs but the spatial difference of ET is insignificant. Higher ET of approximately 450 mm is mainly distributed in the northern part of Jiangsu, the eastern Jiaodong Peninsula and most parts of the Hebei Plain. However, in the low hilly region in the western part of Henan Province, because of weak soil water retention capacity and poor vegetation, the ET is low at only about 320 mm. ET is between 360 and 420 mm in most parts of North China.

Overall there is a surfeit of annual precipitation to the south of the Yellow River, but for most parts of the northern regions ET is greater than precipitation (Fig. 5.5). The difference to the south of the Yellow River mainly comes from the upstream mountain reservoirs and groundwater irrigation supplement. Irrigation supplement is variable among regions in this area. The highest irrigation supplement is distributed in the piedmont in the Taihang and Yan mountains, which is about 150–200 mm, accounting for one-third to one-fourth of the total supplement. This indicates that the water use in this area is seriously higher than in the local water resources. In the wheat-growing period, the amount of farmland irrigation is approximately about 200 mm, and the largest irrigation amount comes out in the piedmont of the mountain, which is more than 200 mm. The rain-fed farmland practised some soil water conservation. In the maize-growing season, precipitation is 100–250 mm higher than the ET to the south of the Yellow River. The precipitation can supply enough water resources for the soil water, groundwater and surface water. But to the north of the Yellow River, there needs to be an addition of about 50 mm irrigation water in the piedmont of the Taihang and Yan mountains. In other areas, soil water storage can barely meet the needs for growing maize. Overall, the water supply and demand have a seasonal imbalance in the NCP. The water used for agriculture in spring is heavily dependent on the upstream mountain reservoir storage, runoff from the Yellow River and the deep groundwater.

5.5 Water Resources under Climate Change Scenarios in the 3H Region

According to the Fourth Assessment Report of the Intergovernmental Panel on Climate Change of 2007, it is a fact that the increase in the concentration of greenhouse gases in the atmosphere is a cause of global climate change. The global average surface

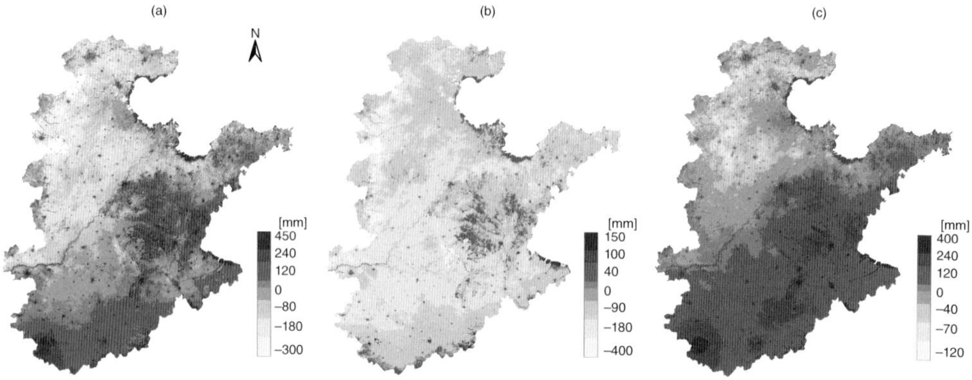

Fig. 5.5. Decadal averages of the differences between precipitation and evapotranspiration (a) on annual total, (b) in winter wheat-growing period and (c) in summer maize-growing period, during 2000–2009.

temperature has increased by around 0.56 to about 0.92°C in the past 100 years (1906–2005). Results have shown that the recent 100-year warming is jointly caused by natural climate fluctuation and anthropogenic activity, but most of the recent 50-year climate change is caused by anthropogenic activity (Ding, 2008).

Many studies have shown that regional warming will be the main feature in the 3H region in the 21st century. By 2030, an increase of 1°C in average temperature in the Yellow River Basin will lead to a 1.3% increase of annual precipitation variability; an increase of 0.9°C in Huai River Basin will lead to an increase of 1.5% precipitation variability. Runoff from rivers in the 3H region will reduce by 4–6% due to augmented evaporation associated with increased temperature. Temperature will increase by 1.1–1.4°C in the 3H region by 2030. There is the possibility of having extremely strong precipitation occurring only every 30–50 years, which will be more frequent in the 2080s and the 2090s in the 3H region. According to forecasts, temperature will significantly increase in the 3H region. The increase will be 1.0°C and 1.2°C under A2 and B2 models in the 2120s. But precipitation is extraordinarily complicated. In the long run; it increases as a whole in the 3H region, but it declines before the 2120s in part of this area and tends to fluctuate more due to temperature rise. Water demands will increase due to higher surface evaporation and crop transpiration.

Climate change is considered as posing the greatest threat to agriculture and food security in the 21st century, particularly in many of the sub-humid and semi-arid regions. As the main food supply area contributes approximately 41% of the total wheat yield and more than 30% of the total maize yield in China, the NCP is vulnerable to climate change (Guo *et al.*, 2010). Also, research showed that by 2030, climate change and increasing water demand will bring about increasing vulnerability to this region, especially for the Hai and Yellow river basins, aggravating the water shortage problems in the future (Xia *et al.*, 2012a, b). Therefore the issues of how to improve the efficiency of water use, how to adapt the crop systems to climate change and what measures should be taken to decrease the uncertainty of agricultural water resources need to be considered. Under future climate change scenarios in China, possibly the nutrient supply in the soil can be improved to increase crop yield by taking green water management measures to decrease the soil evaporation, thus improving the agricultural WUE to deal with the water resources vulnerability of this region.

5.5.1 The hydrologic responses under climate change scenarios

Under the framework of the Variable Infiltration Capacity Model (VIC), 12 climate scenarios were designed to account for possible variations in the future with respect to the baseline of historic climate patterns (Dan *et al.*, 2012). Results from the six representative types of climate scenarios (+2°C and +5°C warming, and 0%, +15%, −15% change in precipitation) show that rising temperatures for normal precipitation and for wet scenarios (+15% precipitation) yield greater increased ET in the south than in the north, which is confirmed by the remaining six scenarios described below. The largest increase or decrease of ET occurs for a 15% change in precipitation. Rising temperatures can lead to a south-to-north decreasing gradient of surface runoff. The six scenarios yield a large variation of runoff in the southern end of the 3H, which means that this zone is sensitive to climate change through surface runoff change.

5.5.2 The response of crop evapotranspiration under the Special Report on Emission Scenarios A2 and B1

The climate change projections from the runs of the GCM HadCM3, archived by the British Atmospheric Data Centre (http://badc.nerc.ac.uk/home/index.html), for A2 and B1 scenarios developed for the Third Assessment Report (Nakicenovic and Swart, 2000) of IPCC Special Report on Emission Scenarios (SRES) are used to simulate the responses of crop yield, ET and WUE to climate changes in the 21st century for the NCP. The A2 scenario describes a very heterogeneous world of high population growth, slow economic development and strong regional cultural identities. B1 is a rather optimistic scenario assuming a 'convergent world' and putting emphasis on global solutions to economic, social and environmental sustainability. The B1 scenario also assumes high economic growth but with a substantial shift to nuclear energy. The data for the A2 scenario include monthly values of maximum, minimum and mean temperatures, precipitation, relative humidity, wind speed and short-wave radiation. So do the data for B1 scenario, but maximum and minimum temperatures are missing. According to the projections, for example, in the 2090s, atmospheric CO_2 concentration, precipitation and daily mean air temperature will increase by 280 ppm, 16% and 2.8°C for B1 and 470 ppm, 48% and 4.5°C for A2, respectively.

Crop water use (ET) under SRES A2 and B1 scenarios is predicted with the Vegetation Interface Processes (VIP) model (an eco-hydrological model) (Mo and Liu, 2001; Mo *et al.*, 2005). Cumulative ET in the growing stage of winter wheat seems to be affected only slightly by climate change. The cumulative ET amounts gently increase for both A2 and B1 scenarios by less than 6%. As is known, the air warming will intensify ET, whereas both lower stomata conductance resulting from higher CO_2 concentration and growing period shortened by warming will mitigate the rising of total ET amount. As a consequence, the change of ET is not remarkable. Cumulative ET in the maize-growing period will significantly increase over 10% after the 2050s. At the end of the 21st century, the cumulative ET amounts under A2 and B1 scenarios will be 37% and 20% higher than the current values over the maize-growing period, respectively (Mo *et al.*, 2009).

5.6 Adaptive Strategies of Sustainable Agricultural Water Resources Utilization

After several decades of past water policies that focused on increasing water supply by constructing more canals and larger reservoirs (Ross, 1983), China's leaders have started to recognize the need to stem the rising demand for water (Boxer, 2001). In particular, under the pressure of climate change, it has become an urgent task for policy makers to change water management from supply side to demand side. That is, transferring from meeting demand with new resources or 'supply side' to managing the demand itself to postpone or avoid the need to develop new water resources. In order to successfully realize this transfer of management strategies, in 2006 the Chinese government set the target for water saving for the 12th 5-Year Plan period (2006–2010); water use per GDP should be reduced by 20% relative to the 2005 level (National Development and Reform Commission *et al.*, 2006). In 2009, the Ministry of Water Resources clearly proposed to implement water demand management through setting up 'Three Red Lines'[2] for the most stringent water management institution. In January 2013, the State Council in China issued the *Assessment Method for Implementing the Most Stringent Water Management Institution.*

As the main water-using sector in China, a large portion of the water saving has been slated to come from the agriculture sector. Therefore, in order to mitigate the impacts of climate change on agricultural water use and realize its sustainable utilization, the key adaptive water management strategy in the agriculture sector is how to improve agricultural WUE through reforming agricultural policies. This section discusses the reforming of agricultural water management and policies.

5.6.1 Improving the performance of participatory irrigation management reform

From the 1980s, China began to attach importance to enhancing the construction of the water resources management system and developing relevant policies. Thus, in 1981, China made an important decision to shift the focus of water conservancy work to management and proposed 'enhancing operation and management and putting focus on economic benefits' as the main policy of water conservancy in 1983. The *Water Law of the People's Republic of China* promulgated by the State in 1988 has defined the significance of water resources management, and the relevant State policies focus more on water resources management. In 2003, the Ministry of Water Conservancy issued *Suggestions on the Implementation of Small Rural Water Conservancy Engineering Management System Reform* and proposed a plan to establish a management system orienting on encouraging the development of participatory irrigation management. In 2005, the Ministry of Water Resources, the National Development and Reform Committee and the Ministry of Civil Affairs jointly released *Opinions on Enhancing Farmer Water User Association (WUA) Building.* These policy documents provided an important policy guarantee for driving water users to participate in irrigation management. Up to 2011, more than 30,000 water user associations had been established in China.

However, some scholars found that not all the participatory irrigation management was successful (Huang *et al.*, 2010a). Wang *et al.* (2010) found that due to implementation of the Five Principles of WUA,[3] the World Bank-funded participatory irrigation management performed very well in improving WUE. However, based on a large field survey and empirical analysis in four large irrigation districts in the Yellow River Basin, Wang *et al.* (2005) pointed out that the transfers of the irrigation management from collective to WUAs or contracting management are mostly nominal reforms, and such reform had not realized the reform purpose of improving WUE. Whether the reform realized the reform purpose depended on the establishment of the water saving incentive mechanism. If an effective water saving incentive mechanism was established after reform, the irrigation water utilization efficiency increased by 40%.

Therefore in the future implementation process of irrigation management reform, governmental authorities must not only push forward effective irrigation management patterns but also pay attention to the reform pattern of internal system building (e.g. as a water-saving incentive mechanism of administrative supervisors) to guarantee the sustainable and effective exertion of reform performance. If focus is not put on the internal system building of management patterns, the reform performance may be one-off and short-term, and it is hard to promote the sustainable development of water resources and social economy. Thus, government authorities must pay close attention to how to establish a long-term and effective water-saving incentive mechanism to proceed with the improvement of reform performance.

5.6.2 Establishing a water rights system

While reforming irrigation management is one effective way to increase WUE, one key problem of such a reform is its sustainability. The local governments (especially in the upstream areas) do not seem to receive any benefits from reducing water use through reforming irrigation management. The major reason is that there are no mechanisms to compensate local governments that have succeeded in reforming irrigation management and have increased WUE. Their saved water is reallocated to other regions without any benefit for them. As a result, local governments in the Yellow River Basin (YRB) have no incentive to push management reforms; and reforms are unlikely to be sustainable in the long run if they lack local government support. In order to allow regions in the upper reaches of the YRB to receive compensation when they save water, one possible solution is to establish a water rights system.

Since 2000, the Yellow River Conservation Commission (YRCC) has begun to promote the establishment of water rights systems through conducting demonstration projects aimed at reducing water competition among sectors. With increasing water shortages, water becomes insufficient to support industrial development, especially the energy industry, in the upstream provinces such as Ningxia and Inner Mongolia. In order to solve this problem, in 2003 the YRCC established some water rights demonstration sites in the upstream reaches of the YRB. The purpose of these demonstration sites is to reallocate water from agriculture to industry through increasing irrigation efficiency (Li, 2007; Molden *et al.*, 2007; Wang, 2007). In 2004, in order to promote the water transfer work in the YRB, the MWR issued the *Guidance on Water Rights Transfer Demonstration Works in Inner Mongolia and Ningxia Provinces*. The YRCC also released two regulations titled *Management and Implementation Measures on Water Rights Transfer in the Yellow River Basin* and *Management Regulation on Water-Saving Engineering*. These regulations have provided the legal foundation for water rights transfers in the YRB.

Despite some progress on water rights transfer, there are still considerable challenges facing both central and local governments regarding the implementation of water rights transfers (Yang *et al.*, 2006; Jiang *et al.*, 2007). The first is the engineering problem. YRCC considers the construction of water-saving infrastructure to be very slow, thus constraining the progress of water rights transfers. In addition, management of this new infrastructure remains a challenge. The second constraint relates to water rights. While some water rights transfer projects have been established, a general water rights system has not been developed. Water users still have no clear ideas about how many water rights they have. Water rights transfers still depend on administrative power, not on developed water markets. Water rights transfers are still a function of the central and local governments, and are not adjusted by market signals or economic measures. In fact, China still has a long way to go in establishing a real water rights system and water markets. How to effectively promote the system of water rights is still hotly debated by many policy makers and researchers.

5.6.3 Reforming agricultural water price

Since the 1990s, China's water officials have begun to consider reforming the pricing of irrigation water as a key policy instrument for dealing with the nation's water scarcity problem. The objective of the reform is to provide agricultural users with economic incentives to save water through higher water prices. Relying on a set of household-level data, Huang *et al.* (2010b) examined the potential for conserving water through water pricing reform. Their study shows that the water pricing policy has the potential to resolve the water scarcity problem in China. However, because the current cost of water is far below the true value of water in many regions, a large increase in the price of water from the current level is required. Some scholars have also recognized the difficulties of implementing water pricing (e.g. Sampath, 1992; Dinar, 2000). Importantly, Huang *et al.* (2010b) revealed the costs associated with higher water prices. Higher irrigation costs will lower the production of all crops, in general, and that of grain crops in particular. This may hurt the nation's food security goal of achieving 95% self-sufficiency for all major grains in the short run (such as within 5 years). Furthermore, when facing higher irrigation costs, households suffer income losses, although income distribution does not deteriorate.

The goal of the water pricing policy, which is to manage water resources in a sustainable way, does not conflict with the long-run goal of the nation's food security policy. On the other hand, dealing with the negative production and income impacts of higher irrigation cost will pose a number of challenges to policy makers, at least in the short run. One possible solution is to set a water saving target in the agriculture sector to be below the national target of 20% (Huang *et al.*, 2010b). In addition to setting a modest water-saving target, if China's leaders plan to increase water prices to address the nation's water crisis, an integrated package of policies will be needed to achieve water savings without hurting rural incomes or national food security. One solution is to develop a subsidy programme in tandem with the water pricing policy that transfers income to households. A subsidy programme is a realistic solution in China's political economy environment. China's agricultural policy has gradually switched from taxing farmers to directly subsidizing them. The tax-for-free reform that targets at eventual elimination of taxation on rural households has been implemented over recent decades (Brandt *et al.*, 2005). Hence, a subsidy programme is well in line with the government's policy agenda. In order to play the effective role of subsidy policy on saving water, instead of a grain subsidy, China should consider giving a 'decoupled water-price reimbursement payment' or 'unconditional payment' to farmers.

5.6.4 Promoting the adoption of agricultural water saving technology

In order to increase WUE, promoting water saving technology has been highly addressed by policy makers in China. The Chinese government stated that the promotion of water saving technology is one of the priorities in its water conservancy reforms. Issued in March 2011, the rural and agricultural parts of the 12th Chinese 5-Year Plan highlight the importance of efficiency and technological innovation. In addition, the Chinese government has announced an expenditure of RMB4 trillion (over US$600 bn) on water conservation over the next 10 years and a specific investment of US$6.03 bn to support the adoption of water saving technology on 2.53 million ha.

Existing literature tells us that policy support is one important driving factor that affects farmers' decisions on adopting water saving technology. Policies promoting adoption of water saving technology often aim to overcome farmers' economic and technical constraints. In order to overcome these constraints, direct provision of subsidy has proven to be one important policy measure in increasing the adoption rate of agricultural water-saving technology, especially when the adoption rates are low. Based on an econometric analysis, Liu *et al.* (2008) confirmed the significant positive relationship between subsidies and adoption of some kind of water saving technology in rural China. For technical constraints, providing knowledge and technical advice through extension service activities is an effective way to increase the adoption rate of agricultural water-saving technology. Therefore, in the future, in order to increase the adoption rate of agricultural water-saving technologies it is necessary to establish rational policy support systems, in addition to which, setting up a rational water price policy will also play an important role in encouraging farmers to adopt water-saving technologies.

5.7 Conclusions

In order to explore the regional crop response to climate change, this study analysed the spatial variability and evolution of crop yield, ET and WUE with a process-based crop model in the NCP and identified the contribution of climate change to their enhancement. The impacts of future climate changes under A2 and B1 scenarios on the wheat–maize double-cropping system were also assessed.

The results show that crop production has increased rapidly over the past decade in the NCP. Accompanying production improvement, crop ET has also risen significantly. There exist spatial patterns of crop yield stemming mainly from soil quality and irrigation facilities. Under IPCC SRES A2 and B1 scenarios, production of winter wheat will increase with slightly intensified ET; in contrast, summer maize production will slightly decline with a significant increase of ET. Also, with agricultural management, maize is more productive than wheat, in that wheat relies more on irrigation than maize, yield level of maize is higher than that of wheat, the water consumption of maize is lower, and the response of maize yield is larger than that of wheat yield to agricultural management. However, the simulation also suggests that wheat is more resilient to climate change than maize. If wheat or maize is to be more favourable in the NCP will depend on the conditions in the future. None the less, our results provide a scientific basis and reference for governments' decision making from the perspective of regional climate change response.

In order to mitigate the impacts of climate change on agricultural water use and realize its sustainable utilization, the key adaptive water management strategy in the agriculture sector is how to improve efficiency of agricultural water use through reforming agricultural water management and policies. The following measures can be implemented to reform agricultural water management and policies:

- Improving the performance of participatory irrigation management reform and paying close attention to the reform pattern of the internal system building in establishing the water rights system to proceed with the improvement of reform performance.
- Establishing a water rights system. In order to allow regions in the upper reaches of the YRB to receive compensation when they save water, one possible solution is to establish a water rights system. Establishing a real water rights system and water markets to promote the system of water rights effectively needs further research.
- Reforming agricultural water price. The current cost of water is far below the true value of water in many regions so that a large increase in the price of water from the current level is required. A subsidy programme developed in tandem with the water pricing policy that transfers income to households can achieve water savings without hurting rural incomes or national food security.
- Promoting the adoption of agricultural water saving technology. Establish rational policy support systems, which are necessary to increase the adoption rate of agricultural water-saving technologies. Setting up a rational water price policy will also play an important role in encouraging farmers to adopt water-saving technologies.

Acknowledgements

This study was supported by the National Basic Research Program of China (2010CB428406/2012CB956204) and the Natural Science Foundation of China (No. 51279140/41071025).

Notes

[1] The Ministry of Water Resources (MWR) completed a nationwide assessment of water resources in 2000, which summarizes the significant challenges China is facing in ensuring the security of its water resources (see Qian and Zhang, 2001; Xia *et al.*, 2011).

[2] 'Three red lines' refer to the total water use, efficiency and dirt holding capacity, with particular emphasis on the 2030 total water, which is controlled at 700 Bm^3.

[3] Principle 1 is adequate and reliable water supply: a WUA is organized only to see that adequate and reliable water supply is available and whether on-farm delivery infrastructure is in good condition and can be properly maintained by its members. Principle 2 is legal status and participation: a WUA should be the farmers' own organization, a legal entity and have a leadership elected by its members. Principle 3 is that WUAs are organized within hydraulic boundaries: the jurisdiction of a WUA should be the hydraulic boundaries of the delivery system. Principle 4 is that water deliveries can be measured volumetrically: a WUA should be able to receive its water under contract from its water suppliers and it should be able to measure the water volumetrically. Principle 5 is that a WUA equitably collects water charges from members: a WUA should equitably assess and collect water charges from its members and make payment for the cost of water.

References

Boxer, B. (2001) Contradictions and challenges in China's water policy development. *Water International* 26, 335–341.

Brandt, L., Rozelle, S. and Zhang, L. (2005) *Tax-for-Fee Reform, Village Operating Budgets and Public Goods Investment.* Report to the World Bank, Beijing.

Cao, J.T., Li, Y.Y., Shen, F.X., Chen, Y.W. and Xia, J. (2012) Drawing down our resources: estimating the total appropriation of water in China. *Water International* 37, 512–522.

Dan, L., Ji, J., Xie, Z., Chen, F., Wen, G. and Richey, J.E. (2012) Hydrological projections of climate change scenarios over the 3H region of China: a VIC model assessment. *Journal of Geophysical Research* 117, D11102. DOI: 10.1029/2011JD017131.

Dinar, A. (ed.) (2000) Political economy of water pricing reforms. In: *The Political Economy of Water Pricing Reforms.* Oxford University Press, New York, pp. 1–27.

Ding, Y.H. (2008) Human activity and the global climate change and its impact on water resources. *China Water Resources* 2, 20–27.

Guo, R.P., Lin, Z.H., Mo, X.G. and Yang, C.L. (2010) Responses of crop yield and water use efficiency to climate change in the North Plain. *Agricultural Water Management* 97, 1185–1194.

Huang, Q., Wang, J., Easter, K. and Rozelle, S. (2010a) Empirical assessment of water management institutions in northern China. *Agricultural Water Management* 98, 361–369.

Huang, Q., Rozelle, S., Howitt, R., Wang, J. and Huang, J. (2010b) Irrigation water demand and implications for water pricing policy in rural China. *Environment and Development Economics* 15, 293–319.

Jiang, B., Zhang, B. and Li, E. (2007) Analysis of monitoring effects of water rights transfer in the Yellow River Basin. *China Water Resources* 19, 47–48.

Li, G. (2007) Exploring and practice of water rights transfer in the Yellow River Basin. *China Water Resources* 19, 30–31.

Liu, Y., Huang, J., Wang, J. and Rozelle, S. (2008) Determinants of agricultural water saving technology adoption: An empirical study of 10 provinces of China. *Ecological Economy* 4, 462–472.

Mo, X. and Liu, S. (2001) Simulating evapotranspiration and photosynthesis of winter wheat over the entire growing season. *Agricultural and Forestry Meteorology* 109, 203–222.

Mo, X., Liu, S., Lin, Z., Xu, Y., Xiang, Y. and McVicar, T. (2005) Simulation of winter wheat production and water use efficiency with GIS and RS Data in the NCP. *Ecological Modelling* 183, 301–322.

Mo, X., Liu, S. and Lin, Z. (2009) Regional crop yield, water consumption and water use efficiency and their responses to climate change in the North China Plain. *Agriculture, Ecosystem and Environment* 134, 67–78.

Mo, X., Liu, S., Lin, Z. and Qiu, J. (2011) Patterns of evapotranspiration and GPP and their responses to climate variations over the North China Plain. *Acta Geographica Sinica* 66(5), 589–598.

Molden, D., Dong, B., Loeve, R., Barker, R. and Tuong, T.P. (2007) Agricultural water productivity and savings: policy lessons from two diverse sites in China. *Water Policy* 9, 29–44.

Nakicenovic, N. and Swart, R. (eds) (2000) *Special Report on Emissions Scenarios.* Cambridge University Press, Cambridge, UK, 612 pp.

National Development and Reform Commission, MWR and Ministry of Housing and Urban-Rural

Development of the People's Republic of China (2006) *Water-Saving Society-Building during the 11th Five-Year Plan*. China Water & Power Press, Beijing.

Qian, Z.Y. and Zhang, G.D. (2001) *Strategic Studies of Water Resources for Sustainable Development in China*, Vol. 1. Chinese Water Resources and Hydropower Press, Beijing.

Ross, L. (1983) Changes in water policy in the People's Republic of China. *Water Resources Bulletin* 19, 69–72.

Sampath, R.K. (1992) Issues in irrigation pricing in developing countries. *World Development* 20, 967–977.

Shi, Y.L. (2008) *Study on Reasonable Allocation of Agricultural Resource and Improvement of the Comprehensive Productivity.* China Agricultural Press, Beijing.

Wang, J.X., Xu, Z.G., Huang, J.K. and Rozelle, S. (2005) Incentives in water management reform: assessing the effect on water use, productivity and poverty in the Yellow River Basin. *Environment and Development Economics* 10, 769–799.

Wang, J.X., Huang, J.K., Zhang, L.J., Huang, Q.Q. and Rozelle, S. (2010) Water governance and water use efficiency: the five principles of WUA management and performance in China. *Journal of the American Water Resources Association* 46, 665–685.

Wang, L.H., Zhao, Z.X., Huang, Z.F., Zhang, C. and Yan, D.H. (2009) Agricultural water use in Huang-Huai-Hai region: problems and countermeasures. *Chinese Journal of Ecology* 28, 2094–2101.

Wang, Y. (2007) Comments on the reform of water price, water rights and water markets in China. *China's Population, Resources and Environment* 5, 153–158.

Xia, J., Liu, C.M., Ding, Y.J., Jia, S.F. and Lin, C.H. (2011) *Water Issues Vision in China*, vol. 1. Science Press, Beijing.

Xia, J., Qiu, B. and Li, Y.Y. (2012a) Water resources vulnerability and adaptive management in the Huang, Huai and Hai river basins of China. *Water International* 37, 523–536.

Xia, J., Qiu, B., Pan, X.Y., Weng, J.Y., Fu, G.B. and OuYang, R.L. (2012b) Assessment of water resources vulnerability under climate change and human activities. *Advances in Earth Science* 27, 51–59.

Yang, X., Zheng, C., Chen, H. and Yang, Z. (2006) Constraints analysis of water rights transfer in China. *Journal of Economics of Water Resources* 4, 65–66.

Zhang, G.H., Lian, Y.L., Liu, C.H., Yan, M.J. and Wang, J.Z. (2011) Situation and origin of water resources in short supply in North China Plain. *Journal of Earth Sciences and Environment* 33(2), 172–176.

6 Climate Change Impacts and Adaptation in Agricultural Water Management in the Philippines

Felino P. Lansigan* and Amparo C. Dela Cruz
University of the Philippines Los Baños, Laguna, Philippines

Abstract

Climate change is now a reality and is expected to have profound impacts on agricultural production systems, thus threatening food security. Local observational evidence of changing climate based on available historical weather data-sets is presented for selected locations in the Philippines. In this study, plausible climate scenarios and downscaled climate projections for representative crop production areas are discussed. Results of downscaling of climate projections for 2020, 2050 and 2080 using a dynamic downscaling procedure and a statistical downscaling method are presented and compared. Impacts of climate change and variability on crop productivity and production systems in selected crop-growing areas in the country are presented. Crop yields using a calibrated crop simulation model for a standard rice variety (IR-64) and a local maize variety (IPB-911) for projected climate were determined. Results indicate a reduction from 8% to 14% in crop yields per 1°C temperature increase depending on season and location. Some adaptation strategies related to agricultural water management to minimize the adverse effects and impacts of climate change are described. These location-specific measures based on best practices include adjusting the planting calendar, improving water use efficiency and irrigation water management, water impoundment, planting stress-tolerant varieties, and weather index-based insurance for crop production.

6.1 Climate Change and Climate Variability in the Philippines

6.1.1 Introduction

Global climate variability influences the local weather systems that exhibit spatial and temporal variation in specific areas. Local weather systems and climate affect the hydrologic regimes in watersheds, agricultural production systems, livelihoods and other socio-economic activities. In recent years, it has been demonstrated that long- and short-term climate variabilities, such as the global El Niño Southern Oscillation (ENSO) episodes, are highly correlated with seasonal climate variability in an area. ENSO events as represented by the sea surface temperature anomaly (SSTA) over the Niño 3.4 Region (i.e. bounded by the 5°N–5°S latitude, and 120°W–170°W longitude), for example, have been associated with the seasonal climate in some countries in Asia such as Indonesia and the Philippines (Naylor *et al.*, 2001; SEARCA and FAO, 2010). The degree of relationships of SSTA with rainfall for different locations in the country ranges from weak to moderate during different periods within the year. Nevertheless, the degree of correlation warrants consideration of using climate information for advanced planning of hydrological and agricultural activities.

Since the ENSO signals at the Niño 3.4 Region are highly correlated with the local weather systems prevailing in the Philippines

* Corresponding author, e-mail: fplansigan@up.edu.ph

3–6 months ahead, advanced climate information can be used to predict the expected seasonal climate in the Philippines. Agricultural crop production in the Philippines is highly correlated with the climate variability as reflected by the high correlation of SSTA with crop production area planted and yield (SEARCA and FAO, 2010). Thus, agricultural production activities such as determining what crops to grow, when and where to best plant the crops, the frequency and timing of irrigation, and also fertilizer application can be planned.

Climate change is now a reality. There is now growing observational evidence at the global, regional and local levels to support that climate has changed (Alcamo *et al.*, 2005; IPCC, 2007; Lansigan, 2009; Comiso *et al.*, 2014). With climate change are found the shifts not only in the mean but also in the variance of climate variables such as temperature and precipitation as well as in the corresponding magnitudes and occurrences of extreme events such as dry and wet episodes. Climate change and climate variability have profound effects and impacts on food production systems, ecosystem stability and socio-economic activities. Responses to climate change in the context of water management in crop production range from growing stress-tolerant rice varieties, adjusting the planting calendar and adoption of weather insurance. Thus, this chapter presents some local observational evidence of changing climate in the Philippines, describes the effects and impacts of climate change on crop production, and also discusses some climate adaptation measures related to agricultural water management in crop production.

6.1.2 Local observational evidence of changing climate in the Philippines

There is increasing local evidence based on historical weather and climate data that climate change is now occurring in the Philippines. A recent study of PAGASA (2011) analysed the trends in climate data based on historical records in a number of weather gauging stations in the country. Figure 6.1 shows the general trends in extreme daily temperature in a number of locations in the Philippines from 1951 to 2008. Warm nights and hot days are generally increasing. Climate data show that over the span of 60 years the mean temperature in the archipelago has increased by 0.65°C from 1951 to 2010. Climate change has also brought erratic and changing rainfall patterns characterized by more intense extreme rainfall events. Figure 6.2 shows the general observed trends in extreme rainfall intensity in a number of locations based on the amount of rainfall exceeding the highest four rain events in the year. While the increases in frequency as well as in the intensity of extreme rainfall events are already being experienced in many areas, the observed changes are not statistically significant (PAGASA, 2011). In recent years, however, changes in rainfall patterns in many locations, particularly in crop-growing areas, have been observed necessitating the adjustments in planting dates and other farm operations.

Using the historical records of annual maximum daily rainfall in Los Baños, Philippines (for location see Fig. 6.3) for two time periods, 1959–1978 and 1979–2006, the frequency distribution as shown in Fig. 6.4 has changed. During the earlier period (1959–1978) annual extreme daily rainfall events tended to be uniformly distributed. The later period (1979–2006) shows increased frequency of average maximum daily rainfall plus the occurrence of the more recent extreme daily rainfall event (348 mm), which led to floods resulting in loss of lives and damage to property and livelihoods. Historical data also in Los Baños indicated that minimum daily temperature has significantly increased by 1.27°C over the span of 27 years although the maximum daily temperature has not shown a significant trend. The increase in minimum temperature is a significant factor in the reduced rice yields in the area since the photosynthetic activity of the crop will be affected.

Moreover, the analysis of historical climate data available in Legazpi City, Albay Province (for location see Fig. 6.3) also shows that monthly mean minimum

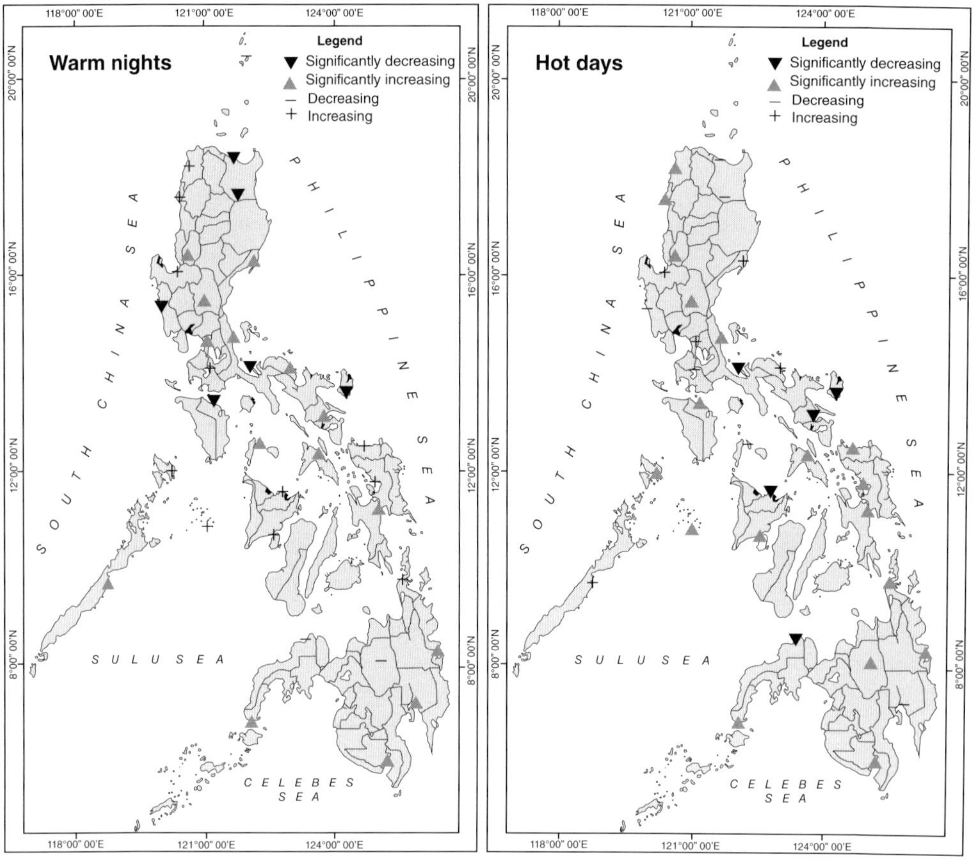

Fig. 6.1. Observed trends in extreme daily temperature in the Philippines based on historical weather records from 1951–2008 (from PAGASA, 2011).

temperature had increased significantly as presented in Fig. 6.5. Similar trends in the analysis of climate variables such as the number of wet days and dry days, the number of days exceeding rainfall threshold levels, the number of hot days and warm nights, etc. have also been observed in other areas in the country with reasonably adequate historical records (Lansigan, 2011a).

6.1.3 Climate change scenarios and projections

General circulation models (GCMs) have been used to project future climate conditions in terms of changes in precipitation and temperature patterns at the global level (IPCC, 2001). Future climates were predicted based on a number of assumptions in terms of the pathways of greenhouse gas (GHG) emissions. These pathways are defined by projected economic development, population growth and human activities. It should be noted that while the Special Report on Emissions Scenarios (SRES) climate scenarios have been widely used, in recent years there has been a new initiative by climate scientists to produce more plausible future climate scenarios based on the so-called Representative Concentration Pathways (RCPs) that determine GHG emissions from human activities (Moss *et al.*, 2010). Producing future climate projections involves integrated assessments considering the main features and facets of human systems, climate and Earth-system models, and impact assessments with a focus on adaptation and

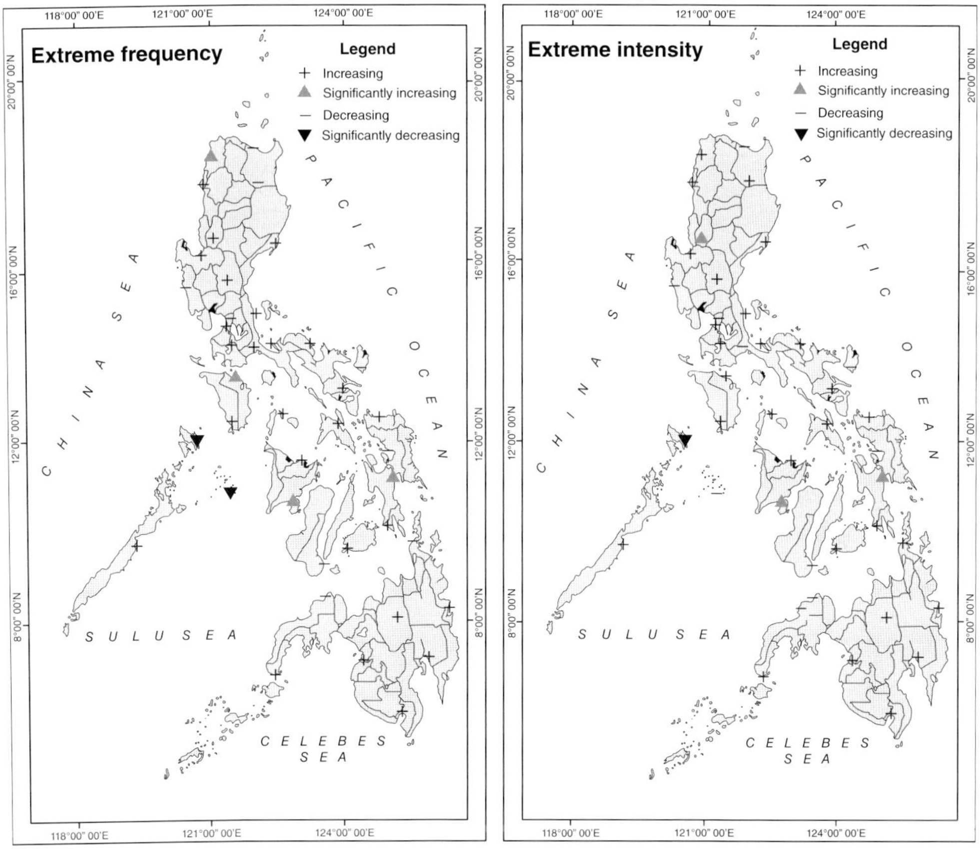

Fig. 6.2. Observed trends in extreme daily rainfall intensity in the Philippines based on historical weather records from 1951–2008 (from PAGASA, 2011).

vulnerability to climate change (Gaffney, 2010). However, there is uncertainty on which model and climate scenario would adequately represent the future climate.

Figure 6.6 shows the observed and the projected climate change in the Philippines in terms of mean temperature for the baseline period (1971–2000), for 2020, 2050 and 2100 for two SRES climate scenarios, namely the medium-range emission scenario (A1B), and the high-range emission scenario (A2) from PAGASA (2011). There is already an observed change of 0.65°C in mean temperature over the span of 60 years from 1951 to 2010. Increase in average temperature is expected to range from 1.0 to 3.1°C and from 0.7°C to 3.4°C for the period 2020 to 2100 under the A1B and A2 scenarios, respectively.

On the other hand, seasonal change in mean rainfall in the Philippines shows wide spatial variability across locations. The dry months are expected to be drier and the wet months wetter. This has important implications in terms of availability and dependability of water from rainfall, particularly for rainfed agriculture.

6.1.4 Downscaled climate projections in the Philippines using PRECIS and SDSM

GCM climate projections, however, are not directly useful for national and local studies such as vulnerability assessments, hydrologic frequency and risks analyses, and impact assessments (IPCC, 2012). The spatial resolution of GCM data is too coarse

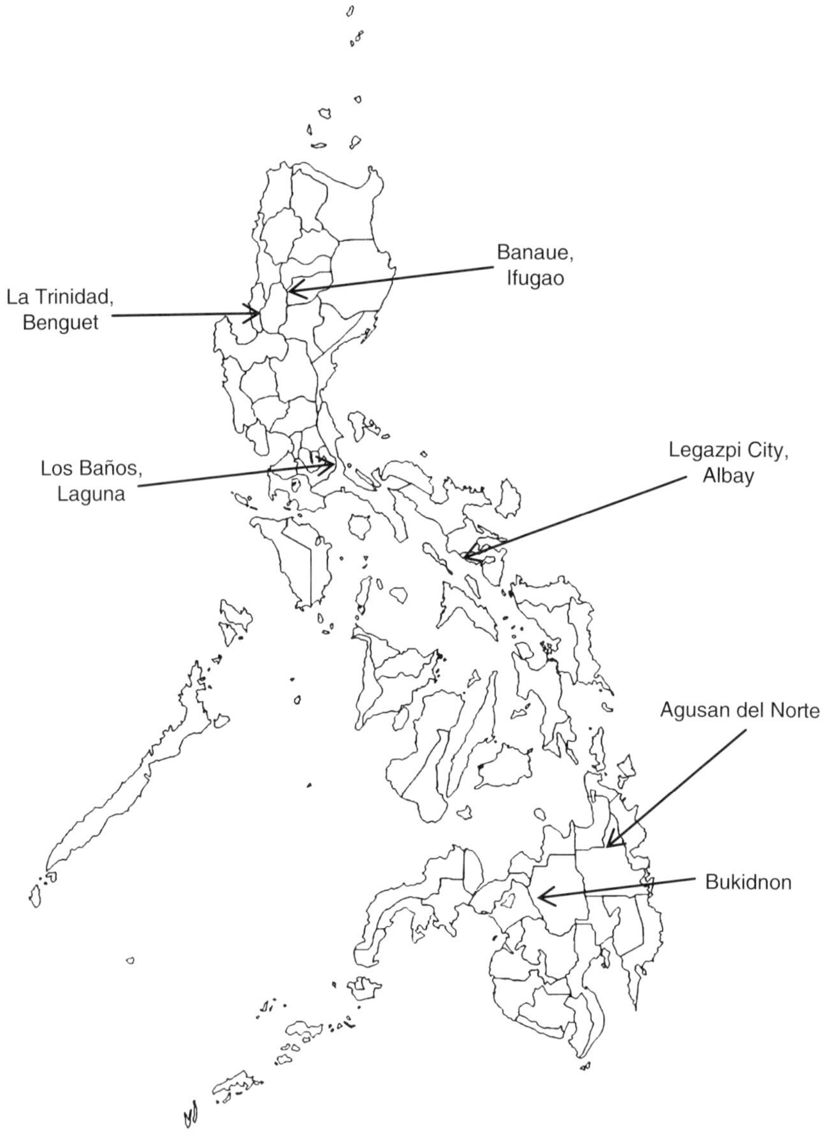

Fig. 6.3. Location map of the different sites in the Philippines considered in the study.

(typically 50,000 km^2) and does not reflect the spatial variability across different locations within the country. While the Philippines has different climatic types, the GCMs considering their coarse resolution, however, provide only one climate projection for the entire country ignoring the local climate variability, i.e. the country is treated as one homogeneous spatial entity.

Thus, downscaling procedures are needed to derive the local-scale surface weather conditions given the regional-scale atmospheric predictor variables used in the global and regional scale models (Wilby and Dawson, 2007). That is, downscaling climate projections is a method used to obtain high-resolution climate information at the scale of 50 km × 50 km or less from the relatively

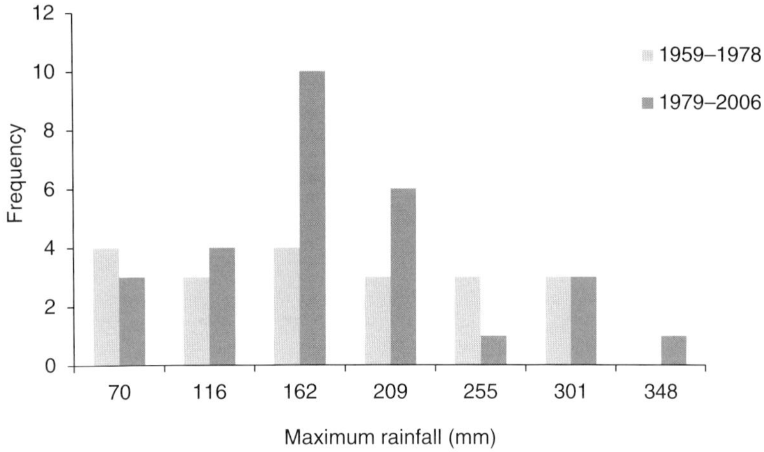

Fig. 6.4. Distribution of annual maximum daily rainfall in Los Baños, Laguna, Philippines during two time periods, 1959–1978 and 1979–2006 (from Lansigan, 2009).

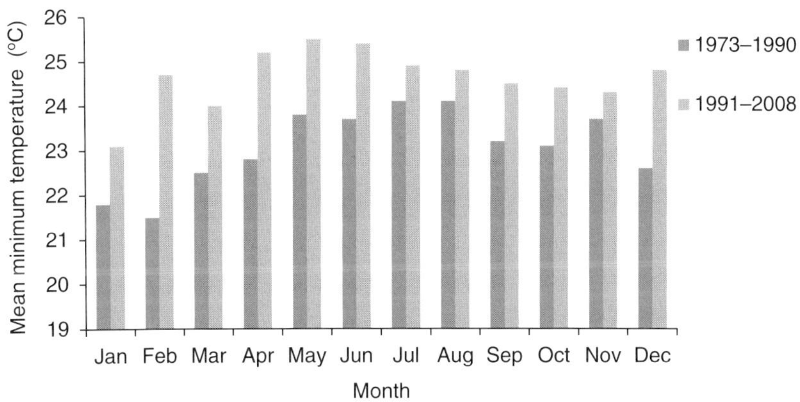

Fig. 6.5. Monthly mean minimum temperature in Legazpi City, Philippines from 1973–1990 and 1991–2008 (from PAGASA, 2011).

coarse-resolution global models such as the GCMs (Anderson, 2008).

There are two types of downscaling procedures, the dynamical downscaling method and the statistical downscaling technique. Dynamic approaches such as regional models (e.g. PRECIS, Providing Regional Climate for Impacts Studies) simulate the interactions between the atmosphere, oceans and land processes at the regional level. It also requires high speed and large memory computing facilities not commonly available in developing countries as well as the need for more intensive training of model users. On the other hand, statistical techniques are based on sound and robust statistical theories and methods that try to capture and mimic the observed relationships between the prediction and the predictors for a given location. Each of these groups of procedures has inherent advantages and limitations with respect to data and information requirements, skills and expertise needed, and facilities to be used.

Dynamic downscaling method uses a limited-area, high-resolution model such as regional climate models (RCMs) driven by boundary conditions from a GCM, to derive

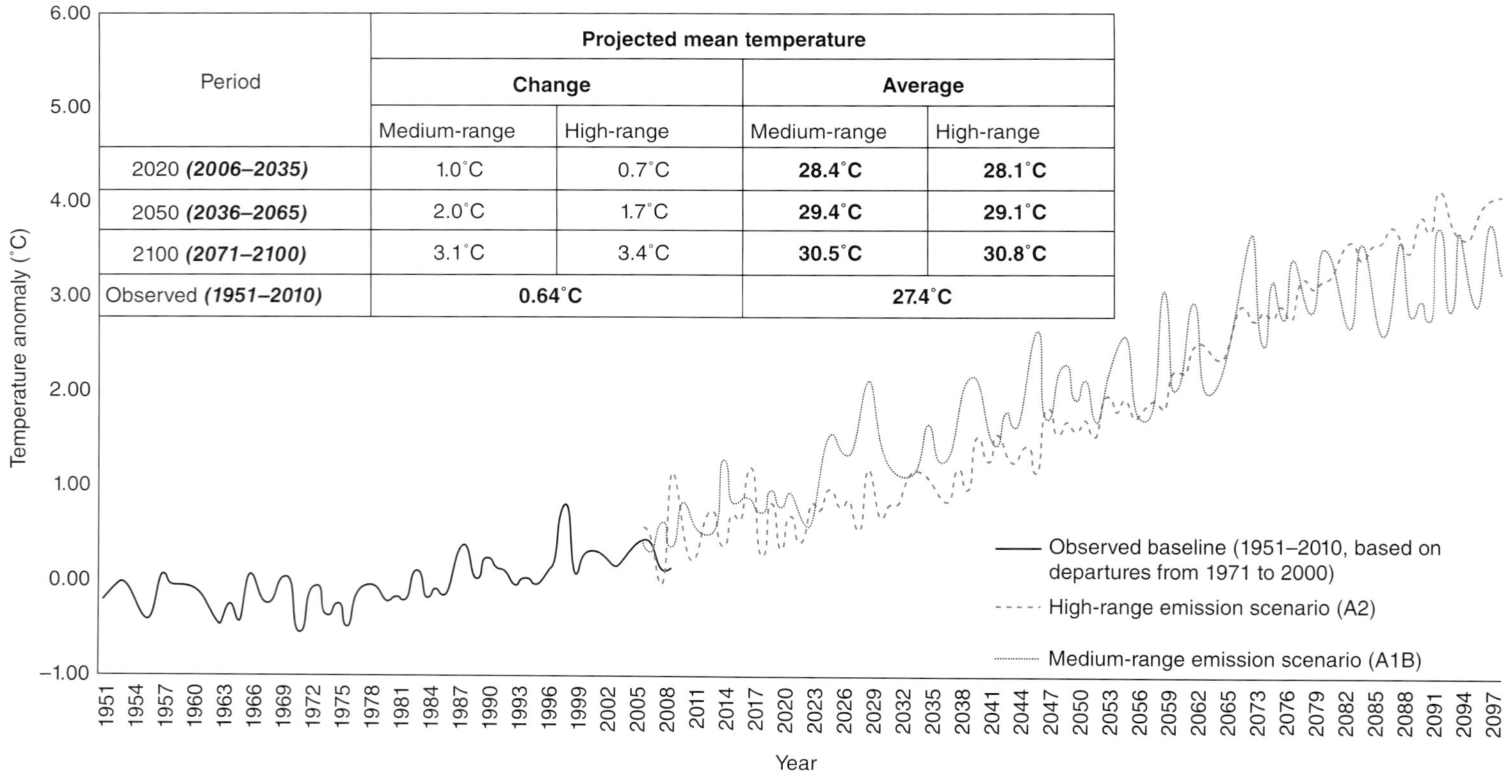

Period	Projected mean temperature			
	Change		Average	
	Medium-range	High-range	Medium-range	High-range
2020 ***(2006–2035)***	1.0°C	0.7°C	**28.4°C**	**28.1°C**
2050 ***(2036–2065)***	2.0°C	1.7°C	**29.4°C**	**29.1°C**
2100 ***(2071–2100)***	3.1°C	3.4°C	**30.5°C**	**30.8°C**
Observed ***(1951–2010)***	**0.64°C**		**27.4°C**	

Fig. 6.6. Historical (observed) and projected annual mean temperature anomalies for the Philippines from 1951 to 2100-based departures from 1971 to 2000 normal values (from PAGASA, 2011).

smaller-scale information. RCMs generally have a domain area of 106 to 107 km^2 and a resolution of 20 to 60 km. An example of a dynamic downscaling approach is the PRECIS (Providing Regional Climates for Impact Studies) model used by PAGASA (2011) to generate climate projections for different provinces in the Philippines. However, regional climate models such as PRECIS are computationally demanding and require intensive training on the use of the model.

Recent studies (e.g. Wilby and Dawson, 2007; Lansigan *et al.*, 2013) using a statistical downscaling method to generate finer resolution climate projections for 2020, 2050 and 2080 show that results of statistical downscaling are quite comparable with the downscaled climate projections generated by PAGASA (2011) using the PRECIS model as shown in Fig. 6.7. Statistically downscaled information was used in the Philippines for impact assessments on the hydrology of the watershed (Delfino *et al.*, 2012), on the frequency analysis of episodes of wet and dry days, agricultural planning, and on impact assessment on crop productivity (Lansigan and Dating, 2012; Lansigan *et al.*, 2013). Results obtained are comparable in terms of average values, variance and extreme events considering the trade-offs in terms of computing facilities needed and training required to apply the dynamic downscaling approach.

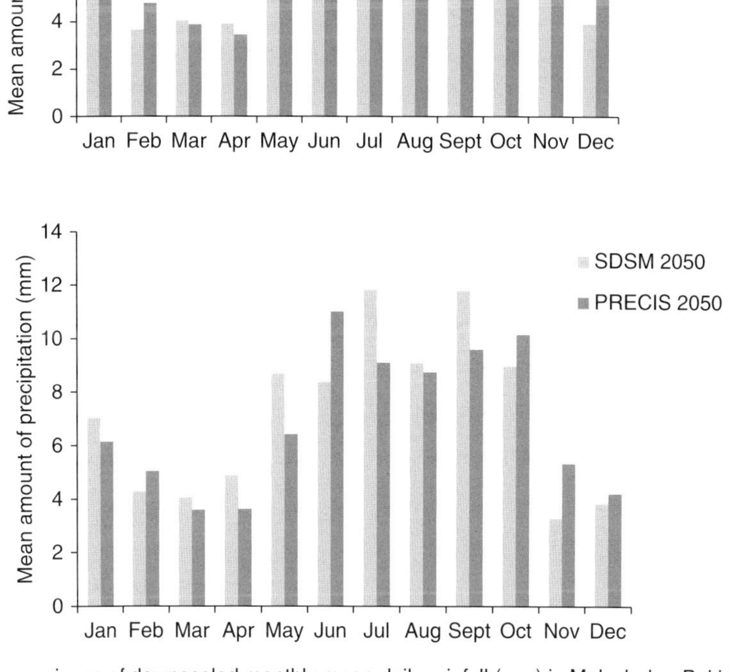

Fig. 6.7. Comparisons of downscaled monthly mean daily rainfall (mm) in Malaybalay, Bukidnon, Philippines for 2020 and 2050 using SDSM and PRECIS (from Lansigan and Dating, 2012).

6.2 Climate Change Effects and Impacts on Agricultural Production

6.2.1 Increased temperature due to global warming

Changes in distributions of climate variables such as rainfall and temperature have profound impacts on agricultural production systems. Climate change affects agricultural crop production through the following processes: (i) increased atmospheric temperature or global warming; (ii) erratic and changing rainfall patterns; (iii) occurrence of extreme events such as more intense rainfall, and typhoons with strong winds; and (iv) sea-level rise.

Temperature gradient tunnel studies, field experiments and crop simulation studies have shown that increased temperature in crop production in the Philippines as well as in many areas in South-east Asia will reduce crop yields. Studies have shown that potential crop yields will be reduced from about 8% to 14% for every 1°C increase in ambient temperature depending on location due to specific climate type. While increased carbon dioxide (CO_2) concentration in the atmosphere will enhance the photosynthetic activity of the plant, and therefore increases crop yields, studies have shown that CO_2 enrichment cannot compensate for the increased maintenance of respiration due to increased temperature, resulting in decreased grain yields.

6.2.2 Erratic and variable rainfall patterns

Changing rainfall distributions have altered the planting calendar in many areas. In certain areas in the country, summer months (March, April, May) have become wetter and wet season (June, July, and August) has become drier. Planting date has to be adjusted considering the variability in rainfall in the area. For example, cropping calendar or planting date may be determined based on rainfall patterns. Optimal planting can be determined using rainfall probabilities, crop yield probabilities determined by crop models, or the use of the modified Penman–Monteith equation considering rainfall distribution and evaporative demand for crop growth. Using the downscaled climate projections for different periods in selected locations, optimal planting dates based on high rainfall probabilities may be derived. Crops have different water requirements for each stage of crop growth and development (Penning de Vries *et al.*, 1989) to give optimal yields. For rainfed crop production systems, water is mainly from rainfall whose availability and temporal variability determine the optimal planting date. Location-specific probabilities are considered, namely: P_w, probability of wet soil ready for planting; P_{200}, probability of at least 200 mm cumulative rainfall during flowering stage; and P_d, probability of a dry period during harvesting (Lansigan, 2010b). Combining these probabilities of meeting the water requirements at different stages of growth by taking their product is the Q probability (i.e. $Q = P_w \times P_{200} \times P_d$) of satisfying the rainfall requirements. Thus the period during which Q probability is high is the best planting date.

Figure 6.8 shows the combined probability Q for Malaybalay, Bukidnon (for location see Fig. 6.3) for different time periods based on the downscaled climate projections using SDSM. The plot shows that the duration of optimal planting date from the baseline period (1971–2000) to 2020 is shifted and shortened. The low probabilities in 2050 indicate the high risk of planting in the periods during which rainfall may not be sufficient to meet the water demand for crop growth. The low probabilities in 2050 may be attributed to the expected large variability of rainfall compared to baseline and 2020. Thus, there is low probability of meeting the water requirements. On the other hand, Fig. 6.9 shows the date and duration of optimal planting for Iloilo, which remain within the same period but the combined probability Q ($P_w \times P_{200} \times P_d$) has increased under the projected climate. The best planting period remains the same but with higher probability of meeting water requirements during different stages of crop growth than for 2020. Thus, the erratic rainfall patterns in many crop-growing areas remain a challenge to farmers in deciding the best time to plant.

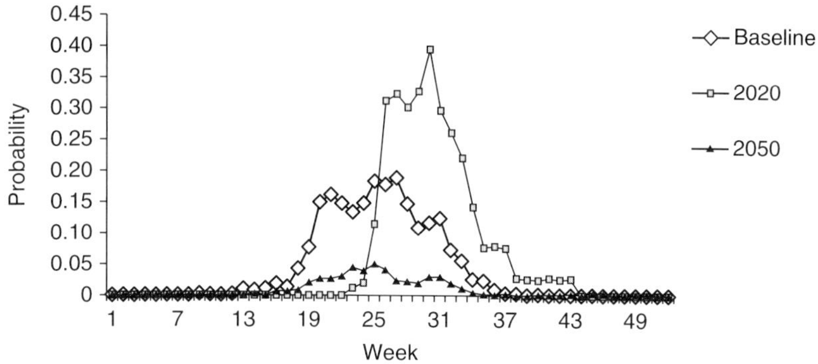

Fig. 6.8. Comparison of the combined probability Q of meeting rainfall requirements for the different crop growth and development stages for the baseline, 2020 and 2050 in Malaybalay, Bukidnon, Philippines.

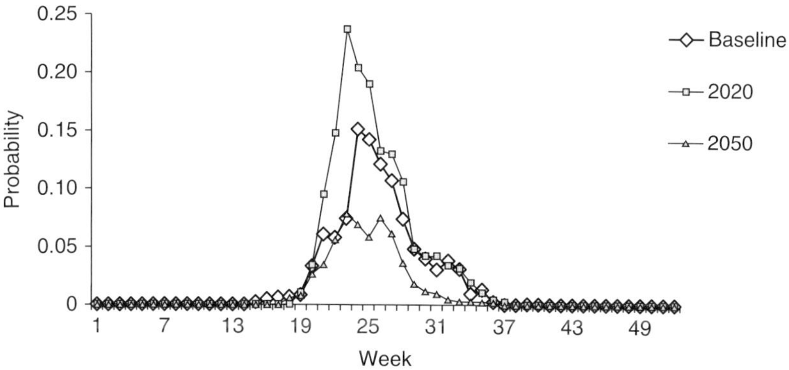

Fig. 6.9. Comparison of the combined probability Q of meeting rainfall requirements for the different crop growth and development stages for the baseline, 2020 and 2050 in Iloilo, Philippines.

6.2.3 More intense extreme climate events

Climate change is expected to result from the disproportionate increase in the mean and variance of climate variables. The percentage change in the mean as well as in the variance of climate variables such as rainfall and temperature are not the same. More intense extreme events, such as more intense rainfall episodes, and hotter days and warmer nights are expected to occur more frequently. However, some downscaled seasonal climate variables do not show significant increase, which may be attributed to the different climate types in the Philippines. Typhoons characterized by more intense and heavy rains and strong winds with shorter recurrence intervals are also expected to occur (IPCC, 2007; PAGASA, 2011; IPCC, 2012). These often lead to floods that result in significant damage to crops and livestock, livelihoods, and even to loss of properties and lives. Damage to crops and livestock, livelihoods and properties exceed 100 million pesos (Php). In recent years, damages and losses have reached more than Php 1.0 bn per occurrence of a strong typhoon.

Historically, an average of 20 typhoons enter the Philippine Area of Responsibility (PAR), although between seven and eight typhoons make landfalls yearly (Yumul *et al.*, 2011). Typhoons are observed to have a 27–35% chance of passing through central to northern Philippines and about a 5–7% chance of passing through the southern part of the country. A recent study (Schellnhuber *et al.*, 2013) on climate scenarios and impacts reported that more intense typhoons are expected in southern Philippines. Thus, crop losses and damage are expected to be high in areas that will be visited by more intense typhoons. Damaging typhoons are expected to significantly affect the gross domestic product (GDP) of the country, which is estimated to average about 4.7%.

6.2.4 Sea-level rise and saltwater intrusion

As high temperature warms the oceans, the sea level rises, inundating the crop-growing areas in the coastal regions. More intense and heavy rainfall events, in combination with high tides and sea-level rise, enhance saltwater intrusion in coastal areas. Thus, submergence and salinity stresses will affect crop growth and development and will reduce yields. The problem of salinity is expected to become even more significant in coastal and deltaic regions affected by sea-level rise. These areas are more exposed and vulnerable to sea-level rise associated with climate change. Salinity-tolerant crop varieties are needed in these coastal areas, since salinity affects the critical stages of crop growth and development especially during seedling and reproductive stages of the crops.

A vulnerability study of South-east Asian coastal areas (David *et al.*, 2008) reported that up to 45,000 ha year^{-1} of land are lost due to submergence, up to 7.7 million people will actually face floods every year by 2100, and there will be a net loss of wetland areas of up to 32,000–435,000 ha due to sea-level rise. In the Philippines, coastal areas account for about 34,000 km^2, covering 804 cities and municipalities and 23,492 *barangays* (political villages). Thus, climate change will not only reduce crop productivity but decrease the areas planted to the crops due to inundation of coastal areas.

6.3 Assessing Impacts of Climate Change on Agricultural Crop Production

6.3.1 Crop model-based evaluation of climate change effects and impacts

The effects and impacts of climate change may be evaluated objectively using eco-physiological or process-based crop simulation models that can estimate crop yields under different climate conditions. These crop models require as inputs the crop genetic coefficients, crop management data and weather data for the specific location. A number of crop models such as DSSAT (Decision Support System for Agrotechnology Transfer) are available, which can be used provided the relevant assumptions are satisfied. The assumptions include the homogeneity of environmental conditions under a specific type of production situation being considered, e.g. potential production, water limited production, nutrient-limited production, etc.

6.3.2 Effects of climate change on major crops

Earlier studies (Matthews *et al.*, 1997; Matthews and Stephens, 2002; Lansigan, 2003; Lansigan and Salvacion, 2007; Centeno and Wassmann, 2009) based on the use of crop models quantified the effects of climate change on crop yields. Figure 6.10 shows the changes in crop yields of major crops such as rice, maize, tomato and groundnut for varying temperature regimes in selected locations in the Philippines. The figures also show that reductions in yields vary depending on locations. Some crops such as C3-crops like rice (Lansigan and Salvacion, 2007) are more vulnerable to climate change than others.

Climate change is expected to reduce crop yields such as in rice crops. Figure 6.11 shows the yield probabilities for IR-64 rice variety planted on 8 May in Malaybalay, Bukidnon (for location see Fig. 6.3) for the baseline period and A1B projected

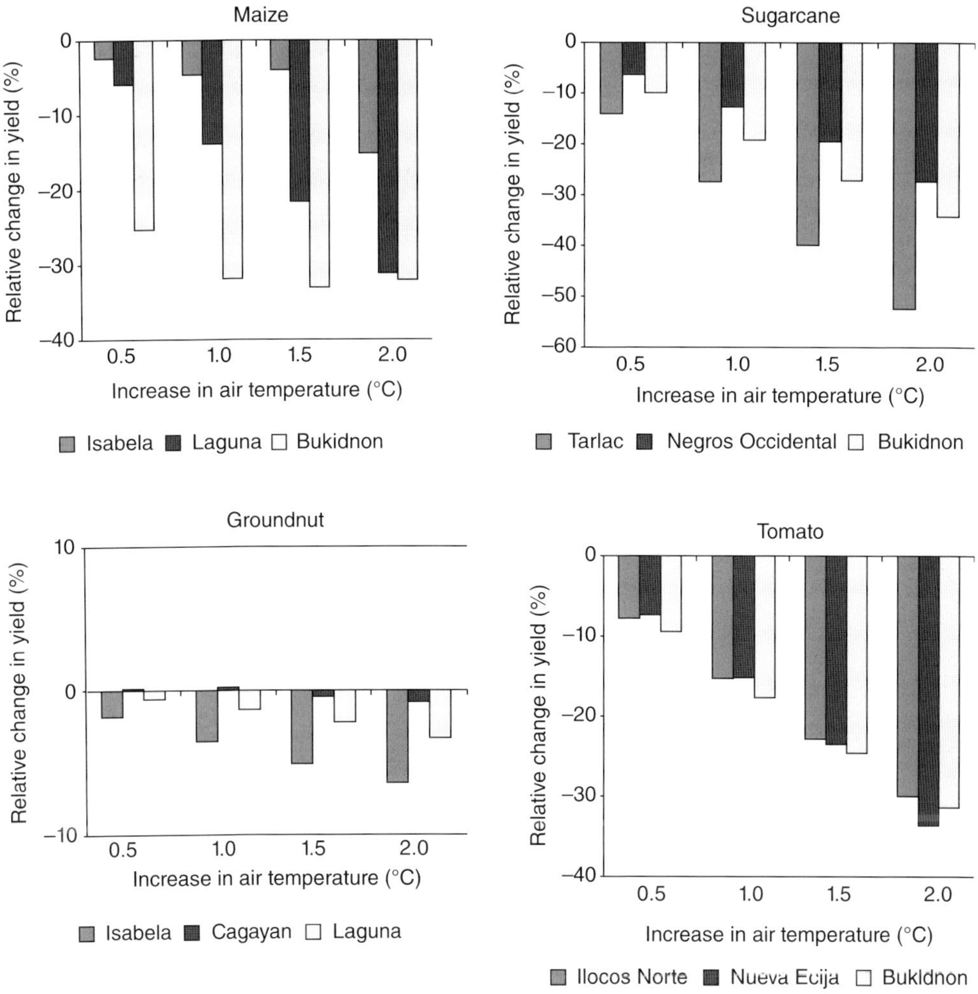

Fig. 6.10. Reduction in simulated crop yields of maize, sugarcane, groundnut and tomato for the projected temperature increase in selected locations in the Philippines (from Lansigan *et al.*, 2008).

temperature and rainfall in 2020 and 2050. The probability of exceeding particular rice yields decreases under projected climate change with the difference being very pronounced in the variety grown on 8 May. This shows the vulnerability of the rice crop to climate change as temperature increases.

Figure 6.12 shows the average rice yields in three locations (Agusan del Norte, Bukidnon and Benguet) in the Philippines (for location see Fig. 6.3) with different climate types and elevation for the baseline period (1971–2000) and under the climate projections for the periods centred on 2020 and 2050 (Lansigan, 2010a). Using the downscaled temperature and rainfall, crop productivity decreases as temperature increases due to climate change except in Benguet Province, which has a higher elevation and therefore a cooler climate. Thus, increase in temperature will benefit rice production in this area.

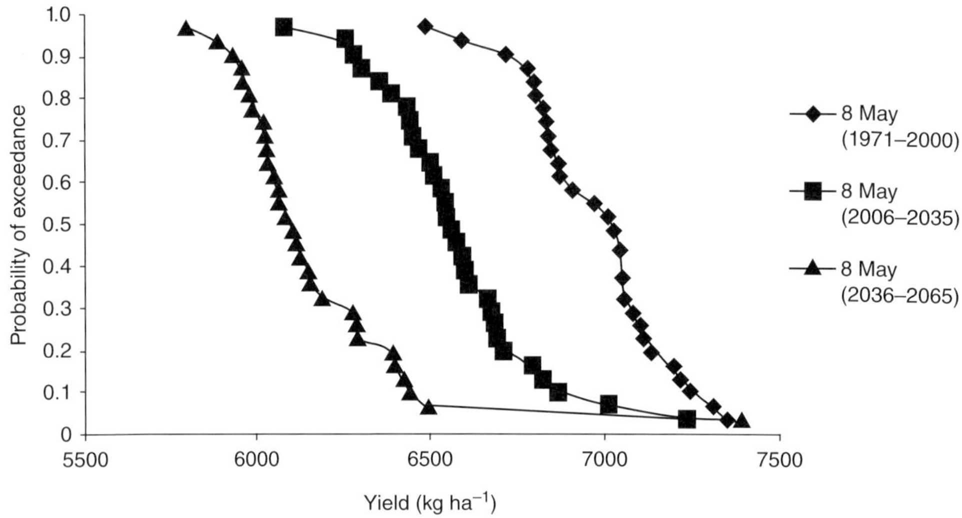

Fig. 6.11. Probabilities of exceedance of rice yields (IR-64 variety) during 8 May planting in Bukidnon, Philippines for the baseline period (1971–2000) and projected climate in 2020 (2006–2035) and 2050 (2036–2065).

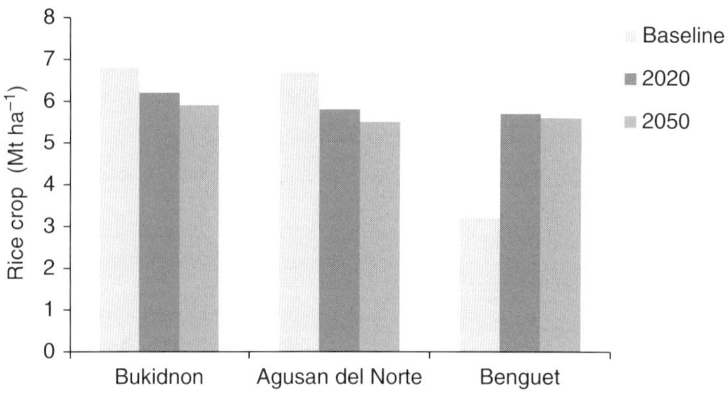

Fig. 6.12. Mean rice crop productivity in three locations in the Philippines (Bukidnon, Agusan del Norte and Benguet) under baseline (1971–2000), 2020 and 2050 projected climate.

6.4 Adaptation Strategies in Agricultural Water Management

There are a number of location-specific adaptation strategies and measures that may be applied to increase climate resilience. These measures are based on best practices in the area, which may be modified to suit local conditions. In crop production, for example, water-related adaptation strategies that may be used to reduce vulnerability to the adverse effects and impacts of changing climate include the following:

- Adjusting cropping calendar based on medium-range weather forecasts and seasonal climate outlook. The best planting window for the major crops may be determined using empirical procedures to estimate yield probabilities, and also frequencies of receiving rainfall threshold values critical in crop growth and development. A dynamic cropping calendar can be developed based on downscaled climate projections. Probabilities of receiving specified cumulative rainfall during the critical stages of crop growth can be determined (see Figs 6.8 and 6.9).

- Water impoundment. Storing water for use during periods with less rainfall is a common best practice in many rainfed areas in the Philippines. The impoundment structure can be made up of natural water impounding, or earth materials, or concrete. This is shown in Fig. 6.13 where rainwater during the wet season is harvested and impounded through a series of farm ponds, such as the example shown in Ifugao Province, Philippines. Stored water is used during the dry periods.
- Improving water use efficiency of irrigation and water management at the farm level. This can be achieved through intercropping or multiple cropping, and crop intensification to maximize the use of available soil moisture, and improved small-scale irrigation systems using shallow tube well (STW) technology. In recent years, it has been demonstrated that alternate wetting and drying (AWD) technology in rice production can improve water use efficiency and can also serve as a mitigating measure since it can reduce GHG emissions from the rice farms when fertilizers are applied in the field (Wassmann, 2010).
- Modifying crop management to reduce water use. This can be achieved through modifications of farm activities such as direct seeding, minimum tillage, and even growing an early-maturing variety which will consume less water. This also involves a shorter management period and less labour.
- Planting stress-tolerant crops or varieties. The adverse effects and impacts of climate change and variability can be reduced by planting crops or using varieties tolerant to environmental stresses such as submergence due to flooding, water stress due to droughts, salinity due to saltwater intrusion and heat stress due to temperature extremes. For example, the Sub-1 rice variety developed recently has been demonstrated to be flood- or submergence-tolerant. Likewise, rice varieties with the

Fig. 6.13. Rainwater harvesting and water impoundment through a series of farm ponds for fish and crop production in Lamut and upland areas of Alfonso Lista municipalities of Ifugao Province, Philippines.

Saltol salt-tolerant gene can withstand salinity.
- Terracing in steep and high-elevation areas. Rice terraces have been practised in many areas, particularly in high elevation areas to maximize area for crop production, reduce soil erosion and maximize water storage. Labour availability and environmental impacts are also considered, since these are important factors in the region where the rice terraces are found. Fig. 6.14a and Fig. 6.14b show the terraces for agroforestry and vegetable crops in Kiangan, and rice terraces in Banaue, Ifugao Province.
- Implementation of innovative agri-insurance for crop production. Crop insurance is one risk transfer mechanism that can be used to reduce vulnerability and risk due to climate-related hazards. They have to be made attractive and affordable to crop growers. Weather index-based insurance (WIBI) has been introduced recently. But development of WIBI products for different climate hazards requires the objective estimation of risks associated with different weather variables based on available historical records, which are often very limited and short, and with missing values. Moreover, an optimal network of weather gauging stations is needed for implementation of WIBI products. For example, Fig. 6.15 shows the risk associated with rainfall deficit for the different stages of growth for the 110-day rainfed rice crop variety PSB Rc14 grown in Iloilo, Philippines (Lansigan, 2013). The plot indicates the low risk-period during which the crop may be grown meeting the required cumulative rainfall for crop growth and development. This information can be used as the basis for developing an objective WIBI product, based on rainfall and also in estimation of premium, based on risk.

In practice, a suite of climate adaptation measures (Tibig and Lansigan, 2007; Asia Rice Foundation, 2010; Lansigan, 2011b) is being implemented in many areas in the Philippines to minimize the adverse effects and impacts to agricultural production systems and related activities. The best strategy is to reduce the vulnerability of the biophysical subsystems to climate-change hazards, and also to increase the resilience of the different stakeholders through a combination of adaptation options.

6.5 Concluding Remarks

Climate and weather are important factors in agricultural crop production. Climate change will lead to increased temperature and CO_2 concentration, sea-level rise and more intense extreme events. Climate change is also expected to alter the hydrologic regime affecting the availability of water supply. Thus, climate change reduces crop productivity and threatens food security.

Assessment of effects and impacts of climate change require the use of GCM climate projections, but they have to be downscaled to be useful for impact assessments using either a dynamic approach such as a regional climate model (e.g. PRECIS), or a statistical downscaling method (SDSM) including weather generator, historical analogue, etc. The use of statistical downscaling techniques gave reasonably adequate results compared to the more computationally demanding dynamic downscaling.

The use of a validated crop simulation model facilitates the objective evaluation of effects of climate change on rice crop yields. Crop variety-specific genetic coefficients, soils and weather data and cultural management practices are model inputs. The model can estimate (simulate) crop yields under any climate condition (e.g. projections). Simulation studies in selected locations show that rice yields are expected to decrease in low elevation areas but increase in high elevation (cool) rice-growing areas like Benguet (and Ifugao). Moreover, the magnitudes of rice yield decrease (in low-lying areas) or increase (in high-elevation areas) vary across locations and also with season (i.e. varying planting dates). Thus, the effects are location-specific and time-dependent.

Fig. 6.14. Terracing in a combination of agroforestry (upper level), vegetable (middle level) and rice (lower level) production systems in Kiangan, Ifugao Province, Philippines. (b) Newly transplanted seedlings in rice terraces of Banaue, Ifugao, Philippines.

Differences in effects and magnitude of impacts of climate change on crop productivity suggest that location-specific adaptation measures are needed to reduce the adverse consequences of changing climate. There are a number of climate-change adaptation strategies and measures based on best practices as well as on recent scientific breakthroughs. What is needed is a suite of adaptation measures that are scientifically sound, cost-effective, economically efficient and socially and culturally acceptable.

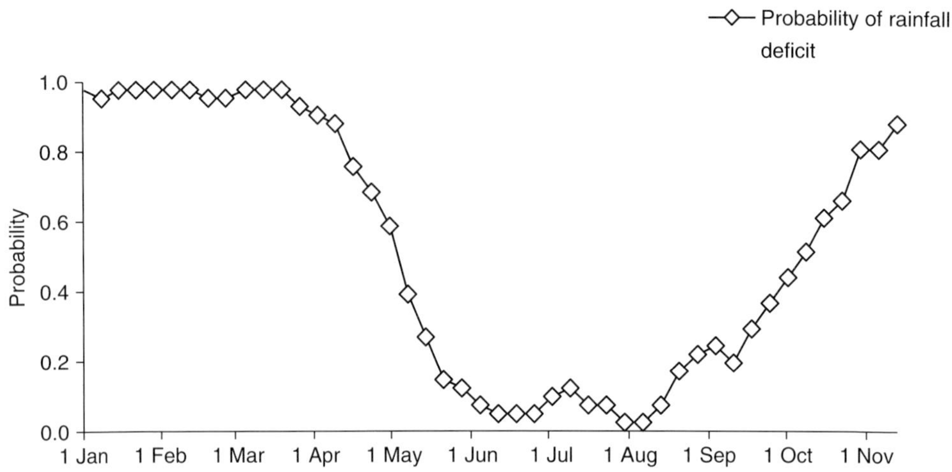

Fig. 6.15. WIBI-based risk of rainfall deficit for different stages of crop growth for rainfed rice production in Iloilo Province, Philippines (from Lansigan, 2013).

References

Alcamo, J., Grassl, H., Hoff, H., Kabat, P., Lansigan, F., Lawford, R. *et al.* (2005) *The global water system project: Science framework and implementation activities.* Earth System Science Partnership (ESSP), Bonn, Germany.

Anderson, C.J. (2008) An overview of downscaling techniques used in climate science. Santa Clara University and Iowa State University, USA. Available at: http://www.colorado.edu/climate_chnage/docs/Anderson_downscaling.pdf (accessed 1 June 2013).

Asia Rice Foundation (2010) *Adaptation to Climate Variability in Rice Production.* Asia Rice Foundation, Los Baños, Laguna, the Philippines, 61 pp.

Centeno, H.G.S. and Wassmann, R. (2009) Regional climate models in predicting Philippine rice production. *Philippine Journal of Crop Science* 25(2), 79–85.

Comiso, J.C., Blanche, C., Espaldon, M.V., Sarigumba, T. and Lansigan, F.P. (eds) (2014) *Changing Philippine Climate: impacts on agriculture and natural resources.* UP Press, Diliman, Quezon City, the Philippines.

David, L.T., Maneja, R.H., Goh, B., Lansigan, F.P., Sereywath, P. *et al.* (2008) Sea level rise vulnerability of Southeast Asia coasts. *LOICZ InPrint* December, 3–6.

Delfino, R.J., Herath, S., Lansigan, F.P., Pulhin, J.M. and Lasco, R.D. (2012) Assessing the impacts of climate change on water resources and rice production in Pantabangan-Carranglan Watershed, Philippines. Paper presented at the National Climate Research Conference, 27 2012. Manila, the Philippines.

Gaffney, O. (2010) One planet, four futures. *IGBP Global Change* 75, 12–15.

IPCC (2001) *Special Report on Emission Scenarios.* Intergovernmental Panel on Climate Change, Geneva, Switzerland.

IPCC (2007) Climate change 2007: The physical basis. Summary for policy makers. *IPCC Fourth Assessment Report.* Intergovernmental Panel on Climate Change, Geneva, Switzerland.

IPCC (2012) Special report on managing the risks of extreme events and disasters to advance climate change adaptation (SREX). Intergovernmental Panel on Climate Change. Available at: http://ipcc-wg2.gov/SREX/report (accessed 4 October 2012).

Lansigan, F.P. (2003) Assessing impacts of climate variability on crop production, and developing coping strategies in rainfed agriculture. *UN ESCAP CGPRT Center Monograph No. 43.* Bogor, Indonesia, pp. 21–36.

Lansigan, F.P. (2009) Frequency analysis of extreme hydrologic events and assessment of water stress in a changing climate in the Philippines. In: Taniguchi, M., Burnett, W.C., Fukushima, Y., Haigh, M. and Umezawa, Y. (eds) *From Headwaters to the Ocean: Hydrological Changes and Watershed Management.* CRC Press, Taylor & Francis, Leiden, the Netherlands, pp. 497–501.

Lansigan, F.P. (2010a) Determining the seasonal climate-based optimal planting dates for rainfed rice in selected locations in the Philippines. *National Convention on Statistics, 4–5 October 2010*, Mandaluyong City, the Philippines.

Lansigan, F.P. (2010b) Managing climate risk to enhance crop productivity for food security: dynamic cropping calendar for farm operations. *University of the Philippines in Los Banos Centennial Professorial Chair Lecture Paper, 11 August 2010*. University of the Philippines in Los Baños, Laguna, the Philippines, 30 pp.

Lansigan, F.P. (2011a) Adaptation to changing climate in rice production in the Philippines. In: *Asia Rice Foundation, Adaptation to Climate Variability in Rice Production, 2011*. Asia Rice Foundation, Los Baños, Laguna, the Philippines.

Lansigan, F.P. (2011b) Responding to changing climate on rice production in the Philippines. In: *Adaptation to Climate Variability in Rice Production*. Asia Rice Foundation , Los Baños, Laguna, the Philippines, pp. 32–46.

Lansigan, F.P. (2013) Climate risk for weather index-based insurance for crop production in the Philippines. *Proceedings, 12th National Convention on Statistics*, Mandaluyong City, the Philippines.

Lansigan, F.P. and Dating, M.J. (2012) Statistical downscaling of climate projections and assessment of impacts on crop productivity in the Philippines. Paper presented at the National Academy of Science and Technology (NAST) Conference on Climate Change and Variability, 27 September 2012, Manila, the Philippines.

Lansigan, F.P. and Salvacion, A.R. (2007) Assessing the effect of climate change in rice and corn yield in selected provinces in the Philippines. *Proceedings, 10th National Convention on Statistics*, Mandaluyong City, the Philippines.

Lansigan, F.P., Salavacion, A.R. and Garcia, A.G. (2008) SEARCA Seed Fund for Research and Training (SFRT). Southeast Asian Regional Center for Research and Agriculture (SEARCA), Los Baños, Laguna, the Philippines.

Lansigan, F.P., Dating, M.J.P. and Guiam, E.G.A. (2013) Comparison of statistical downscaling methods of climate projections in selected locations in the Philippines. *National Convention on Statistics*. October 1–2, 2013, Mandaluyong City, the Philippines.

Matthews, R.B. and Stephens, W. (2002) *Crop-Soil Simulation Models: Applications in Developing Countries*. CAB International, Wallingford, UK.

Matthews, R.B., Kropff, M.J., Horie, T. and Bachelet, D. (1997) Simulating the impact of climate change on rice production in Asia and evaluating options for adaptation. *Agricultural Systems* 54(3), 399–425.

Moss, R.H., Edmonds, J.A., Hibbard, K.A., Manning, M.R., Rose, S.K., van Vuuren, D.P., Carter, T.R., Emori, S., Kainuma, M., Kram, T., Meehl, G.A., Mitchell, J.F.B., Nakicenovic, N., Riahi, K., Smith, S.J., Stouffer, R.J., Thomson, A.M., Weyant, J.P. and Wilbanks, T.J. (2010) The next generation of scenarios for climate change research and assessment. *Nature* 463, 747–756.

Naylor, R.L., Falcon, W.P., Rochberg, D. and Wada, N. (2001) Using El Niño/Southern oscillation climate data to predict rice production in Indonesia. *Climatic Change* 50(3), 255–265.

PAGASA (2011) *Climate change in the Philippines*. Philippine Atmospheric, Geophysical and Astronomical Services Administration (PAGASA), Department of Science and Technology (DOST). Millennium Development Goal-Fund (MDGF) and DOST, June 2011, Quezon City, the Philippines, 86 pp.

Penning de Vries, F.W.T., Jansen, D.M., ten Berge, H.F.M. and Bakema, A. (1989) *Simulation of Ecophysiological Processes of Growth in Several Annual Crops*. Simulation Monograph. PUDOC, Wageningen, the Netherlands and International Rice Research Institute, Los Baños, Laguna, the Philippines, 271 pp.

Schellnhuber, H.J., Hare, B., Serdeczny, O., Adams, S., Baarsch, F., Schwan, S., Coumou, D., Robinson, A., Schaeffer, M., Vieweg, M., Piontek, F., Donner, R., Runge, J., Rehfeld, K., Rogelji, J., Menon, A., Perrette, M., Schleussner, C.-F., Bondeau, A., Svirejeva-Hopkins, A., Schewe, J., Frieler, K. and Warszawski, L. (2013) *Turn Down the Heat II: global hotspots and regional cases*. World Bank, Washington, DC, 236 pp.

SEARCA and FAO (2010) *The Impacts of El Niño Southern Oscillation (ENSO) Events in Cereal Production, Area and Yield in Asia. Project Report*. Southeast Asian Regional Center for Graduate Study and Research in Agriculture (SEARCA) and Food and Agriculture Organization of the United Nations (FAO), Los Baños, Laguna, the Philippines.

Tibig, L.V. and Lansigan, F.P. (2007) Coping strategies to agro-meteorological risks and uncertainties in crop yields. In: Sivakumar, M.V.K. and Motha, R.P. (eds) *Managing Weather and Climate Risks in Agriculture*. Springer-Verlag, Berlin, pp. 209–235.

Wassmann, R. (ed.) (2010) *Advanced Technologies of Rice Production from Coping with Climate Change: 'no regret' options for adaptation and mitigation and their potential update*. International Rice Research Institute (IRRI), the Philippines, 76 pp.

Wilby, R.L. and Dawson, C.W. (2007) SDSM 4.2 – A decision support tool for the assessment of regional climate change impacts. Available at: http://co-public.lboro.ac.uk/cocwd/SDSM (accessed 5 October 2012).

Yumul, G.P., Cruz, N.A., Servando, N.T. and Dimalanta, C.B. (2011) Extreme weather events and related disasters in the Philippines, 2004–2008: a sign of what climate change will mean? *Disasters* 35(2), 362–382.

7 Adaptation Strategies to Address the Climate Change Impacts in Three Major River Basins in India

Krishna Reddy Kakumanu,[1] Kuppannan Palanisami,[1*] Pramod Kumar Aggarwal,[1] Coimbatore Ramarao Ranganathan[2] and Udaya Sekhar Nagothu[3]

[1]*International Water Management Institute, Hyderabad, India;* [2]*Tamil Nadu Agricultural University, Coimbatore, India;* [3]*Norwegian Institute for Agricultural and Environmental Research-Biforsk, Ås, Norway*

Abstract

The agriculture sector is highly vulnerable to climate change in many parts of the world. There is an increasing concern among farmers, researchers and policy makers about the potential impacts of climate change on food security and livelihoods. Researchers are using several climate change models to make an assessment of the impacts and identify adaptation strategies. The present chapter reviews the current state of understanding of the climate change impacts on irrigation water in South Asia and specifically on the crop yield and relevant adaptation measures in three major river basins (Godavari, Krishna and Cauvery) in India. Optimization model was used to evaluate the different adaptation practices and their potential to maximize rice production and income, and minimize water use for the mid- and end-century climate-change scenarios. Adaptation practices such as systems of rice intensification, machine transplantation, alternate wetting and drying (AWD) and direct seeding could reduce the water and labour use by 10–15% and stabilize rice production in the long term. The study suggests the need for technology upscaling, which should be backed up with well-planned capacity-building programmes for the farmers.

7.1 Introduction

Climate change is a complex subject that requires an interdisciplinary approach needing an impact assessment to develop corresponding adaptation measures. Therefore, multiple-level assessment and data sets are required to effectively capture the current and future situations. Data on climate, soils, water, crop pattern, crop productivity and socio-economic variables mainly contribute to model estimation. A large and growing body of research shows that socio-economic factors can be as important as the magnitude of a climatic event in determining the impact of environmental change on the agriculture sector (Patt and Gwata, 2002; Fraser *et al.*, 2003). However, there are no clear-cut procedures to characterize human coping and adaptation mechanisms as these vary from place to place (Elisabeth *et al.*, 2010). Where climate change affects yields, impact models should integrate environmental factors, such as available water and temperature that directly affect yield, with socio-economic factors that encourage pro-active adaptation

* Corresponding author, e-mail: k.palanisami@cgiar.org

and policy factors that support adaptation. On the other hand, climate change not only affects yield but also its variability (Barnwal and Kotani, 2010). However, little is known about how the available water resources in the future can be effectively used under the changing climate. Therefore, the economics of climate change impacts as well as adaptation to climate change through optimization of the available resources is a challenging area of research.

In this chapter, optimal allocation of resources viz., land, labour and water, to the changing climatic conditions under different irrigation projects of the Godavari, Krishna and Cauvery river basins is discussed. Section 7.2 deals with the background followed by Section 7.3, which comprises the methodology used and a review of understanding the climate change impacts in South Asia. In Section 7.4, the model used for the optimal use of resources is explained along with the study area and projects. Section 7.5 focuses on the results and discussion and finally recommendations. By comparing the results from the individual river basins, the chapter thus draws some lessons for overall management of the water in the river basins.

7.2 Review on Understanding the Climate Change Impacts

South Asia is the home for one-fifth of the world's population and is the most disaster-prone region in the world (UNEP, 2003). Climate change is affecting a large number of people across South Asia in different ways, which includes variability in the monsoons, increase in average temperature, warm winters, increased salinity in coastal areas, reduced discharges from rivers, etc. The Intergovernmental Panel on Climate Change (IPCC) has also projected that the mean annual temperatures of South Asia will increase by 0.5–1.2°C by 2020, 0.88–3.16°C by 2050 and 1.56–5.44°C by the end of the century (IPCC, 2007). High temperatures are likely to reduce the yields of different crops and increase the proliferation of weeds and pests, thus providing new challenges for agricultural scientists (Cruz *et al.*, 2007; Nelson *et al.*, 2009). In tropical parts of South Asia, temperature rise will negatively impact rice and wheat yields as they are already being grown close to their threshold (Kelkar and Bhadwal, 2007). In sub-humid, semi-arid and arid regions wheat yields are predicted to decline by 6%–9% with a 1°C rise in temperature (Sultana and Ali, 2006). Cash crops such as cotton, mango and sugarcane will be severely impacted with a decadal rise of 0.3°C (MoE, 2003). Thus, the overall impacts of temperatures on agriculture are expected to be negative, threatening global food security.

Droughts or floods are destructive but when they last for longer periods then the effects can be devastating or irreversible (Conway, 2009). Widespread flooding is seen in many small island and delta regions, for example the Mekong Delta. The floods in Myanmar during 2008 devastated 1.75 million ha (Mha) of rice land while in Bangladesh it caused a production loss of about 0.8 t of rice during 2007 (Craufurd *et al.*, 2011). In India, 70% of the arable land is prone to drought and 20% to floods and cyclones. Of the total precipitation of around 4000 km^3 in the country, availability of surface water and replenishable groundwater is estimated at 1893 km^3. But due to the variations of topography and uneven distribution of rain over space and time, only about 1123 km^3, including surface water and groundwater resources, can be put to beneficial use (Aggarwal *et al.*, 2012).

As a result, water scarcity is expected to become an ever-increasing problem due to the changing climate in India. The water balance will change due to the accelerated rate of evaporation from soil and water bodies and transpiration from plants. Several studies have shown that unless we adapt, there is a probability of a 10–40% loss in crop production in India by the end of the century owing to global warming (Aggarwal, 2009; Knox *et al.*, 2011).

Among other things, the river basins are going to be highly affected by sedimentation, reducing water storage, water availability (due to poor monsoon) and area under production. The per capita water resource availability of the basins in India also varies from a low of 240 m^3 (Sabarmati Basin) to a high of 17,000 m^3 (Brahmaputra Basin) (Amarasinghe *et al.*, 2005). Authors also reported that many river basins record significantly lower per capita water availability in terms of total renewable water resources, thus increasing the demand for water resources. There are several factors influencing water supply and demand in the basins such as population growth, urbanization and income, changes in dietary preferences, irrigation expansion and environmental flow requirements.

Though there is much concern about reduced water supplies and the substantial impact of climate change in crop production in South Asia, still there is limited understanding of the adaptation strategies needed. Adaptations are adjustments or interventions which take place in order to manage the losses or take advantage of the opportunities presented by a changing climate (IPCC, 2001). Adaptation occurs at two levels: (i) farm-level adaptation, which mainly focuses on farming-related interventions or adjustments and are related to short-term periods and influenced by seasonal climate variations and local agricultural cycles; and (ii) the regional- or national-level adaptation, which focuses on the agricultural production at macro-level linking domestic and international policies (Kandlinkar and Risbey, 2000; Bradshaw *et al.*, 2004).

Several adaptation measures have already been examined by researchers. For example, Palanisami *et al.* (2011) have examined the adaptation measures such as systems of rice intensification, machine transplanting, alternate wetting and drying and maize water management options and incorporated them in the optimization model that could help minimize the incidence of climate change impacts on crop yields, labour use and water use. In addition, there are various other adaptation practices, such as direct seeding of rice and drum seeding of rice, which are gaining acceptance among farmers (Gurava Reddy *et al.*, 2013). Abraha and Savage (2006) studied the potential impact of climate change on maize yield at Cedara, KwaZulu-Natal, South Africa with different planting dates, such as normal, 15 days earlier and 15 days later. The farm management components appear to be prominent in the literature (Till *et al.*, 2010) where most of the adaptation practices included the adjustments in farm management and technology, followed by knowledge management, networks, governance, diversification, government interventions and farm financial management.

7.3 Methodology

7.3.1 Selection of river basins

Godavari, Krishna and Cauvery are some of the major river basins covering central and southern parts of India (Fig. 7.1). The highlights of each river basin are given in Table 7.1. The rainfall pattern ranges from humid to arid and the basins have diversified cropping patterns (rice, cotton, chilli, banana, sesame, maize, groundnut, pulses), indicating water variability over the years due to climate change. The river basins are also more adverse to climate variability in their hydrological regime, which affects irrigation to a large extent. However, there is a need to adapt to new management strategies and practices to overcome water scarcity and achieve food security in the irrigation projects throughout the basins. Results from a study conducted by Gosain and Rao (2012) on impact assessment of water resources from the Godavari Basin show that the water balance, mean annual precipitation, water yield and sedimentation are likely to increase along with temperatures in the mid- and end-centuries compared to the baseline scenario. The simulations of rainfed maize, sorghum and rice yields also indicated the adverse effect due to increases in temperatures, although increased rainfall and change in management practices can partly offset these effects (Kattarkandi *et al.*, 2010; Srivastava *et al.*, 2010).

7.3.2 Model

Palanisami *et al.* (2011) have estimated production functions focusing on the relation between yield and its variability in the context of climate change. The authors have estimated the Just–Pope production function using the maximum likelihood method by assuming the relation between yield of a crop and climate variables (temperature and

Fig. 7.1. Map showing the projects in the respective basins in India.

Table 7.1. Details of the Godavari, Krishna and Cauvery river basins (from http://www.india-wris.nrsc.gov.in).

Details	Godavari	Krishna	Cauvery
No. of states covered	6	3	3
Catchment area (km^2)	312,812	258,948	81,155
Length (km)	1,465	1,400	800
Annual rainfall (mm)	1000–3000	784	956
Average water resource potential (Mm^3)	110,540	78,120	21,358
Utilizable surface water resource (Mm^3)	76,300	58,000	19,000
No. of hydrological observation stations	17	53	34
No. of flood forecasting stations	77	9	0
Major crops	Rice, wheat, maize, sugarcane, cotton	Rice, cotton, chilli, maize, sugarcane, groundnut, millet and horticultural crops	Rice, sugarcane, maize, groundnut, banana, turmeric, sesame oil, etc.

precipitation) for different districts (Just and Pope, 1978). In the present study, functions developed by the authors were fitted for an irrigation project in Godavari, Krishna and Cauvery river basins. A quadratic form was assumed for the mean function (Isik and Devadoss, 2006; Ranganathan, 2009), which ensures positive output variance. In addition, the riskiness of an input variable was also derived from the sign of the coefficient. The mean function was used to study the maximum and minimum possible yields and also the impact of climate change on the crop yield.

The first and second order conditions were derived by assuming that precipitation and temperature will vary and technology will be held at the current level. None the less, accurate region-specific predictions for changes in temperature and rainfall are needed to capture the impact of climate change. Gosain and Rao (2012) have predicted the season-wise changes in the Godavari River Basin for baseline period (1960–1990), mid-century period (2021–2050) and end-century period (2071–2098).

Two scenarios were formulated based on the mid-century and end-century periods (Table 7.2). The mid-century scenario for kharif[1] season showed an increase of 1.93°C and an overall increase of 13.6% in precipitation. This scenario is denoted by 1.93°C/13.6% and for the rabi[2] season the scenario is 2.22°C/13.6%. Similarly, the end-century scenarios for kharif and rabi are respectively 4.03°C/17.8% and 4.28°C/17.8%. In all these scenarios, only the annual change in precipitation (and not seasonal changes) is considered, as the annual precipitation reflects inter-seasonal water accumulation. These predicted changes were used in the mean and variance functions to predict the average yield and variability in yield induced by climate change.

The precipitation is increasing in both scenarios but the climate models do not have information on the rainfall distribution pattern. There can be a sudden deluge where the present storage reservoirs are not sufficient to meet the demand. The Assessment Report 4 (AR4) by IPCC noted that the frequency of more intense rainfall events in many parts of Asia has increased, causing severe floods, landslides and mud flows. At the same time, the number of rainy days has decreased. Analysis of rainfall data for India highlights the increase in the frequency of severe rainstorms over the last 50 years. The number of storms with more than 100 mm rainfall in a day is reported to have increased by 10% per decade (UNEP, 2007). However, due to storage-related issues, the actual irrigation water availability can be decreased in contrast with the projected increased rainfall, which would increase the demand for food grain production.

The projected changes from Table 7.2 were assumed to be same for all the three river basins for modelling purposes as the estimates for Krishna and Cauvery basins are not available.

The parameters were estimated by using the following equation with the assumption that $\omega_{it} \sim N(0,1)$, the likelihood function is given by:

Table 7.2. Projected changes in climatic variables during kharif and rabi seasons (calculations based on figures from Gosain and Rao, 2012).

Change in mean daily average temperature (°C)	Kharif (June to November)	Rabi (December to April)
Change from baseline to mid-century	1.93	2.22
Change from baseline to end-century	4.03	4.28

Change in mean precipitation (%)	Kharif (June to November)	Rabi (December to April)	Overall
Change from baseline to mid-century	12.5	17.6	13.6
Change from baseline to end-century	13.0	53.4	17.8

$$L=\left[\frac{1}{2\pi}\right]^{N/2}\prod_{t=1}^{T}\prod_{i=1}^{R}\left[\frac{1}{h(x_{it};\delta)}\right]^{1/2}\exp\left[-\{y_{it}-f(x_{it};\beta)\}^2/2h(x_{it};\delta)\right] \quad (7.1)$$

where, R is the number of districts and T is the number of time periods and $N=RT$. So the log likelihood function is given by:

$$\ln L=-\frac{1}{2}\left[N\ln(2\pi)+\sum_{t=1}^{T}\sum_{i=1}^{R}\ln\left(h(x_{it};\delta)\right)+\sum_{t=1}^{T}\sum_{i=1}^{R}\frac{\{y_{it}-f(x_{it};\beta)\}^2}{h(x_{it};\delta)}\right] \quad (7.2)$$

This was then maximized to estimate the parameter vectors β and δ. STATA software package has inbuilt *ml* command and it was used to maximize the log likelihood function.

The climate change scenarios and the most likely management options considered in the study are given in Tables 7.3 and 7.4.

The scenario S1 reflects the current state of affairs. Scenario S2 corresponds to 'near future' status. This scenario assumes a 10% likely reduction in water availability for agriculture and a 5% reduction in labour availability. The reduction in water availability is based on the assumption that, in spite of increased and uneven precipitation predicted by climate change scenarios, the share available for agriculture will be reduced due to increased population, domestic consumption demand and industrial use. The Krishna and Cauvery river basins also fall under arid and semi-arid regions (Table 7.1). Similarly, reduced labour supply is due to migration from rural areas and labour demand met by agricultural mechanization. In the mid-century scenario, S3, the productivities of the crops during kharif and rabi seasons are considered. The last scenario, S4, uses the end-century predictions of yield levels. The eight management options are based on promising technologies for rice and maize (Table 7.4).

Table 7.3. Climate scenarios and resource availability considered in the study.

Scenario symbol	Description of the scenario
S1	Current levels of yield, water availability and labour availability
S2	Current level of yield and 10% reduction in water availability and 5% reduction in labour availability, as observed from the past data in the basin (near future)
S3	Projected mid-century yield levels and 10% reduction in water availability and 5% reduction in labour availability
S4	Projected end-century yield levels and 10% reduction in water availability and 5% reduction in labour availability

Table 7.4. Management options considered in the study.

Management option – symbol used	Description of the option
M1	Current management intervention
M2	System of Rice Intensification (SRI) is an improved rice cultivation practice, which could save 20% irrigation water
M3	SRI + Machine Transplanting (MT) will result in a 15% reduction in labour use for rice
M4	Alternate Wetting and Drying (AWD) will result in reduction of water use for rice by 10%
M5	AWD + MT
M6	Maize Water Management (MWM) will result in a 10% reduction in water use for maize
M7	AWD + MT + MWM
M8	SRI + MT + MWM
M9	Direct Seeding of Rice (DSR) will result in a 20% reduction in water use, 10% reduction in labour and 10% reduction in yield

For example, system of rice intensification (SRI) is the most recommended technology for rice cultivation and maize water management (MWM) for maize in the same season. The SRI and machine transplanting as modified SRI (MSRI) with specified spacing, has been implemented in the Krishna and Cauvery river basins of Andhra Pradesh and Tamil Nadu, which resulted in a 10% reduction in water and a 15% reduction in labour (Kakumanu *et al.*, 2011; Lakshmanan *et al.*, 2012). Similarly, reduction was seen in the alternate wetting and drying, and direct seeding of rice (Gurava Reddy *et al.*, 2013). Rice and maize were considered in the same season under the command area for the optimum allocation of resources, based on the historic cropping pattern.

For minimizing water use, four targets for rice production and income were fixed, as given in Table 7.5. The targets are desired for rice production as the basins are known for rice production and hence stabilizing rice production under the climate change scenarios is important.

The target T1 refers to the current situation. The maximum possible rice production and maximum income achievable under eight management options are the targets to be met while optimizing water usage. For the second target, T2, the maximum rice production and income levels possible are derived when available water and labour are reduced by 10% and 5%, respectively. The same target levels are also used in T3 and T4, where the productivity is reduced by climate change.

The details of the optimization framework developed are given in Appendix 7.1. In this optimization framework, the objectives are first prioritized as stated above. First priority is given to maximum rice production as food security is of primary importance for society.

The second priority is given to maximum income, as it will ensure a better livelihood for the farmers. A new constraint is used while running the linear programs to meet the second objective, which will ensure that the rice production will be at least equal to the maximum level as dictated by the first objective. Thus, the results of the first two objectives will ensure maximum food production and a better livelihood.

The third objective is to minimize water usage in agriculture. This objective is important because, as shown by historical data, the share of water use for agriculture is declining over the years as demand increases for other sectors. The results of the first two objectives are incorporated as constraints in meeting the third objective. Thus, the results of all the three optimization models will provide a complete framework to plan for optimum land and water use for the triple targets: increased food production, increased income and reduced water use.

7.3.3 Study area for model application

One project case from each of the three river basins, Godavari, Krishna and Cauvery, were selected for the study, namely the Sri Ram Sagar Project (SRSP) and Nagarjuna Sagar Project (NSP) from Godavari and Krishna river basins, respectively, and the Lower Bhavani Project (12 districts located within the Bhavani Sagar) from the Cauvery Basin (Fig. 7.1). The details of the projects are

Table 7.5. Target levels with minimum water use.

Target	Description
T1	Current maximum rice production and maximum income
T2	Near future[a] maximum rice production and maximum income
T3	Near future maximum rice production and maximum income with mid-century climate change-induced productivities
T4	Near future maximum rice production and maximum income with end-century climate change-induced productivities

[a]Near future represents the next 20 years.

summarized in Table 7.6 and discussed in the following sub-sections.

Sri Ram Sagar Project

The Sri Ram Sagar Project (SRSP) is a multi-purpose project, located across the Godavari River near Pochampad of Nizamabad District in Andhra Pradesh. The project was cleared in 1946 for utilization of 1869 million m^3 (Mm^3) of water from the Godavari River. As a result of an inter-state accord, the allocation was increased from 1869 Mm^3 to more than 5664 Mm^3 (http://www.aponline.gov.in; accessed 4 January 2012), but later limited to 3455 Mm^3 due to capacity constraints. The catchment area upstream of the dam site is 91,751 km^2 and the surface area of the reservoir is 453 km^2. The reservoir water irrigates 0.39 Mha of land through three canals (Kakatiya and Laxmi on the right bank and Saraswathi on the left bank), and supplies nearby areas with drinking water and water for hydropower generation. The water used for hydropower production is later released for irrigation in the Kakatiya Canal, 146 km long, from the SRS Dam. The Kakatiya Canal crosses the Manair River and water is also stored at the Lower Manair Dam. A flood flow canal discharges excess water towards the right bank also flowing into the Lower Manair Dam when the reservoir level exceeds 326 m. In addition, water from the groundwater is also an important source supplementing irrigation in this region.

In addition to irrigation and hydropower, SRSP also provides drinking water to urban and rural areas along the canal system, particularly to the towns of Karimnagar and Warangal located in Andhra Pradesh. The drinking water allocation for rural areas ranges from 55–100 l per capita day^{-1} (l cap^{-1} day^{-1}), whereas for urban areas it is 70–120 l cap^{-1} day^{-1}. The population served by SRSP is approximately 12.6 million.

The cropping pattern proposed for SRSP in the early stages was to grow only irrigated dry crops such as maize, groundnut, sorghum and pulses. However, the cropping pattern has changed over the years into more water-demanding crops, and rice, together with maize and groundnut, are now the main crops (SRSP, 2009) in both kharif and rabi seasons, which poses a major challenge for water distribution and management. The cropping pattern changes with the variation in rainfall. The mean annual rainfall in the area of SRSP is 900 mm of which more than 75% is received in the south-west monsoon period (kharif season). In the last decade, the driest year was

Table 7.6. Highlights of the three irrigation projects considered in the study.

	Sri Ram Sagar	Nagarjuna Sagar	Lower Bhavani
Name of the basin	Godavari	Krishna	Cauvery
No. of districts covered	4	5	12
Catchment area (km^2)	91,751	215,185	4,200
Project type	Multi-purpose	Multi-purpose	Multi-purpose
Project capacity (TMC)[a]	112	408	70
Length (km)	364.5	382	200
Command area (ha)	387,456	896,000	84,000
Annual rainfall (mm)	878	785	730
On-farm application efficiency – wet crops (%)	34.5	33.10	38
Average on-farm application efficiency (%)	57.28	38.93	48
Overall project efficiency (%)	44.66	21.8	52
Soil type	Black clay to red soils	Black clay to red coarse soils	Red loamy soils
Major crops	Rice, maize and groundnut	Rice, cotton, chilli, maize, groundnut and pulses	Rice, maize, sesame, turmeric, sorghum and pulses

[a]TMC = Thousand million cubic feet.

recorded during 2004/2005, where only 55% of the mean annual rainfall was received and no water was available through canals to the command area downstream (SRSP, 2009). Hence there is a need for optimization of resource use by maximizing the food production and income with minimum water use through improved water management practices.

Nagarjuna Sagar Project

The Nagarjuna Sagar Project (NSP) is one of the largest and highest masonry dams (125 m) in the world. It is situated downstream of the Srisailam Reservoir on the main Krishna River in Andhra Pradesh. It is a multi-purpose project with irrigation, hydropower and flood-control components.

The catchment area of the dam is 215,193 km^2; the annual rainfall in the catchments is 889 mm, the maximum observed flood is 30,050 $m^3 s^{-1}$ (cumec) and the design flood (return period 1000 years) is 58,340 cumec. NSP complex has a substantial capacity for hydropower generation. It has one conventional and seven reversible units, each with 110 MW capacity. The right bank canal powerhouse has three units of 30 MW each and the left canal powerhouse has two units of 20 MW each.

NSP annually provides 7465 Mm^3 water on average to a command of 0.89 Mha. The project was completed in 1974 and comprises a dam with two canals taking off on either side. The Nagarjuna Sagar Right Canal (NSRC) is 203 km long and creates an irrigation potential for 0.47 Mha in Guntur and Praksam districts, while the Nagarjuna Sagar Left Canal (NSLC) is 179 km and creates irrigation potential for 0.42 Mha in Nalgonda, Khammam and Krishna districts. Of the five districts under NSRC and NSLC, Guntur District has the highest command area of 284,000 ha covering 39 *mandals* (blocks) in the district.

The command area under NSP is designed for a mixed cropping pattern, that is, one-third wet and two-thirds irrigated dry (ID). The areas close to the head reaches were localized as ID rabi (December to April) and the command in the lower reaches were localized as kharif wet (June–November). Water is supplied for a single crop, either only for kharif wet or rabi ID. Rice is the major crop grown in the kharif season and ID crops are sown after October in rabi. The important ID crops grown in the project area are chilli, cotton, pulses and groundnut. The cropping pattern observed from 1995/1996 to 2004/2005 has not had much variation (I&CAD, 2009). Whatever changes that have occurred in the cropping pattern are due to the availability of water in the reservoir, coupled with the seasonal rainfall pattern. Most of the cropping takes place during the kharif season. Though the upper reaches are designed for ID crops, wet crops are generally sown, resulting in excess use of water compared to its original design, depriving tail-end users. The tail-end users are thus compelled to cultivate ID crops or supplement irrigation of wet crops with groundwater. The water in the canal flows continuously in kharif and on and off during the rabi season. The crops grown also depend on the soils, climatic conditions, irrigation facilities, and market and price conditions. Rice is the most favoured crop for the farmers as it is a staple food. In the rabi season, farmers cultivate pulses or vegetables and ID crops such as groundnut and maize.

In NSP, problems regarding availability of water at the tail-end exist due to excess use of water at the head regions. The average on-farm efficiency is 39% and the project efficiency is 21.8% (I&CAD, 2009). The roles and responsibilities of the water user associations are not clearly known and defined. Water management options are also poorly disseminated to the farming community. Hence, there is a need to address some of these challenges by adapting to water management practices to improve water use efficiency at the farm and project level.

Lower Bhavani Project

The Lower Bhavani Project (LBP) is the major project in the Cauvery River Basin (CRB), which lies in the eastern part of Tamil Nadu. CRB covers 12 districts of Tamil Nadu.

The Erode District makes up about 16% followed by Coimbatore with 14% of the total basin area (Government of Tamil Nadu, 2006). The region experiences rainfall during both the summer south-west (June through September) and early winter north-east monsoons (October through December), with the peak rainfall during the north-east monsoonal season. While most parts of central and northern India experience decreasing rainfall in both seasons, the peninsular parts of India, particularly the region 9–16°N encompassing the CRB, shows a tendency for increasing rainfall. This increase is particularly strong during the north-east monsoonal season. While part of this trend may be due to multi-decadal monsoonal variability, a potential impact of anthropogenic greenhouse gas (GHG) forcing cannot be ruled out (Annamalai and Nagothu, 2009). The forced emerging signal due to increase in GHG concentrations, or climate change, is manifested as a long-term trend. At any rate, the observed long-term changes in rainfall may have already been influencing the agriculture sector in the CRB.

The total cropped area of the CRB, Tamil Nadu is about 1.8 Mha. Rice is the major food crop in all the districts and is predominantly grown in Thanjavur, Thiruvarur and Nagapattinam, with 66.78% of the total rice production in the basin. Sorghum and pulses make up the second and third most important crops in the basin, with 16.25% and 15.13% shares, respectively.

The project is facing water shortages due to reduced inflows over years and increasing withdrawal for the domestic sector. Groundwater use is increasing to offset the reduction in canal supplies. Water usage for agriculture is also high with poor water use efficiency, especially in the case of the rice crop. Improving the water use efficiency at system level will help to address the water shortages that will increase in the future.

7.4 Results and Discussion

As water use in these projects is mostly during the kharif crop season (June–November months), optimization of the resources was carried out for kharif season for all three river basins. The mean functions, variance functions, standard errors of the coefficients and log-likelihood function of all the three projects for different crops are presented in Appendix 7.2. The maximum and minimum possible yields and also the impact of climate change on the crop yield for all the three projects are presented in Appendix 7.3. The project-wise optimization results are presented below. The results have highlighted only the significant different management options.

7.4.1 Godavari river basin – Sri Ram Sagar Project

Current rice production in the kharif season is 570,000 t, with a predicted production of 500,000 t in each of the next 20 years. The corresponding predictions for the mid- and end-century are 480,000 t and 360,000 t, respectively (Fig. 7.2). Adoption of SRI in the current and future scenarios increases the yield by 13%, AWD by 11%, and combination of SRI–machine transplantation with MWM by 23%.

Total gross income from crop production during the kharif season using the current market price of the crops is at present Rs2.9 bn and will reduce to Rs2.67 bn over the next 20 years assuming that the same prices would prevail. At mid- and end-century, the gross income will reduce to Rs2.52 and Rs2.37 bn, respectively (Fig. 7.3).

In order to maintain the current level of production and gross income 3868 Mm^3 of water is required, and this will decrease to 3479 Mm^3 in the near future. However, during the mid- and end-century, the water required for maintaining the current level of production and gross income will be 3679 and 4794 Mm^3, respectively (Fig. 7.4). This can be reduced by adapting management practices like SRI with machine transplantation, MWM and AWD. The water use with the adaptation of SRI and AWD reduces by 17% and 9.5%, respectively. A similar trend was noticed for the near future, and mid- and end-centuries.

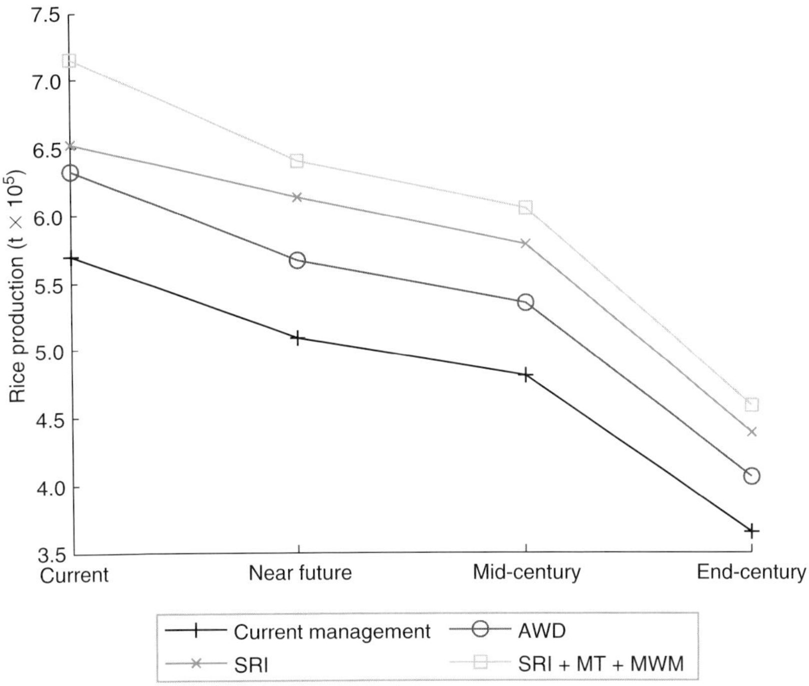

Fig. 7.2. Rice production under different scenarios and management options in the kharif season.

7.4.2 Krishna basin – Nagarjuna Sagar Project

Current rice production in the kharif season from NSP is 380,000 t and will be about 300,000 t, each season, in the near future (next 20 years) and mid-century. The production will be 250,000 t at the end of the century (Fig. 7.5). Adding water-saving technologies such as SRI and AWD will increase rice production by about 50,000 to 100,000 t, which is about 13–26% of current production. With the addition of machinery transplantation (MT) along with the SRI and AWD options, the production levels remain the same but the labour use increases for each option. MT is addressing the issue of labour scarcity both in present and future time periods. Direct seeding of rice (DSR) is also showing better yields compared to the current practices by improving water productivity, except in the near-future scenario. DSR reduces water use without any variation in the production levels. The system is presently practised at the tail-ends of the canal systems. This practice is expected to be increasingly adapted during the water stress years at mid- and end-century periods.

The second objective is to maximize income from crop production during the kharif season. Total gross income from crop production during the kharif season with the current level of production is Rs15.7 bn and will reduce to Rs15.4 bn in the next 20 years. During the mid- and end-century, the gross income will reduce to Rs12.7 and Rs11.8 bn, respectively (Fig. 7.6). The SRI and AWD options improve the gross income by Rs0.42 and Rs0.18 bn with the current practices. MT does not influence the production levels and income levels with the SRI and AWD options for the near future and for the mid- and end-century periods, but will reduce the scarcity of labour and minimize water use.

In order to maintain the current level of production and gross income, 8384 Mm^3 of water are required. The water requirement decreases with change in technologies. SRI

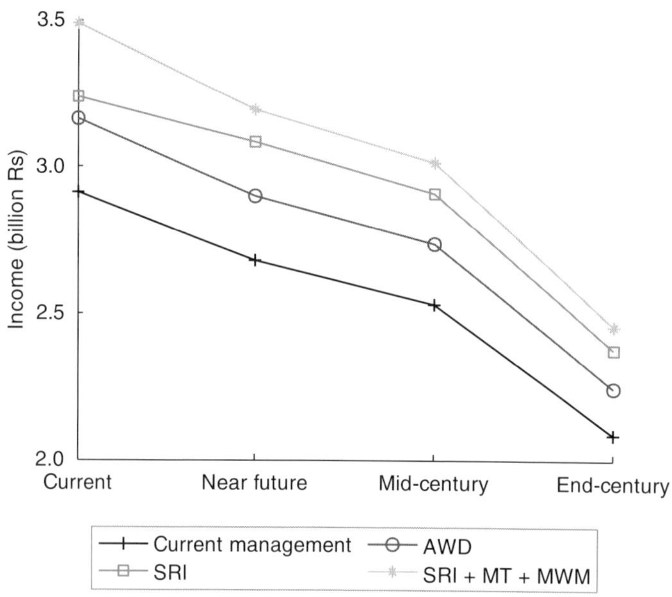

Fig. 7.3. Income under different scenarios and management options during the kharif season.

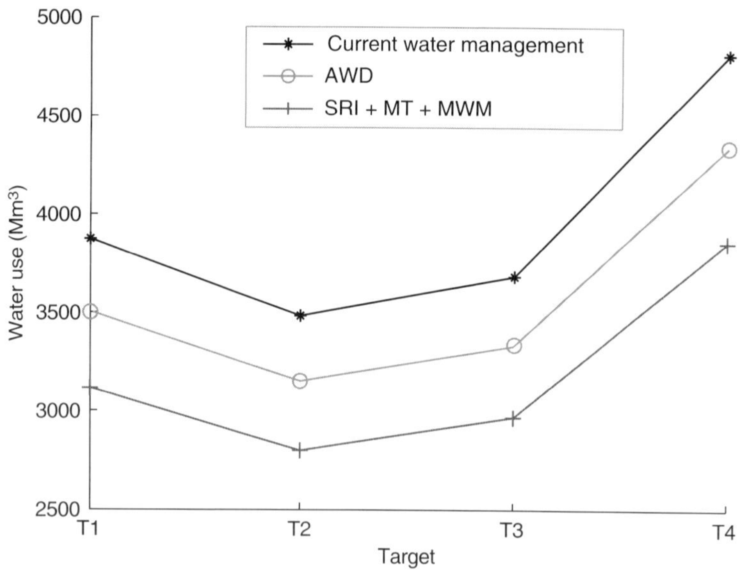

Fig. 7.4. Water use under different scenarios and management options, kharif season.

requires 7067 Mm^3 (15.7% less), AWD requires 7726 Mm^3 (7.8% less) and DSR requires 8149 Mm^3 (2.8% less). The minimum water requirement to maintain the current level of production and gross income is more or less similar during the near future, and mid- and end-century periods.

It is therefore important to see how production and income can be maintained in the future using various adaptation strategies. As indicated earlier, water- and labour-saving technologies will help to minimize the production and income losses due to climate change impacts, and will also reduce water

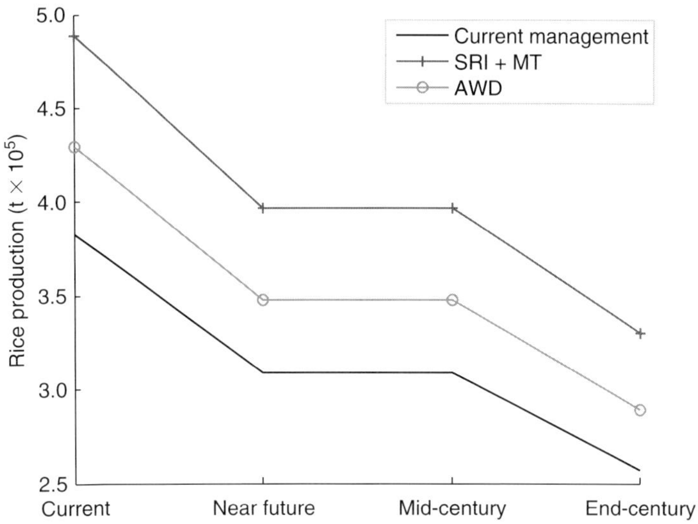

Fig. 7.5. Rice production under different scenarios and management options during the kharif season.

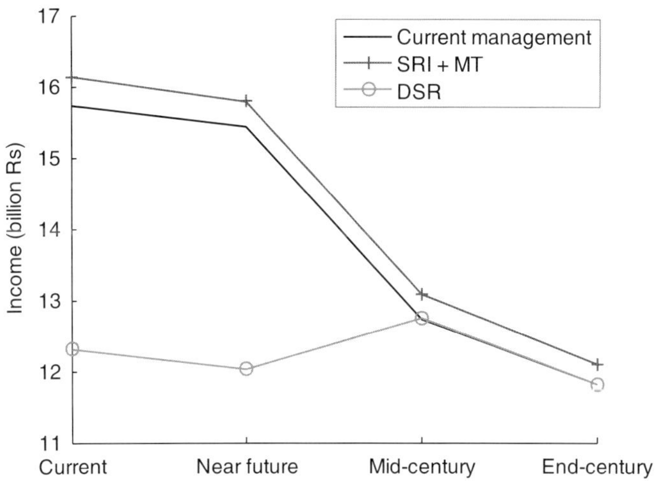

Fig. 7.6. Income for different scenarios and management options – kharif season.

use. From the optimization results it was observed that when the water- and labour-saving technologies are adopted in crop cultivation, the current rice production will be 710,100 t (24.5% increase), 640,000 t (12.2% increase) in the next 20 years, 600,000 t (5.2% increase) at mid-century and 460,000 t at end-century (19% decrease), respectively. This end-century result of 19% reduction in rice production can be compared to 37% reduction if no technological interventions are made. Water management and labour-saving technologies will help to address the negative impact of climate change in rice production in the project area. A similar trend is seen in the case of gross income and water use during different periods.

7.4.3 Cauvery river basin – Lower Bhavani Project

Current rice production in the kharif season is 118,000 t and is predicted to be 105,000 t each season in the next 20 years (near future). SRI with machine transplantation and maize water management has increased the yield by 17%, SRI 14% and AWD 11%. As per the mid-century predictions, the production will be 91,000 t (a decrease of 22.9%) and it will be 80,000 t (a decrease of 32.2%) at the end of the century (Fig. 7.7).

The major crops from which farmers derive income during the kharif season are rice and maize. At present, the total gross income from crop production during kharif is Rs709.8 million and is predicted to reduce to Rs633.7 million, every season, in the next 20 years. At mid- and end-century, the gross income is predicted to reduce to Rs541.7 million (23% reduction) and to Rs474.0 million (33% reduction), respectively (Fig. 7.8). But with the adoption of SRI, MT, AWD and MWM practices the income can also be increased by 11%–17%.

In order to maintain the current level of rice production of 118,000 t and gross income of Rs709.8 million during the kharif season, the amount of water required is 480 Mm^3. However, if SRI techniques are used for rice and MWM is used for maize, the same targets can be achieved with 394 Mm^3 of water, saving 86 Mm^3. In the near future, the total water available will be 432 Mm^3. With this amount of water, and the current water management options, the maximum achievable rice production is 105,000 t and maximum income will be Rs633.7 million. Application of SRI to rice crop will decrease the need for water by 86 Mm^3, improving water productivity with the decrease in water consumption by 9%–18% in all the scenarios except in the end-century.

However, at mid- and end-century, due to the negative impacts of climate change on crop productivities, there will be a greater water demand to maintain the current level of productivity and income (target T1) or at least near-future productivity and income (T2). To achieve this, we set the targets T3 and T4 listed in Table 7.5. These targets are not possible to be met with the current availability of water and current management interventions. Table 7.7 provides the predicted quantity of water needed to attain the targets under various management options.

It is evident from the Table 7.7 that SRI is a very promising management option to minimize water use. The excess demand for water during mid-century can be nullified by applying this technique. However, at the end of the century, the effect of climate change will be much more severe, and more water will be needed to maintain target T2. Even with SRI intervention there will be a 24.6 Mm^3 deficiency of water.

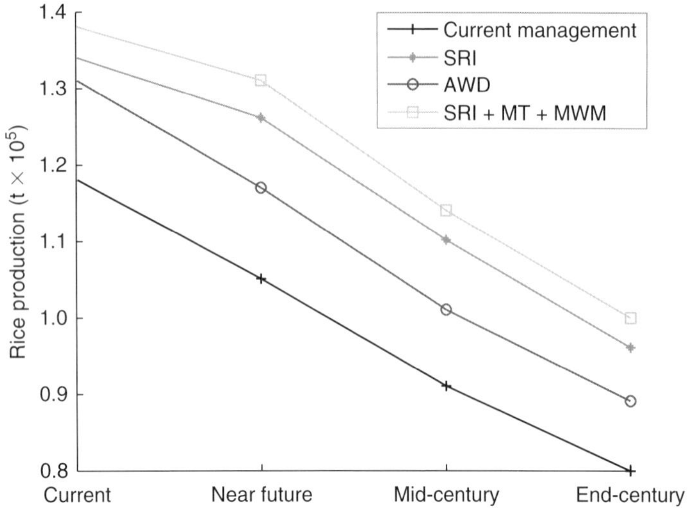

Fig. 7.7. Rice production for different scenarios and management options during the kharif season.

Adoption of water- and labour-saving technologies contribute to rice production in the project area. In all the cases, SRI resulted in higher production, gross income and water saving, compared to MT and AWD. MT helped the rice production by releasing labour to cultivate additional rice areas. In the future, labour scarcity is expected to lead to a reduction of areas under rice cultivation, as it will be a constraint to the transplanting operations. MT helps to ease the labour scarcity by 20–25%. It is clear that the returns from investing in water management technologies for coping with future climate change impacts are high, if the farmers adopt them properly (Palanisami *et al.*, 2011).

7.5 Conclusions and Recommendations

Climate change impacts will, in the long run, reduce rice production in the project areas by 25–30%. Water is the key long-term constraint in rice production and land currently under fallow due to water scarcity will be a key issue to address in the future. By implementing various water- and labour-saving technologies (MT, SRI, DSR and AWD), one can minimize the reduction in rice production by 20–25% during the mid- and end-century periods, and these technologies will also help to minimize water use as well.

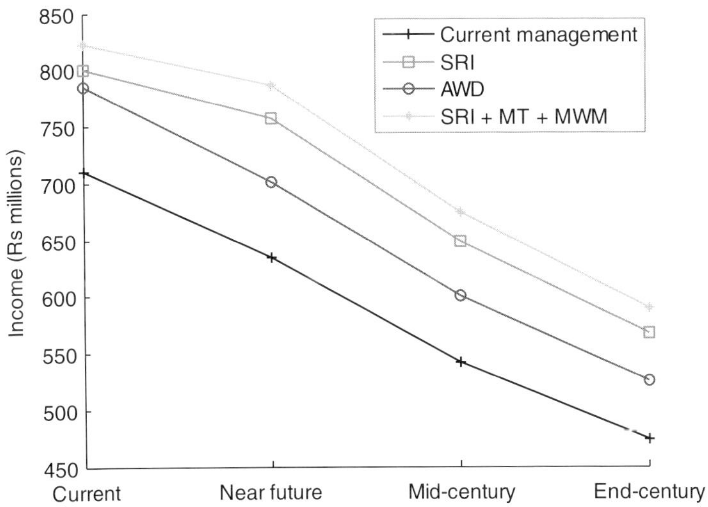

Fig. 7.8. Income under different scenarios and management options during the kharif season.

Table 7.7. Water use for different scenarios and management options during the kharif season.

Management option	Target			
Rice production (t)	118,000	105,000	105,000	105,000
Income (Rs million)	709.8	633.7	633.7	633.7
Water available (Mm^3)	480	432	432 and mid-century projected crop productivities	432 and end-century projected crop productivities
Water required (Mm^3)				
Current management	480.0 (0.0)	432.0 (0.0)	493.1 (–61.1)	557.0 (–125.0)
SRI	394.2 (+85.8)	355.8 (+76.2)	405.3 (+26.7)	456.6 (–24.6)
AWD	437.1 (+42.9)	393.9 (+38.1)	449.2 (–17.2)	506.8 (–74.8)
SRI+MT+MWM	393.7 (+86.3)	355.3 (+76.7)	377.9 (+54.1)	450.6 (+18.6)

+, excess water availability; –, deficit.

The results from the three basins have shown that adoption of various water management technologies improves the water productivity and income from the projects. However, the performance of the technologies varied across the basins, indicating their mixed performance. The factors contributing for the successful adoption have to be studied and follow-up actions can be initiated in other basins.

However, in general, the level of technology adoption is currently poor in all the basins due to poor access to the technologies, lack of skills in handling the improved technologies and the recurring costs (Palanisami *et al.*, 2014). A recent study conducted in these basins also indicated that upscaling the technologies will help address the climate change impacts and hence promotion of the water management technologies is the key intervention to be targeted at basin level (Nagothu *et al.*, 2012). These technologies need to be disseminated and up-scaled with a capacity-building framework considering their impacts on the production, income and conservation of water resources. As piloting the technologies on individual farms will not have a major impact, a cluster approach (covering a group of villages in a location for each technology) will be more useful in up-scaling these management technologies.

Notes

[1] Kharif is the wet season, which covers the months June/July–November.

[2] Rabi is the dry season, which covers the months December–March/April.

References

Abraha, M.G. and Savage, M.J. (2006) Potential impacts of climate change on the grain yield of maize for the midlands of KwaZulu-Natal, South Africa. *Agriculture Ecosystems & Environment* 115, 150–160.

Aggarwal, P.K. (2009) Global climate change and Indian agriculture. *Case Studies from ICAR Network Project.* Indian Council of Agricultural Research, New Delhi, India, 148 pp.

Aggarwal, P.K., Palanisami, K., Khanna, M. and Kakumanu, K.R. (2012) Climate change and food security of India: adaptation strategies for the irrigation sector. *World Agriculture* 3(1), 20–26.

Amarasinghe, U.A., Sharma, B.R., Aloysius, N., Scott, C., Smakhtin, V. and de Fraiture, C. (2005) Spatial variation of water supply and demand across river basins of India. *Research Report 83.* International Water Management Institute, Colombo, Sri Lanka.

Annamalai, H. and Nagothu, U.S. (2009) Climate change detection and prediction for adaptation decision-making with relevance to Cauvery basin. Technical Brief no.7, May 2009. Available at: http://www.tnau.ac.in/climarice/Technical_brief_7.pdf (accessed January 2013).

Barnwal, P. and Kotani, K. (2010) Impact of variation in climatic factors on crop yield: a case of rice crop in Andhra Pradesh, India. *Economics and Management Series*, 17. International University of Japan, Minamiuonuma, Japan.

Bradshaw, B., Dolan, H. and Smit, B. (2004) Farm-level adaptation to climatic variability and change: crop diversification in the Canadian Prairies. *Climatic Change* 67, 119–141.

Conway, G. (2009) The science of climate change in Africa: impacts and adaptation. *Discussion Paper No. 1.* Imperial College, London.

Craufurd, P.Q., Jagadish, S.V.K. and Padgham, J. (2011) Impacts of climate change on rainfed agriculture and adaptation strategies to improve livelihoods. In: Suhas, P., Wani, J.R. and Sahrawat, K.L. (eds) *Integrated Watershed Management in Rainfed Agriculture.* CRC Press/Balkema, the Netherlands, pp. 421–440.

Cruz, R.V., Harasawa, H., Lal, M., Wu, S., Anokhin, Y., Punsalma, B., Honda, Y., Jafari, M., Li, C. and Huu, N. (2007) Asia climate change, 2007, impacts, adaptation and vulnerability. In: Parry, M.L., Canziani, O.F., Palutikof, J.P., Van der Linden, P.J. and Hanson, C.E. (eds) *Contribution of Working Group II to the Fourth Assessment Report of the Intergovernmental Panel on Climate Change.* Cambridge University Press, Cambridge, UK. pp. 469–506.

Elisabeth, S., Misselhorn, A., Eakin, H., Devereux, S., Drimie, S., Msangi, S. and Stafford, S.M. (2010) Vulnerability to what? In: Ingram, J., Ericksen, P. and Liverman, D. (eds) *Food Security and Global Environmental Change.* Earthscan, UK.

Fraser, E.D.G., Mabee, W. and Slaymaker, O. (2003) Mutual vulnerability, mutual dependence: the reflexive relation between human society and the environment. *Global Environmental Change* 13(2), 137–144.

Gosain, A.K. and Rao, S. (2012) Analysis of climate scenarios in the Godavari river basin. In: Nagothu, U.S., Gosain, A.K. and Palanisami, K. (eds) *Water and Climate Change: An Integrated Approach to Address Adaptation Challenges*. Macmillan Publishers India Ltd, New Delhi, India, pp. 59–79.

Government of Tamil Nadu (2006) *Season and Crop Report 2006. Directorate of Economics and Statistics*. Government of Tamil Nadu, India.

Gurava Reddy, K., Kakumanu, K.R., Tirapamma, S., Palanisami, K. and Uday, S.N. (2013) Partnership as key for enhancing technology transfer: a case of direct sowing rice in Guntur district, Andhra Pradesh, India. Paper presented at the National Seminar on Futuristic Agricultural Extension for Livelihood Improvement and Sustainable Development 19–21 January 2013, Hyderabad, India.

I&CAD (2009) *Water Use Efficiency Studies – Nagarjuna Sagar Project*. Irrigation and Command Area Development Department, Government of Andhra Pradesh, India.

IPCC (2001) *Climate change 2001: Impacts, adaptation, and vulnerability*. Intergovernmental Panel on Climate Change, Cambridge University Press, Cambridge, UK.

IPCC (2007) Climate change 2007: the physical science basis. In: Solomon, S., Qin, D., Manning, M., Chen, Z., Marquis, M., Averyt, K.B., Tignor, M. and Miller, H.L. (eds) *Contribution of Working Group I to the Fourth Assessment Report of the Intergovernmental Panel on Climate Change*. Cambridge University Press, Cambridge, UK.

Isik, M. and Devadoss, S. (2006) An analysis of the impact of climate change on crop yields and yield variability. *Applied Economics* 38(7), 835–844.

Just, R.E. and Pope, R.D. (1978) Stochastic specification of production functions and economic implications. *Journal of Econometrics* 7(1), 67–86.

Kakumanu, K.R., Gurava Reddy, K., Palanisami, K. and Udaya, S.N. (2011) Machinery transplantation: labour and water saving technology in rice cultivation. Technical Brief-8. Available at: http://www.climarice.com (accessed 31 October 2013).

Kandlinkar, M. and Risbey, J. (2000) Agricultural impacts of climate change: If adaptation is the answer, what is the question? *Climatic Change* 45, 529–539.

Kattarkandi, B.S., Naresh, K. and Aggarwal, P.K. (2010) Simulating impacts, potential adaptation and vulnerability of maize to climate change in India. *Mitigation and Adaptation Strategies for Global Change* 15(5), 413–431.

Kelkar, U. and Bhadwal, S. (2007) South Asian Regional study on climate change impacts and adaptation: implications for human development. *Occasional Paper. Human Development Report Office*. United Nations Development Programme, New Delhi, India.

Knox, J.W., Hess, T.M., Daccache, A. and Perez Ortola, M. (2011) What are the projected impacts of climate change on food crop productivity in Africa and South Asia? *DFID Systematic Review, Final Report*. Cranfield University, Cranfield, UK, 77 pp.

Lakshmanan, A., Geethalakshmi, V., Nagothu, U.S., Mary, J.K. and Bhuvaneswari, K. (2012) System of rice intensification (SRI) – an adaptation technology for changing climate. Technical Brief 12. Available at: http://www.climarice.com (accessed 19 August 2013).

Ministry of Environment (MoE) (2003) Pakistan's initial national communication on climate change. Government of Islamic Republic of Pakistan, MoE, Islamabad. Available at: http://unfccc.int/resource/docs/natc/paknc1.pdf (accessed 17 October 2013)

Nagothu, U.S., Gosain, A.K. and Palanisami, K. (2012) *Water and Climate Change: an integrated approach to address adaptation challenges*. Macmillan Publishers India Ltd, New Delhi, India.

Nelson, G., Rosegrant, M., Koo, J., Robertson, R., Sulser, T., Zhu, T., Ringler, C., Msangi, S., Palazzo, A., Batka, M., Magalhaes, M., Valmonte-Santos, R., Ewing, M. and Lee, D. (2009) *Climate Change: impact on agriculture and costs of adaptation*. International Food Policy Research Institute, Washington, DC.

Palanisami, K., Ranganathan, C.R., Kakumanu, K.R. and Udaya, S.N. (2011) A hybrid model to quantify the impact of climate change on agriculture in Godavari river basin. *Energy and Environment Research* 1(1), 32–52.

Palanisami, K., Ranganathan, C.R., Nagothu, U.S. and Kakumanu, K.R. (2014) *Climate Change and Agriculture in India: Studies from Selected River Basins*. New Routledge-Taylor & Francis, New Delhi, India.

Patt, A. and Gwata, C. (2002) Effective seasonal climate forecast application: Examining constraints for subsistence farmers in Zimbabwe. *Global Environmental Change* 12(3), 185–195.

Ranganathan, C.R. (2009) Quantifying the impact of climatic change on yields and yield variability of major crops and optimal land allocation for maximizing food production in different agro-climatic zones of Tamil Nadu, India: an econometric approach. Working Paper. Research Institute for Humanity and Nature, Kyoto, Japan.

Srivastava, A., Naresh Kumar, S. and Aggarwal, P.K. (2010) Assessment on vulnerability of sorghum to climate change in India. *Agriculture, Ecosystems and Environment* 138(3–4), 160–169.

SRSP (2009) *Water Use Efficiency Studies – Sri Ram Sagar Project.* Irrigation and Command Area Development Department, Government of Andhra Pradesh, India.

Sultana, H. and Ali, N. (2006) Vulnerabiltiy of wheat production in different climatic zones of Pakistan under climate change scenarios using CSM-CERES-Wheat model. Paper presented in the Second International Young Scientists Global Change Conference, 7–9 November 2006, Beijing, China.

Till, B., Artner, A., Siebert, R. and Sieber, S. (2010) Micro-level practices to adapt to climate change for African small-scale farmers. *IFPRI Discussion Paper 00953.* International Food Policy Research Institute, Washington, DC.

UNEP (2003) *GEO Yearbook 2003.* United Nations Environment Programme, Nairobi, Kenya, 12 pp.

UNEP (2007) *GEO Yearbook 2007: An Overview of Our Changing Environment.* United Nations Environment Programme, Nairobi, Kenya, 86 pp.

Appendix 7.1. Details of the Optimization Framework

Optimization is done by formulating suitable linear/quadratic programming problems. The objective functions generally included are:

1. Maximize rice production.
2. Maximize farmers' income.
3. Minimize water use in agriculture.

Depending on the project objectives, other goals can also be included, such as maximizing food grain production and minimizing agricultural area.

A set of constraints for various resources like land, water, labour, etc., can be formulated depending on the availability of data for each basin. For example, in the selected basins, the following constraint formulations were made.

Variable	Explanation
x_{sdC}	Area under crop *C* in district *d* during season *s* *C*=Rice (R), Maize (M), Groundnut (G), Cotton (C), . . . *d*=districts *s*=Kharif, Rabi
y_{sdC}	Yield under crop *C* in district *d* during season *s*
A_{sC}	Area under crop *C* in season *s* in all the four districts
P_{sC}	Production of crop *C* in season *s* in all the four districts
AC_s	Total area under cereals in all the four districts in season *s*
w_{sdC}	Water required per hectare for crop *C* in season *s* in district *d*
R_{sdC}	Revenue under crop *C* in district *d* during season *s*

Objective 1: Maximize rice production in kharif

The predominant crops during the season in project areas are, e.g. rice, maize, cotton, chilli, etc.

$$\text{Maximize rice production: } \sum_{d=1}^{4} x_{sdR} y_{sdR}, \ s=Kharif$$

Constraints:

1. Total area under the crops during kharif season should not exceed the available crop area in the command in the districts (let us say, $d=4$)

$$\sum_{d=1}^{4} x_{sdR} + \sum_{d=1}^{4} x_{sdM} \leq AC_s, \ s=Kharif$$

2. Total water required for all crops during kharif in all the districts in the project area is less than or equal to the total water available

$$\sum_{c} \sum_{d=1}^{4} w_{sdC} x_{sdC} \leq W_s, \ s=Kharif$$

3. Total labour required for all crops during kharif in all the districts in the project area is less than or equal to the total labour available.

$$\sum_{c}\sum_{d=1}^{4} l_{sdC}x_{sdC} \leq L_s, s=Kharif$$

4. The normal (average of the last 5 years area) area under rice cultivation during the kharif season in the four districts is approximately in the ratio 1:2.5:1.7:2.5. It is assumed that this area ratio will continue. Hence:

$$x_{sAR}:x_{sKR}:x_{sNR}:x_{sWR}=1:2.5:1.7:2.5$$

The equivalent linear constraints that are included in the model are:

$$2.5x_{sAR}-x_{sKR}=0$$
$$1.7x_{sAR}-x_{sNR}=0$$
$$2.5x_{sAR}-x_{sWR}=0$$

5. The normal (average of the last 5 years area) area under maize (example) during the kharif season in the four districts is approximately in the ratio 1:5.1:2.6:2.4. It is assumed that this ratio of areas will continue. Hence

$$x_{sAM}:x_{sKM}:x_{sNM}:x_{sWM}=1:5.1:2.6:2.4$$

The equivalent linear constraints which are included in the model are

$$5.1x_{sAM}-x_{sKM}=0$$
$$2.6x_{sAM}-x_{sNM}=0$$
$$2.4x_{sAM}-x_{sWM}=0$$

6. The normal area under maize is about 70,000 ha. It is assumed that at least this much area should be allotted to the maize crop.

$$\sum_{d=1}^{4} x_{sdM} \geq 70000$$

Objective 2: Maximize farmers' net income

The objective is to maximize farmers' net income during the kharif season. That is:

$$\text{Maximize net income } \sum_{d=1}^{4}\sum_{C} R_{sdC}y_{sdC}, s=Kharif$$

In addition to the constraints described in Objective 1, an additional constraint is included which guarantees that the rice production during the kharif season will not be lower than the maximum level.

That is, if $R_{maxKharif}$ is the maximum rice production, then the new constraint added is:

$$\sum_{d=1}^{4} x_{sdR}y_{sdR} \geq R_{\max Kharif}$$

Objective 3: Minimize water use

The objective is to minimize water use in agriculture. That is:

$$\text{Minimize water use } \sum_{d=1}^{4}\sum_{C} w_{sdC}y_{sdC}, s=Kharif$$

The constraints on water availability and labour are removed and all other constraints that were included for maximizing income are retained. Two new constraints, one for fixing target for rice production and the other one for fixing target for income, are added. If *T* and *MI* are maximum rice production and maximum income, respectively, the constraints can be written as:

$$\sum_{d=1}^{4} x_{sdR} y_{sdR} \geq T$$

$$\sum_{d=1}^{4} \sum_{C} R_{sdC} y_{sdC} \geq MI$$

Thus the model will estimate the required minimum quantity of water that will ensure the target rice production *T* and maximum income *MI*.

Appendix 7.2. Just–Pope Production Function for the Crops Grown in all Three Projects: Parameter Estimates

	Sri Ram Sagar Project (Godavari)			Lower Bhavani (Cauvery)	
Mean yield	Rice	Maize	Groundnut	Rice	Maize
Precipitation (*R*) (mm)	7.210**	1.458***	–0.140	–0.2329	17.9885***
Temperature (*T*) (°C)	2,245.915**	–1,684.180**	4,621.546**	–190.385***	–15,954.3100***
Trend (year)	42.436***	73.988***	20.096***	42.0260***	2.3598
R^2	–0.001***	–0.001	0.000	0.0003	0.0001
T^2	–40.172	29.238***	–86.010***	0.7508	291.7981***
$R \times T$	–0.151	–0.005	–0.002	–0.0400*	–0.6436***
Adilabad/Nalgonda	–710.666	–443.900***	–237.896***	–	–
Karimnagar/Khammmam	119.036**	317.455***	–40.902	–	–
Nizamabad/Krishna	5.293	168.429	78.230	–	–
Guntur	–	–	–	–	–
Constant	–31,671.7	24,377.59	–61,385.95	7,933.15	219,145.80
Variability in yield					
Precipitation (*R*)	–0.001**	–0.001	0.000	0.0073**	0.0009
Temperature (*T*)	0.629**	0.133*	0.281**	2.8463**	–3.1340***
Trend	0.026**	0.035**	0.029	–0.0522	–0.2699***
Adilabad/Nalgonda	1.074**	0.443	0.656	–	–
Karimnagar/Khammam	0.854**	–0.136	0.347	–	–
Nizamabad/Krishna	1.922***	0.368	1.576	–	–
Guntur	–	–	–	–	–
Constant	–6.100	8.376	2.566	–75.6846	103.1627
Likelihood function	–1,096.8	–1,182.2	–1,081.3	–280.5	–292.1

	Nagarjuna Sagar Project (Krishna)			
Mean yield	Rice	Chilli	Cotton	Groundnut
Precipitation (*R*) (mm)	7.131	1.267	–0.316	–11.764
Temperature (*T*) (°C)	517.565***	–4,550.519***	1,642.900**	2,685.464**
Trend (year)	38.678***	77.024***	7.1934***	25.137***
R^2	–0.001***	0.0002	0.006**	–0.0001
T^2	–9.407***	80.033***	–30.086	–53.971***
$R \times T$	–0.167	–0.065	0.0154	0.425
Adilabad/Nalgonda	–138.399	183.360	–161.045***	–284.919**
Karimnagar/Khammmam	–563.750***	958.678***	–99.833**	–74.356
Nizamabad/Krishna	–155.250	925.118***	–9.513	–16.537
Guntur	206.042***	1,607.058***	107.545***	83.639
Constant	–6,141.211	65,400.22	–22,254.53	–31,988.28
Variability in yield				
Precipitation (*R*)	–0.002**	–0.002***	0.0016***	–0.0009
Temperature (*T*)	0.421	–0.763	0.952***	0.153
Trend	0.022**	0.034*	–0.053**	0.127
Adilabad/Nalgonda	0.030	0.397	–0.793	–0.646
Karimnagar/Khammam	0.289	0.452	–1.080**	–2.490***
Nizamabad/Krishna	0.768*	0.474	–0.433	–1.399**
Guntur	–0.055	1.228***	0.253	–1.252***
Constant	0.501	35.116	–17.845	6.491
Likelihood function	–10,593	–1,167.2	–907.2	–1,026.3

*, Significant at 10% level; **, significant at 5% level; ***, significant at 1% level.

Appendix 7.3. Impact of Climate Change for the Crops Grown in all Three Projects

		SRSP (Godavari)			LBP (Cauvery)	
CC-Scenario		Rice	Maize	Groundnut	Rice	Maize
	Normal yield (kg ha^{-1})	2972	3922	1556	3994	3840
Mid-century (MC) 1.93°C/13.6%	MC-predicted yield (kg ha^{-1})	2747	3708	1254	3469	2182
	Loss (%)	7.6	5.5	19.4	13.2	43
	Standard deviation	575	696	383	225	587
End-century (EC) 4.03°C/17.8%	EC-predicted yield (kg ha^{-1})	2065	3778	338	3033	1990
	Loss (%)	30.5	3.7	78.3	24.1	48
	Standard deviation	1086	789	507	987	684

		NSP (Krishna)			
CC-Scenario		Rice	Chilli	Cotton	Groundnut
	Normal yield (kg ha^{-1})	2944	3568	442	1831
Mid-century (MC) 1.93°C/13.6%	MC-predicted yield (kg ha^{-1})	2923	2947	332	1483
	Loss (%)	0.7	17.4	25.1	19.1
	Standard deviation	452	335	183	647
End-century (EC) 4.03°C/17.8%	EC-predicted yield (kg ha^{-1})	2439	2867	220	913
	Loss (%)	17.1	19.7	50.3	50.2
	Standard deviation	680	145	198	749

8 Water Management for Agricultural Production in a Coastal Province of the Mekong River Delta Under Sea-level Rise

Ngo Dang Phong,[1,4]* Chu Thai Hoanh,[2] Tran Quang Tho,[3] Nguyen van Ngoc,[3] Tran Duc Dong,[3] To Phuc Tuong,[1] Nguyen Huy Khoi,[3] Nguyen Xuan Hien[3] and Nguyen Trung Nam[3]

[1]International Rice Research Institute (IRRI), Los Baños, Laguna, Philippines; [2]International Water Management Institute (IWMI), Colombo, Sri Lanka; [3]Southern Institute for Water Resource Planning, Ho Chi Minh City, Vietnam; [4]Nong Lam University, Ho Chi Minh City, Vietnam

Abstract

Bac Lieu province, a low-lying coastal province in the Ca Mau peninsula of the Mekong River Delta (MRD) of Vietnam, is recognized as an area strongly affected by sea-level rise (SLR) accompanying global climate change. SLR will aggravate inundation and salinity intrusion and hence exert a strong influence on agriculture and aquaculture production in the province. The study presented in this chapter aims to quantify the impacts of SLR on agriculture in this province and to propose adaptive options.

The 'Vietnam River Systems and Plains' (VRSAP) model was used for simulation of water level, flow and salinity in the canal network of the MRD for 3 years of low, average and high water volume from upstream of the MRD and different levels of SLR (12, 17, 30, 50 and 75 cm). Under present sea level conditions, the western part of the province faces the highest flooding risk in October during the rainy season. In the dry season, salinity is high in the western part where farmers grow brackish water shrimp, while it is low in the eastern part where rice is grown. The inundation depth increases with the level of the SLR. For SLR less than 30 cm, salinity is expected to decrease slightly due to more fresh water from the Bassac River. When SLR is higher than 30 cm, salinity in the eastern part will also increase because the saline water intrudes into freshwater intake canals along the Bassac River. In the near future, adjustment of the cropping calendar as well as the operations of existing water control structures will be required. In the distant future, additional structures will be needed to cope with aggravated inundation and salinity.

8.1 Introduction

Rice is the most important food source for half of the world's population and the most important crop in Vietnam. The MRD, with an annual rice production of 16 Mt, accounts for 50% of the national rice production. The MRD, with a total area of 4 Mha and a total population of 17.5 million in 2011, has been identified as a region that will be significantly affected by SLR associated with climate change (MONRE, 2012). The most

* Corresponding author, e-mail: n.phong@irri.org

important impact of SLR is the aggravation of inundation and salinity intrusion that strongly influences agriculture and aquaculture production in the MRD, especially in the coastal provinces (Dasgupta *et al.*, 2007; IPCC, 2007).

An area of 770,000 ha of land (19% of the MRD) may suffer from frequent and severe submergence by the end of the century due to SLR (MONRE, 2012). In addition, SLR also enlarges the areas affected by salinity intrusion. With a 75 cm SLR by 2100 (MONRE, 2012), 17.5% of the population in the MRD will be directly affected. Developing adaptation strategies to SLR for the coastal provinces is crucial to Vietnam's economy, food security and livelihoods of people living in these provinces. It requires quantitative information on impacts of SLR on the salinity and submergence in the affected areas, in particular at the 'hot spots' where impacts are the most aggravated (Wassmann *et al.*, 2004). Bac Lieu, a low-lying coastal province in the Ca Mau peninsula of the MRD, is a representative province for considering influences of SLR. It was estimated that with SLR of 1 m in 2100, 39% of the provincial area and 45% of its population will be affected (Carew-Reid, 2007).

The objectives of the study presented in this chapter are: (i) to evaluate the impacts of SLR on salinity intrusion and submergence in Bac Lieu province under different conditions of upstream flows; and (ii) to recommend the adaptation strategy for agriculture production in the province.

8.2 Methodology

8.2.1 The study area

The study area is Bac Lieu province at the southern extent of the MRD (Fig. 8.1). It is very flat (elevation from 0.2 m to 0.8 m above mean sea level) and has a dense canal network for navigation, irrigation and drainage. The network comprises the Quan Lo Phung Hiep (QLPH) main canal that connects the study area to the Bassac River, a main branch of the Mekong River in the MRD (Fig. 8.1), and a series of canals of different sizes with a total length of 715 km. Four primary canals, typically 30–50 m wide and 4–10 m deep with 10 m-wide embankments, are perpendicular to the QLPH canal at 4–5 km apart. Secondary canals, 10–15 m wide and 1.5–2.0 m deep with 7 m-wide embankments, connect to the primary canals at 1 km spacing. Tertiary canals, typically 5–8 m wide and 1–2 m deep with 5 m-wide embankments, are spaced at 500 m along the secondary canals. Rice and shrimp fields are usually connected to secondary and tertiary canals by tiny channels.

Twenty seven sluices along the national road No. 1 (NR1) and a series of seasonal dams (Fig. 8.1) are operated for controlling saline and freshwater flows in the study area. Among these sluices, the Ho Phong (HP) and Gia Rai (GR), the two largest three-gate sluices 3 × 8 m and 3 × 7.5 m wide, respectively, play an effective role in managing salinity and drainage. However, sluices at the West Sea side have not been built. Therefore the province is still intruded by saline water, although inflow from the West Sea is not as strong as from the East Sea due to smaller tidal amplitude and diurnal tide. Fresh water for rice and upland crops in the eastern part of the province is delivered from the Bassac River through the QLPH main canal. In general, the sluice system and temporary dams divide the province into three different areas (Fig. 8.1): (i) the freshwater area including zones I, V, VI and VII in the east; (ii) the brackish area including zones II, III, IV, VIII and IX in the west; and (iii) the saline area of zone X in the south.

There are two distinct seasons in the province: the rainy season from May to November, accounting for 86% of total annual rainfall, which averages 2300 mm year^{-1}, and the dry season from December to April. Land use in the province is associated closely with the distribution of brackish and fresh water. Farmers grow rice and upland crops on a total area of 72,542 ha (SNIAPP, 2008) in the eastern part comprising of zones I, V, VI and VII. Shrimp culture is mainly practised in the western part (zone IV, VIII and IX) and at the south of NR1 (zone X). In zones II and III, the buffer area

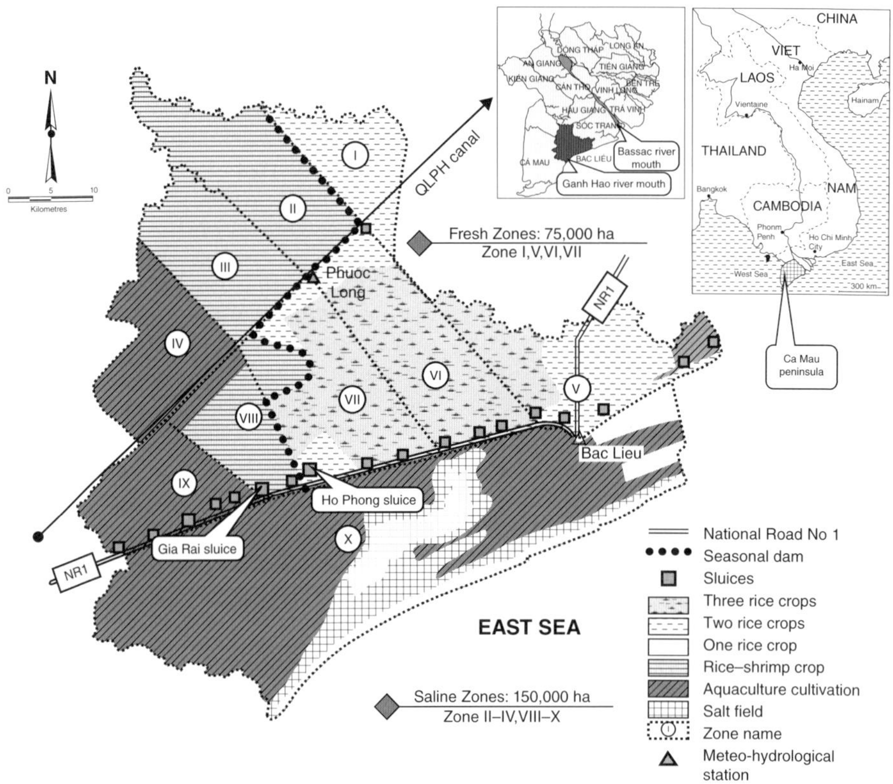

Fig. 8.1. Land use in Bac Lieu province in 2008 (from SNIAPP, 2008).

between fresh and brackish water north of the QLPH canal, shrimp–rice combination is practised with shrimp raised during February–July while rice cropping starts in August–September and is harvested in December–January.

Three rice crops are grown in the province: (i) He Thu (summer–autumn) is often sown in the middle of April and harvested in late July. Summer–autumn rice is mainly rainfed with supplementary irrigation when drought occurs and may suffer from salinity if rainfall in April–May is low; (ii) Thu Dong (autumn–winter) is usually grown from the end of July to early August. Autumn–winter rice is completely rainfed and often suffers from submergence in late September and October; and (iii) Dong Xuan (winter–spring) is sown after the harvest of the autumn–winter crop at the end of November–December. Winter–spring rice is fully irrigated, therefore it can be grown only in areas where salinity intrusion is completely controlled. In some years this crop suffers from shortage of irrigation water, therefore some farmers have replaced it with upland crops.

8.2.2 Setting the Vietnam River Systems and Plains model

The 'Vietnam River Systems and Plains' (VRSAP) model is a hydraulic and water quality model to simulate water flow, salinity and acidity in a complex canal network in the coastal zone influenced by tide (Phong *et al.*, 2007). In this study, VRSAP was used to simulate water level, flow and salinity in the canal and river systems in MRD, bounded by the Mekong river system from the delta boundary at Kratie in Cambodia to the East Sea and the West Sea in the Vietnamese

Mekong Delta. The model scheme includes 5066 segments (canal and river reaches and hydraulic structures), 3457 nodes (junction of two or more segments) and 2882 fields (space bounded by three or more segments). Boundary conditions were set using water level and salinity observations at 11 stations along the East and West Seas, and flow at Kratie station. Daily rainfall at 24 meteorological stations in MRD was also used as input data. Topographical data of mainstream rivers and canals were updated with surveyed data from 1999 to 2006. Topographical data of secondary and tertiary canals in Bac Lieu province were provided by the local authority in 2008. Elevation of ground surface in the MRD was derived from the elevation map provided by MONRE (2010) at the scale of 1:2000. Water demand for all land uses in the study area was calculated from evapotranspiration data for every 10-day interval.

Water level, discharge and salinity estimates from the VRSAP model were calibrated and validated with observed data in 1996 in the study area (Wassmann *et al.*, 2004). The year of 2008 is the most recent year having sufficient hydrological data and sluice operation information for modelling and is used as the 'baseline' year. In that year two main sluices along NR1, the Ho Phong and Gia Rai, were opened a few days in each month of the dry season (November–May) for intake of saline water into zones VIII, IX, IV, III and II for shrimp cultivation, and opened permanently in the wet season from June to October for drainage. The model results showed a comparison between simulated and observed water levels and salinity in 2008 at several locations in the Ca Mau peninsula such as Phuoc Long and Bac Lieu (Fig. 8.2).

8.2.3 Sea-level rise scenarios

Water levels at boundary stations along the East and West Seas were assumed to rise

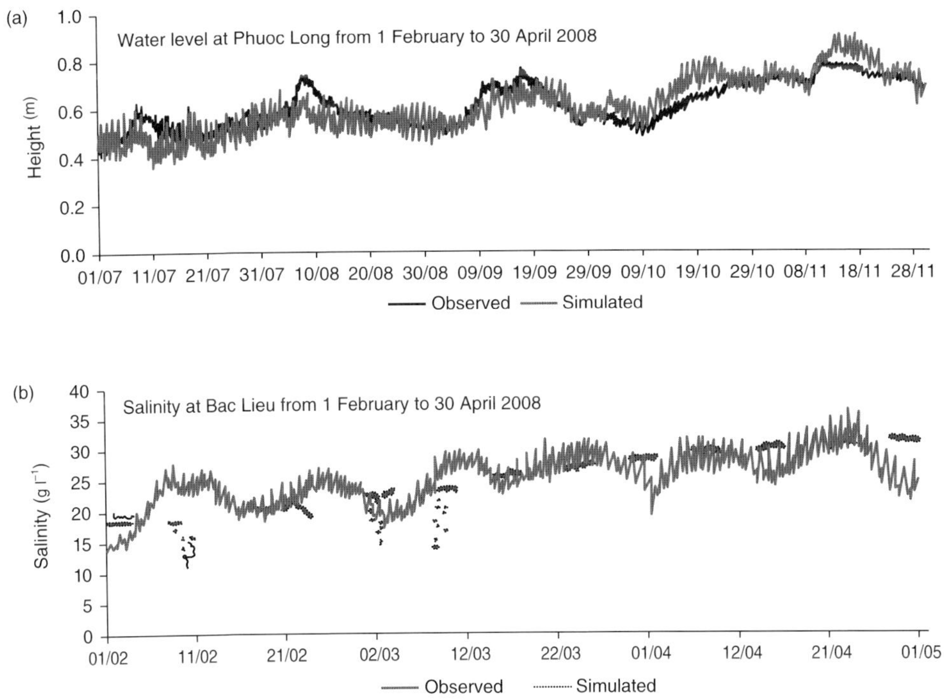

Fig. 8.2. Comparison of observed and simulated water level at Phuoc Long in the wet season and salinity at Bac Lieu in the dry season.

with sea level by 12, 17, 30, 50 and 75 cm SLR in 2030, 2040, 2050, 2075 and 2100, respectively (MONRE, 2012). The impacts of SLR were considered for different upstream flows. From time series data of daily upstream discharge at Kratie from 1985 to 2008 (MRC, 2008), low, average and high water conditions in 1998, 2004 and 2000, with the total annual volumes of 240 Mm^3, 362 Mm^3 and 507 Mm^3, respectively, were selected to compare with the baseline year 2008 with a total annual volume of 404 Mm^3. Impacts of climate change and future upstream hydropower and irrigation development are not included. The sluice operation in 2008 is applied for all cases.

Daily water level and salinity at nodes in the canal network estimated by VRSAP model were used to assess the impacts of SLR. Submergence and drainage capacities in each zone of the province are evaluated by comparing maximum and minimum water level in canals with average ground surface elevation. Salinity levels exceeding a threshold of 4 $g l^{-1}$ are considered deleterious for rice production. From these assessments, adaptation strategies in agriculture production are recommended, for example, application of new rice varieties with enhanced submergence or salinity tolerance.

8.3 Results and Discussions

8.3.1 Submergence and salinity in Bac Lieu province under present sea level

Daily water level is particularly high if heavy rain coincides with high tide and increased flow into the province from the Bassac River, as indicated by the variation of water level at Ganh Hao river mouth (Fig. 8.3, location in inset map of Fig. 8.1). The effects of tide and local rainfall in the province are stronger than upstream flow as reflected by higher water level in the baseline year 2008 than in all other years (Fig. 8.3), although water volume from upstream in 2008 is only slightly higher than in the normal water year 2004 and lower than in the high water year 2000.

Similarly, Fig. 8.4 also shows that maximum inundation depth, which usually occurs in zones II, III and IV during October–November, is rather high in the baseline 2008 compared to other years. These maps reaffirm that for a province close to the coast such as Bac Lieu, the effect of total volume from the Mekong upstream on inundation depth is not as important as the simultaneous occurrence of high tide, local rainfall and high daily upstream flow.

The tide from the East Sea is a dominant influencing factor on salinity intrusion in

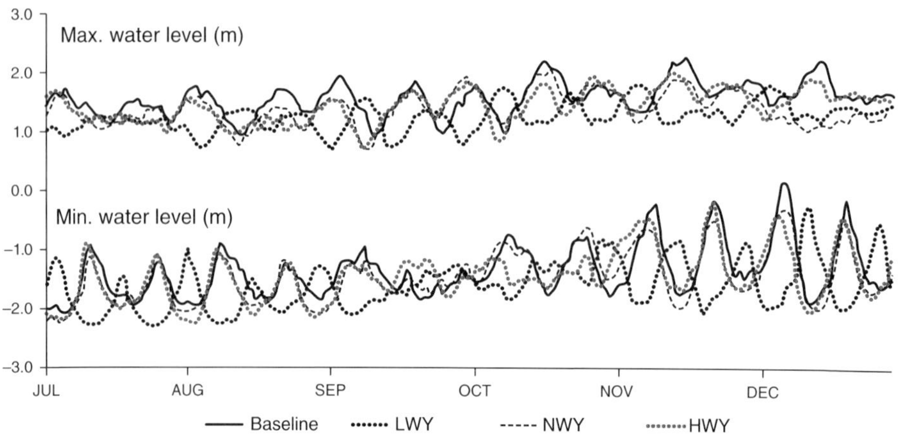

Fig. 8.3. Daily maximum and minimum water levels at Ganh Hao river mouth in low water (LWY, 1998), normal water (NWY, 2004) and high water (HWY, 2000) years compared with baseline (2008).

the eastern zones (Fig. 8.5). Salinity in the unprotected zone X and in the well-protected zones I, V, VI and VII does not vary much year by year. On the other hand, variations of salinity in different hydrological years are clearer in the western zones II, III, IV, VIII and IX with lower salinity and shorter duration in the high water year. In the low water year, the salinity peak in zone IX occurs in April ($32\,g\,l^{-1}$) and extends until May.

8.3.2 Impacts of sea-level rise on inundation

For all SLR scenarios analysed using inland water conditions of the baseline year 2008, both maximum and minimum daily water levels in the canals in the study area will increase. Under the present sea level the drainage capacity in the province, represented by the difference between the daily minimum water level and ground elevation

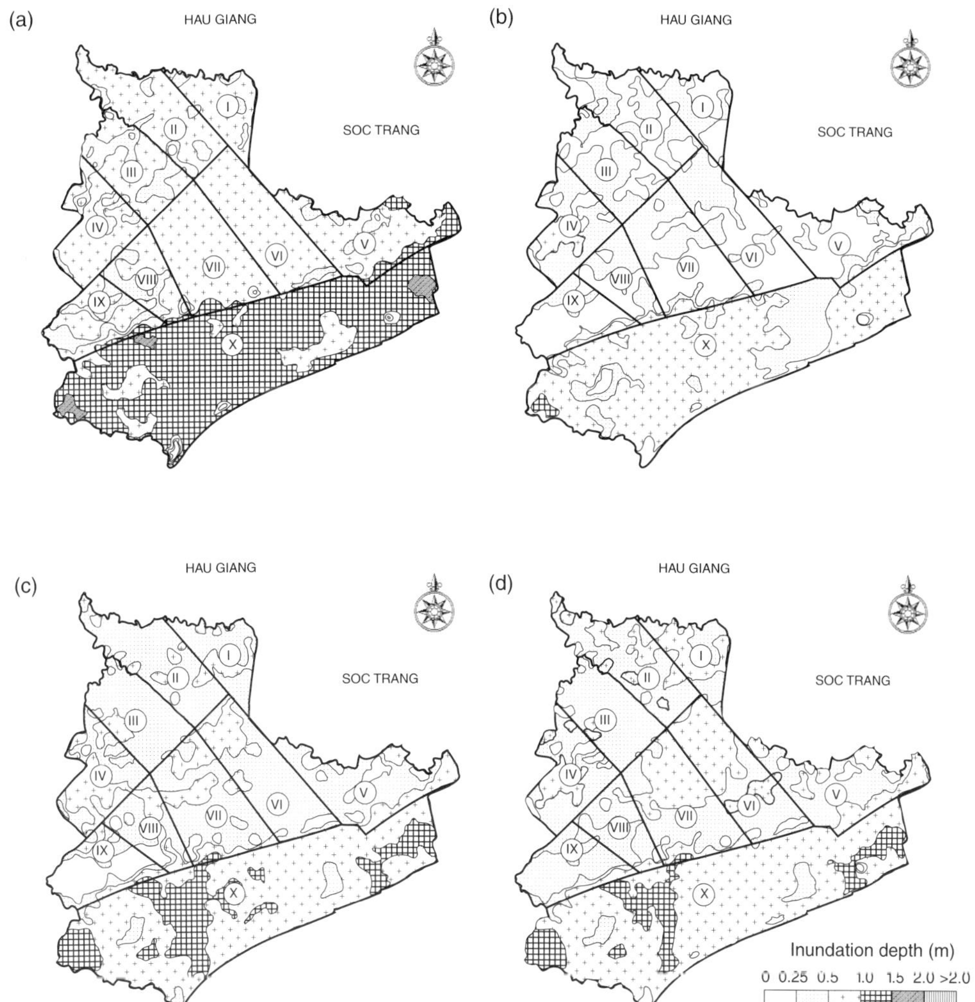

Fig. 8.4. Maximum inundation depth in rainy season for (a) baseline year 2008, (b) low water year 1998, (c) normal water year 2004 and (d) high water year 2000.

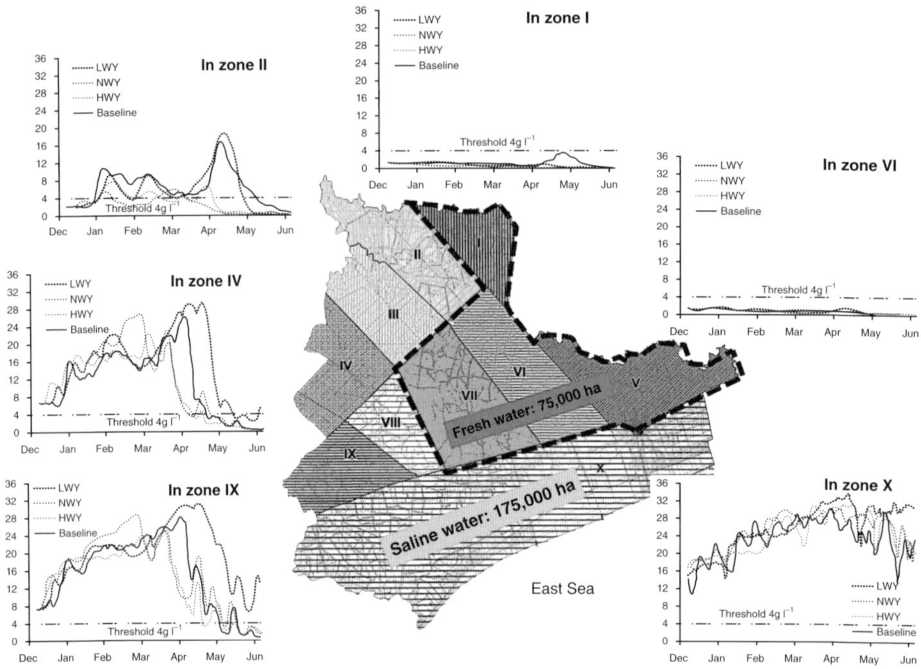

Fig. 8.5. Maximum salinity (g l^{-1}) in dry season in baseline 2008, low water year (LWY, 1998), normal water year (NWY, 2004) and high water year (HWY, 2000).

in Fig. 8.6, is hindered by the high water level at both high and low tides. Water in zones VIII and IX at the south-west of QLPH canal, and in zone X at the south of NR1 can be easily drained out during low tide because the minimum water level in canals is generally lower than the ground elevation. Drainage in the western zones of QLPH canal (zones I, II, III, IV) is poor in the rainy season, therefore inundation will be more serious with SLR.

With a 17 cm or 30 cm SLR, drainage capacity could be maintained in the eastern part of QLPH canal (such as zone V) where sluices are operated for drainage during low tides from July to October. With SLR up to 30 cm, the area with inundation depth of above 0 to 1 m and above 1 m will be 50% and 35% of the total province area, respectively (Fig. 8.7d, e, f).

For SLR higher than 30 cm, inundation will be severe if the water control systems (sluice, dyke and canal) are not improved (Figs 8.6, 8.7e, f). Under a 75 cm SLR scenario, the area with inundation depths of less than 1 m and above 1 m will cover 10% and 90% of the total province area, respectively (Fig. 8.7f).

The effects of a 75 cm SLR on maximum inundation depth in the low and high water years are shown in Fig. 8.8. Compared with the situation in these years under present SLR (Figs 8.4b, d), a larger area will be submerged deeper than 1 m. However, due to a worse combination of high tide, local rainfall and daily upstream flow, the maximum inundation in 2008 (Fig. 8.7f) was as severe as in the high water year 2000 (Fig. 8.8b) as discussed above.

8.3.3 Impacts of sea-level rise on salinity intrusion

In the baseline year 2008 sluices along NR1 and temporary dams were operated to manage satisfactorily saline water intake for shrimp production in western and southern parts of the Bac Lieu province and also to

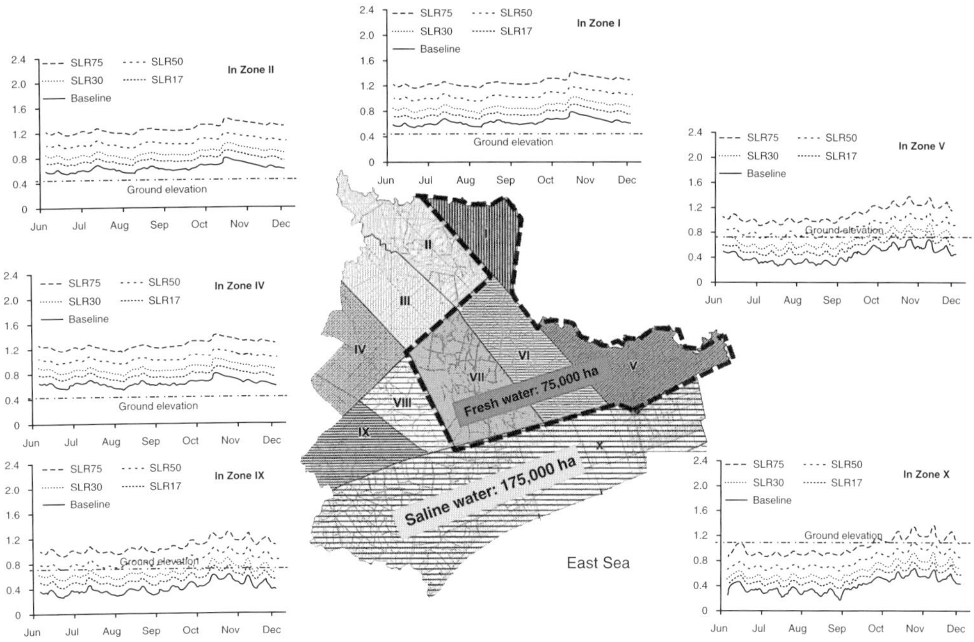

Fig. 8.6. Minimum daily water level (m) in the rainy season under 17, 30, 50 and 75 cm SLR with inland water conditions of the 2008 baseline year.

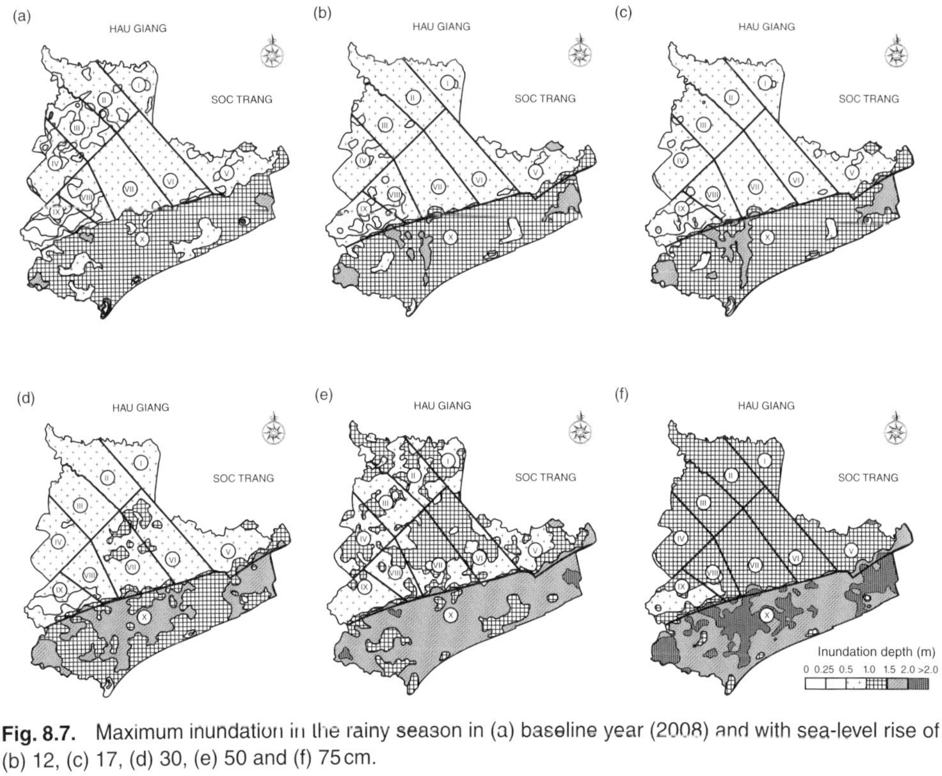

Fig. 8.7. Maximum inundation in the rainy season in (a) baseline year (2008) and with sea-level rise of (b) 12, (c) 17, (d) 30, (e) 50 and (f) 75 cm.

protect the rice production area in the eastern part from salinity intrusion.

SLR will cause higher water level in the Bassac River and consequently increase water level in the QLPH canal and other canals that connect this river to the Ca Mau peninsula. Figure 8.9 indicates the increase of water level at stations along the Bassac River and the up-river extent of the salinity boundary of 4g l^{-1} (a threshold for rice production) for different SLR scenarios. For all SLR scenarios this salinity boundary will not go beyond 50 km from the mouth of the Bassac River. The maximum intrusion is 47.6 km, which is about 7 km further upstream compared with the situation at the present sea level. The projected increase in salinity intrusion up the Bassac River is not expected to affect intake of fresh water into the canal network of the QLPH area as the intake points are from Cai Con station, 60 km from the river mouth, and further upstream. On the other hand elevated water levels upstream of Cai Con station will increase freshwater volume flowing through the upper intake points into the QLPH area. This additional amount will reduce salinity, even in the dry season, at the northern and western parts of Bac Lieu province.

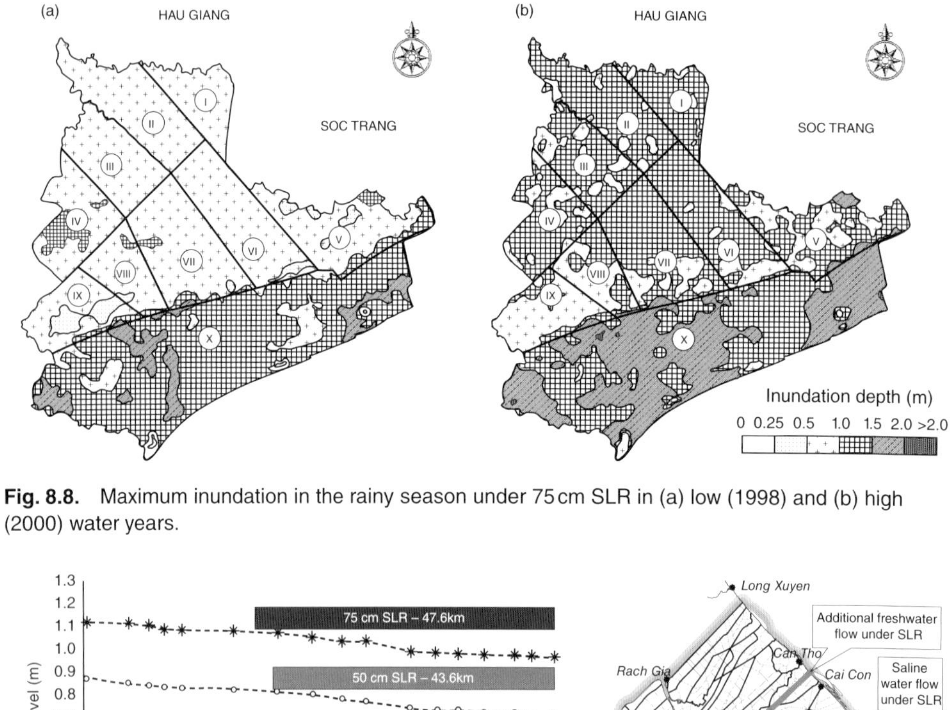

Fig. 8.8. Maximum inundation in the rainy season under 75 cm SLR in (a) low (1998) and (b) high (2000) water years.

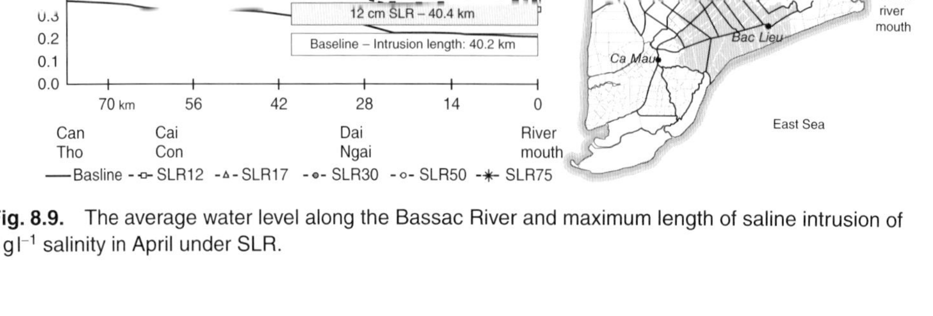

Fig. 8.9. The average water level along the Bassac River and maximum length of saline intrusion of 4 g l^{-1} salinity in April under SLR.

However, SLR above 30 cm will cause high salinity in the intake canals below Cai Con station such as the canal at Dai Ngai which is not sluice-protected. This will cause expansion of the salinity-affected area in the eastern part of Bac Lieu province. Figure 8.10 shows that for all SLR scenarios salinity in zone X (outside of the sluice and dam system) is almost the same as in the baseline. On the other hand, salinity in zone VI is higher than in the baseline, and increases with higher SLR. Variations in other zones (e.g. zones I, II, IV, IX) are complex, with higher salinity in certain months and lower in other months, depending on the tidal flows in the canal system and the operation of the sluice system. In this study the operation in each month is assumed for the same purpose of salinity control for shrimp and rice production as in 2008.

Figure 8.11 shows that for a 75 cm SLR scenario, maximum salinity in the western and southern parts in the dry season will be lower by 4–8 $g l^{-1}$ compared to the baseline while salinity in the eastern part will be higher by 4–8 $g l^{-1}$ (Fig. 8.11f). The mapped results suggest that under high SLR scenarios sluices and dykes are required to protect the rice area from salinity intrusion from the Bassac River into the eastern part of the province.

The effects of SLR on salinity in the low and high water years are shown in Fig. 8.12. Maximum salinities in the low and high water years under present sea level (Fig. 8.12a, c) are respectively higher and lower than in the baseline year 2008 (Fig. 8.11a), indicating that unlike inundation, salinity in the dry season in Bac Lieu province is strongly influenced by upstream flow. Under a 75 cm SLR scenario, salinity variations in these years are similar to the baseline year 2008 with a decrease in the western part and an increase in the eastern part, and salinity

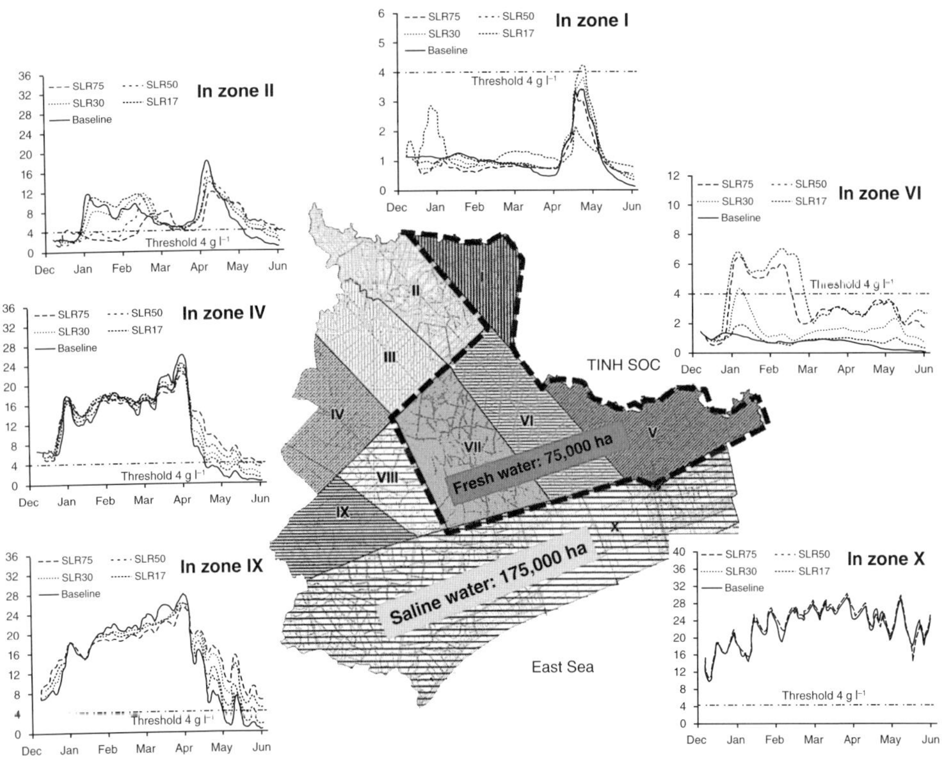

Fig. 8.10. Maximum daily salinity (g l^{-1}) in different zones of Bac Lieu province under SLRs of 17, 30, 50 and 75 cm.

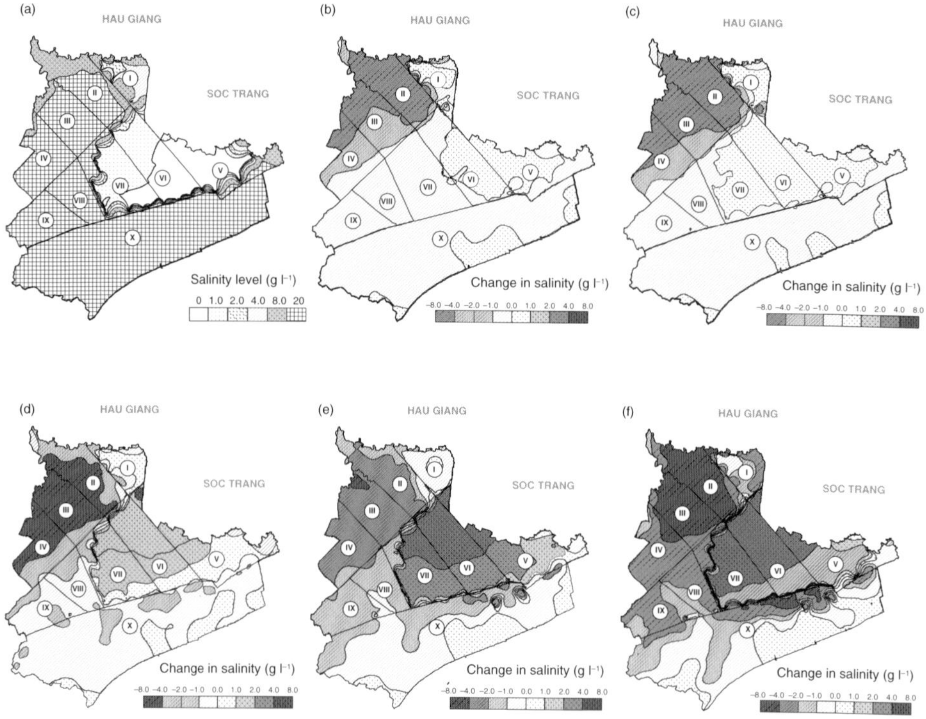

Fig. 8.11. Changes of maximum salinity in Bac Lieu province in the dry season compared with (a) the 2008 baseline under present sea level and under (b) 12, (c) 17, (d) 30, (e) 50 and (f) 75 cm sea-level rise.

in the high water year is consistently lower than in the low water year.

8.4 Adaptation Strategies to Sea-level Rise in Bac Lieu Province

Based on the changes in inundation depth and salinity in three representative zones (Fig. 8.13) where rice and shrimp cropping systems are predominant, the following adaptations to SLR are recommended.

SLR below 30 cm before 2050 will cause a maximum increase in inundation depth of less than 20 cm in all three zones and a slight decrease in salinity, 5 g l^{-1} in zone IV and 1 g l^{-1} in zones I and VII, respectively. In general, infrastructure measures can be applied to reduce the effects of these changes such as: (i) opening and closing all sluices at ebb tide and flood tide, respectively, for improving drainage in the wet season; (ii) elevating the field bunds by 20 cm to avoid over-bund flow into the rice fields; and (iii) opening the Ho Phong and Gia Rai sluices for more days during the dry season for intake of saline water into zone IV for shrimp cultivation. Major adaptation strategies to SLR in different zones are:

- In the freshwater zones I, V, VI and VII, more water flowing into the province under SLR will be favourable as a source for irrigation of the winter–spring rice from December to March. Maximum salinity in these zones will slightly increase but it is still not a major constraint to rice cultivation. On the other hand, new rice varieties with higher submergence tolerance will be required for the autumn–winter rice from August to November.
- In the brackish water zones II, III and IV where shrimp is raised in the dry season and rice is grown in the wet season, slightly lower salinity and longer duration of fresh water will provide more

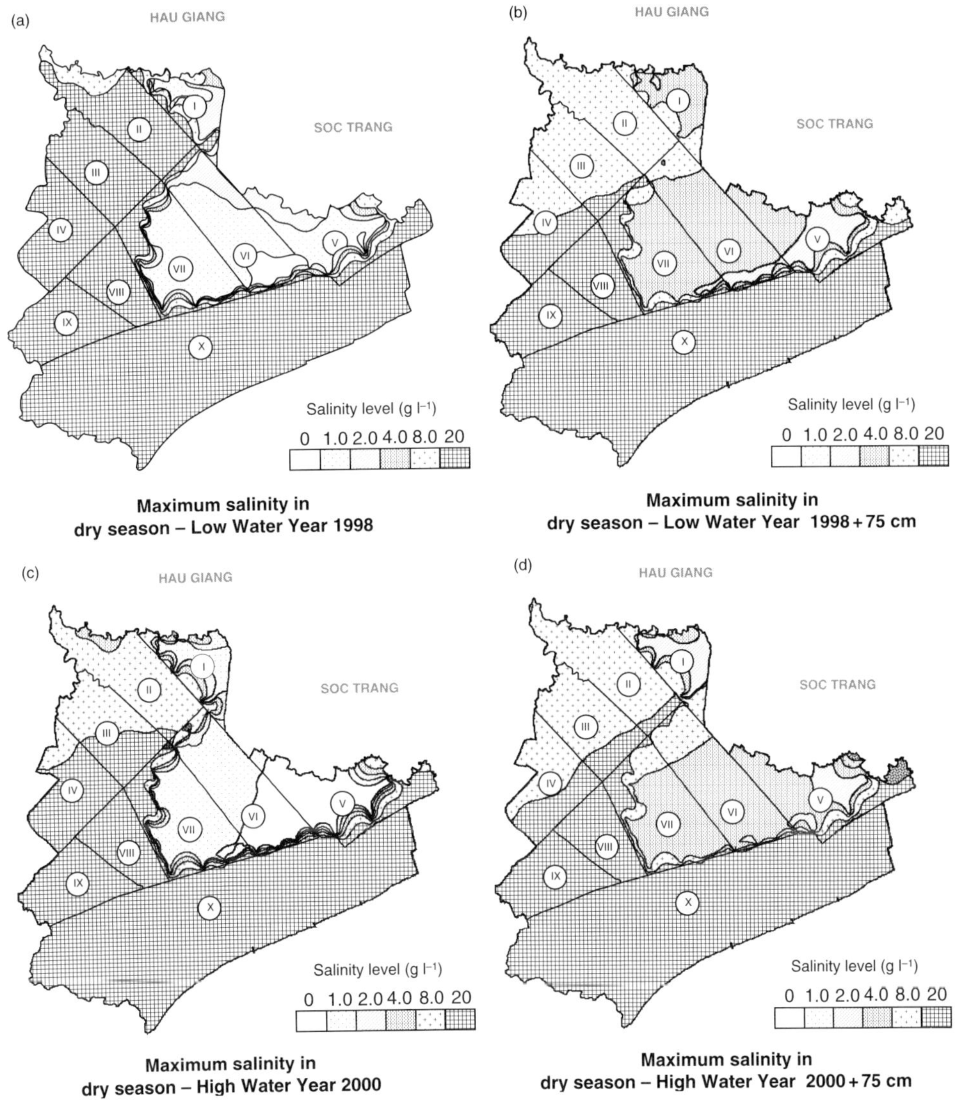

Fig. 8.12. Salinity in the low and high water years under present sea level and 75 cm SLR.

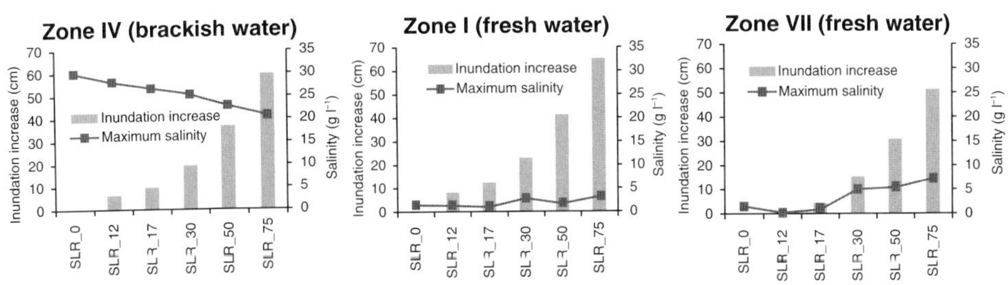

Fig. 8.13. Changes in inundation depth and salinity due to SLR in brackish water zone (IV) and freshwater zones (I and VII)

favourable conditions for the rice crop, although new rice varieties with higher submergence tolerance are also required for the summer–autumn crop.
- In the brackish and saline water zones VIII, IX and X where shrimp-based systems are predominant, adjustment in sluice operation will be needed to ensure suitable salinity for shrimp.

Beyond 2050 when SLR is higher than 30 cm, the daily maximum and minimum water levels in the canal network will increase with about 50–80% of the SLR. These increases will cause deeper inundation and hinder gravitational drainage in all zones. Besides the adjustment in production systems and operations of existing sluice systems, additional hydraulic structures will be needed to cope with these impacts. For example, embankments should be raised to a higher elevation than 20 cm, new sluices should be built in the canals upstream of Dai Ngai and pumping capacity should be improved. Major adaptation strategies to SLR in different zones are:

- In the freshwater zone I at the north of QLPH canal, SLR will cause higher submergence and restrict the drainage by gravity during the peak water level in September–November. The rice-cropping calendar must be adjusted to avoid this peak period. The autumn–winter rice crop will not be suitable unless pumps are deployed for drainage. Rice varieties with enhanced submergence tolerance will be needed for this cropping season, and also for the summer–autumn crop that is harvested in late July. On the other hand, the winter–spring crop in the dry season should be started earlier than under the present sea-level conditions to avoid increased salinity at the end of the crop season. Rice varieties with enhanced salinity tolerance are also needed.
- In the freshwater zones V, VI and VII to the south of QLPH canal, inundation depth will be lower than in zone I, but the increase in salinity will hamper the cultivation of the winter–spring rice crop in the dry season, although varieties with enhanced salinity tolerance (such as OM4900, OM6677, OM10252, etc.) have been applied. Unless new sluices are built upstream of Dai Ngai station to control salinity intrusion into these zones, this rice crop may be replaced by upland crops such as maize or vegetables with less water requirement and shorter crop duration. The summer–autumn and autumn–winter rice crops will also be hampered by increased submergence and reduced drainage capacity, but not as seriously as in zone I.
- In the buffer zones II and III between brackish and fresh water, rice varieties with enhanced submergence tolerance will be needed. With reduction in salinity, the time window for the winter–spring rice crop will be prolonged and the duration of shrimp cultivation at upper locations will be shorter. Even if the Ho Phong and Gia Rai sluices are opened for more days during the dry season, saline water from the East Sea would still not be able to reach these zones.
- In the brackish water zones IV, VIII and IX, lower salinity under SLR will provide more favourable conditions for expanding and extending the rice crop in the rainy season. A conflict in water requirement as in the past (Hoanh *et al.*, 2012) will occur if sluices are opened for more days, for intake of brackish water to meet the demand of shrimp growers, while rice farmers need fresh water for their rice crop. Coordination between sluice operation at regional level and adjustment of cropping calendars at farm level should be strengthened to minimize such conflict.
- In the saline zone X, the slight change of salinity under SLR will not influence significantly the main production of shrimp and other aquaculture production such as crab and fish. However, because this zone is located along the seashore, the increased inundation depth will be close to the level of SLR and therefore embankments should be elevated.

8.5 Conclusions

In contrast with previous global studies (Zeidler, 1997; Nicholls *et al.*, 1999; Dasgupta *et al.*, 2007) whereby digital elevation models (DEM) of coastal zones were used directly to determine the impacts of SLR, this study analysed the hydrodynamics of unsteady flows in a complex canal and river network within a delta that is strongly affected by tide and SLR. The study showed that in a province close to the coast such as Bac Lieu, changes of inundation depth due to the variation in total annual water volume from upstream are not significant, but timing of the peak flow, tidal cycle and local rainfall are more relevant. The effects of SLR on salinity are more complex than on inundation depth. In the wet season SLR causes an increase of maximum and minimum water levels, but this increase gradually reduces inland, about 50%–80% of the increase of sea level. In the dry season, SLR will cause more fresh water from the mainstream flowing into the province at the northern and western parts, hence salinity in these parts will be lower. On the other hand salinity will be higher in the eastern part where saline water will reach the intake canals. Soft measures such as adjustment of cropping calendar or sluice operations are possible to adapt to low SLR. However, beyond 2050 when SLR is higher, hard measures such as new sluices, dykes and pumps will also be needed.

While the study is of a very specific case study area, it illustrates two important general points. The first is the importance of modelling to explore impacts that may not be intuitive (e.g. the decrease in salinity in some areas under SLR), but are significant in management terms. The second is that the study demonstrates very clearly that adaptation responses are not static, but must evolve as the severity of impacts increases; and that planning needs to reflect this strategy.

In this study impacts of climate change and development in the Mekong upstream as well as in the MRD were not taken into account when determining the impacts of SLR on production systems in Bac Lieu province. Nevertheless, if climate change is included, and given the current uncertainty in daily rainfall projection, we cannot expect more accurate results to develop suitable adaptation strategies. Another limitation of this study relates to the nature of SLR. Unlike sudden onset events such as a tsunami, SLR is a slow change amounting to an increase <1 cm year^{-1}. Therefore changes in siltation and streambed configuration of channels and in ecosystems are gradual, and adaptations by local people or heightening of canal embankments are likely to happen incrementally. These adaptations require other detailed investigations that were not considered in this study.

Acknowledgements

The authors highly appreciate the project 'Improving Rice Tolerance of Submergence and Salinity to cope with Climate Change in coastal area of Vietnamese Deltas' funded by DANIDA and the project on Climate Change Affecting Land Use in the Mekong Delta: Adaptation of Rice-based Cropping Systems (CLUES) funded by ACIAR to support this study.

References

Carew-Reid, J. (2007) Rapid assessment of the extent and impact of sea level rise in Viet Nam. *Climate Change Discussion Paper 1.* ICEM – International Centre for Environmental Management, Brisbane, Australia.

Dasgupta, S., Laplante, B., Meisner, C., Wheeler, D. and Yan, J. (2007) The impact of sea level rise on developing countries: a comparative analysis. *World Bank Policy Research Working Paper 4136.* Available at: http://www-wds.worldbank.org/external/default/WDSContentServer/WDSP/IB/2007/02/09/000016406_20070209161430/Rendered/PDF/wps4136.pdf (accessed 9 March 2014).

Hoanh, C.T., Phong, N.D., Trung, N.H., Dung, L.C., Hien, N.X., Ngoc, N.V. and Tuong, T.P. (2012) Modeling to support land and water management in the delta context: experiences from the Mekong Delta, Vietnam. *Water International* 37(4), 408–426.

IPCC (2007) Climate change 2007: the physical science basis. Contribution of Working Group I to the Fourth Assessment. In: Solomon, S., Qin, D., Manning, M., Chen, Z., Marquis, M., Averyt, K.B., Tignor, M. and Miller, H.L. (eds) *Report of the Intergovernmental Panel on Climate Change.* Cambridge University Press, Cambridge, UK, 996 pp.

MONRE (2010) *Digital Topography Map at Scale 1/2000.* Ministry of Natural Resource and Environment (MONRE), Vietnam.

MONRE (2012) *Scenarios of Climate Change and Sea Level Rise for Vietnam.* Vietnam Publisher on Natural Resources-Environment and Maps, Ha Noi, Vietnam (in Vietnamese).

MRC (2008) *Monitored Daily Discharge at Kratie from 1986 to 2008.* Mekong River Commission (MRC), Vientiane, Lao PDR.

Nicholls, R.J., Hoozemans, F.M.J. and Marchand, M. (1999) Increasing food risk and wetland losses due to global sea-level rise: regional and global analyses. *Global Environmental Change* 9, S69–S87.

Phong, N.D., Hoanh, C.T., Tuong, T.P., Malano, H. and Weaver, T. (2007) Modelling acidity propagation in a coastal zone with acid sulphate soils. In: Oxley, L. and Kulasiri, D. (eds) MODSIM 2007 International Congress on Modelling and Simulation. Modelling and Simulation Society of Australia and New Zealand, December 2007, pp. 170–176.

SNIAPP (2008) *Land Use and Cropping Pattern Map of Mekong Delta in Scale 1/250.000.* Sub-National Institute of Agricultural Planning and Projection, Ho Chi Minh City, Vietnam.

Wassmann, R., Hien, N.X., Hoanh, C.T. and Tuong, T.P. (2004) Sea level rise affecting the Vietnamese Mekong Delta: water elevation in the flood season and implications for rice production. *Climatic Change* 66, 89–107.

Zeidler, R.B. (1997) Continental shorelines: climate change and integrated coastal management. *Ocean Coastal Management* 37(1), 41–62.

9 Aquaculture Adaptation to Climate Change in Vietnam's Mekong Delta

Suan Pheng Kam,[1]* Tran Nhuong,[1] Chu Thai Hoanh[2] and Nguyen Xuan Hien[3]

[1] *WorldFish, Penang, Malaysia;* [2] *International Water Management Institute, Colombo, Sri Lanka;* [3] *Southern Institute for Water Resources Planning, Ho Chi Minh City, Vietnam*

Abstract

Most of the aquaculture production in South-east Asia occurs in the floodplains and coastal areas that are highly exposed and vulnerable to climate change impacts and sea-level rise (SLR). This chapter presents an example of economic estimation of autonomous adaptation by shrimp and catfish farms in the Mekong Delta of Vietnam. It illustrates how planned adaptation measures can help defray catfish farmers' escalating costs of raising pond dykes in response to increased flooding in the delta. It also indicates that government policy and public investment into planned adaptation towards climate change impacts, particularly for water resources management, would necessarily take account of socio-economic development targets of the aquaculture industry. From these analyses, broader implications of plans for water resources management in the delta on the prospects and challenges to the aquaculture sector are discussed. In the long term, a 'no-regrets' strategy of reducing the high dependence on shrimp and catfish culture and diversifying into more ecologically oriented production systems can also hedge the aquaculture industry against the increasing risks and uncertainties brought about by climate change.

9.1 Introduction

Aquaculture is one of the fastest-growing animal food-producing sectors in the world. In the last three decades (1980–2010), production of farmed food fish increased 12-fold, at an average annual rate of 8.8% (FAO, 2012). The Asian region accounts for almost 90% of total aquaculture production and one-third of its global seafood export value (estimated from FAO online statistics). Excluding China, over half of the exported aquaculture products from Asia (accounting for almost two-thirds of the continent's global seafood export value) are produced in Bangladesh, Indonesia, the Philippines, Thailand and Vietnam. Most of the aquaculture production in South-east Asia occurs in the floodplains and coastal areas that are highly exposed and vulnerable to climate change impacts – not only the direct climatic parameters, such as temperature and rainfall patterns, but also SLR and consequent flooding and coastal salinity intrusion. These impacts will have significant economic as well as social costs to those who directly and indirectly depend on the aquaculture industry for their livelihoods.

Adaptation is an imperative for enhancing the resilience of this economically important and dynamic sector. Aquaculture operators will, and do, respond to changes in land and water availability, market incentives and commodity prices by changing farm practices or even altering culture

* Corresponding author, e-mail: s.kam@cgiar.org

species and production systems. All of these constitute autonomous adaptation, where actors respond 'spontaneously' to change triggers (Adger *et al.*, 2007). Climate change and its attendant hydrological and coastal phenomena constitute additional environmental triggers to which farmers have to respond and adapt. By implementing these changes at the farm level, individual farmers will have to incur incremental capital investment and operational costs.

In 2010, a study was conducted by a WorldFish team to estimate the costs of autonomous adaptation by shrimp and catfish farms in the Mekong Delta of Vietnam (Kam *et al.*, 2012). This chapter extracts and examines the main findings of the study and discusses broader implications of plans for water resources management in the delta, in light of climate change and future land use, on the prospects and challenges that the aquaculture sector in the delta would face and the possible response strategies.

9.2 Current Status of the Aquaculture Industry in the Mekong Delta of Vietnam

The Mekong Delta of Vietnam, with its intricate system of canals, embankments and water-control structures, provides a mosaic of freshwater, brackish-water and marine environments that accommodate diverse aquaculture systems producing a variety of fish, crustacean and mollusc species (Table 9.1).

Presently the aquaculture industry in the delta is dominated by culture of the brackish-water shrimp (predominantly the black tiger shrimp, *Penaeus monodon* and with an increasing presence of the Pacific white leg shrimp, *Litopenaeus vannamei*) and the freshwater striped catfish (*Pangasianodon hypophthalmus*) – the two most important fish food commodities of Vietnam with a total export value of almost US$4bn in 2012,[1] accounting for about 4% of the national Gross Domestic Product (GDP). The Mekong Delta accounts for about 80% and 98% of the total shrimp and catfish export values, respectively (GSO, 2008).

The spatial distribution of brackish-water shrimp and freshwater fish culture in the delta (Fig. 9.1) is influenced by its hydrological regimes, which result from a combination of local rainfall, seasonal discharges of the Mekong River into the delta and tidal effects from the sea.

Prolonged deep flooding occurs in the upper delta from July to December (Tuan *et al.*, 2007) as a result of the rainy season (May–October) coinciding with the high flow period (June–August), while the coastal zone experiences salinity intrusion from December to May (Tuan *et al.*, 2007) as a result of the dry season (November–April) coinciding with the period of low flows (February–April). High tidal ingress up the river branches of the delta aggravate flooding in the upper delta as well as salinity intrusion in the coastal zone.

Brackish-water shrimp culture in ponds is carried out in the salinity-affected areas in the coastal zone (Fig. 9.1) at improved extensive, semi-intensive and intensive scales, which are differentiated by stocking rates and feeding regimes. Presently the improved extensive system covers the largest area and occurs mainly in the salinity-affected intertidal zone of Ca Mau and Bac Lieu provinces. In areas further away from the intertidal zone there have been shifts to semi-intensive and intensive shrimp culture with mechanical aeration and water recirculation to maintain high stocking densities for high yields. Shrimp–rice aquaculture is practised in areas where a dual brackish and freshwater regime is maintained, particularly in parts of the south-western Ca Mau peninsula where a system of sluice operations and temporary dams allow sea water to intrude during the dry season for shrimp and to be flushed out during the rainy season for rice[2] (Hoanh *et al.*, 2009).

Catfish culture is concentrated in the inland provinces of An Giang, Dong Thap and Can Tho in the upper delta where freshwater supplies are ample (Fig. 9.1). Catfish ponds are located close to rivers and canals for ease of constant pumping to have a high level of water exchange for the intensively cultured system. As these areas are subjected to seasonal flooding, catfish ponds

Table 9.1. Diversity of aquaculture production systems and farmed species in the Mekong Delta, Vietnam.

Production system	Culture species	Extent and location
Brackish-water shrimp-based systems		
Pond monoculture at improved extensive (IE), semi-intensive (SI) and intensive (IN) scale	Hatchery *Penaeus monodon* at IE, SI and IN scale; hatchery *Litopenaeus vannemai* at IN scale	300,000 ha IE in Ca Mau and Bac Lieu provinces; 18,000 ha SI in Bac Lieu, Soc Trang and Tra Vinh; 22,000 ha IN in Soc Trang, Tra Vinh and Ben Tre
Shrimp–rice rotation (dry-season shrimp monoculture)	Hatchery *Penaeus monodon*	100,000 ha in Ca Mau, Bac Lieu, Soc Trang and Kien Giang
Shrimp–crab pond culture	Hatchery *Penaeus monodon*; hatchery *Scylla serrata*	Ca Mau, Ben Tre, Kien Giang and Tra Vinh
Mangrove–shrimp	Wild seeds supplemented with hatchery *Penaeus monodon*	20,000 ha in Ca Mau; other provinces: Ben Tre, Kien Giang, Tra Vinh
Mangrove–shrimp–mud crab; mangrove–shrimp–mud crab–blood cockle	Wild seeds supplemented with hatchery *Penaeus monodon*; hatchery *Scylla* sp.; *Anadara granosa* spats	30,000 ha shrimp–mud crab in Ca Mau; 850 ha of mixed culture with blood cockle
Other brackish-water aquaculture systems		
Cockles and clams: bed culture on intertidal mudflats for blood cockle; sandy beaches for clams	*Anadara granosa*, various clams	Ben Tre, Soc Trang, Tien Giang, Kien Giang and Tra Vinh
Pond or tank monoculture of eel	*Anguilla mamorata*	2000 ha in Ca Mau
Mud-skipper: pond monoculture; in rotation with shrimp, mud-crab and *Artemia*; integrated with shrimp and mud crab	*Pseudapocryptes elongatus*	150 ha in Bac Lieu and Soc Trang
Freshwater catfish culture systems		
Pond monoculture of river catfish at highly intensive scale	*Pangasius hypophthalmus* (cátra)	4500 ha in An Giang, Dong Thap, Can Tho and Vinh Long
Cage and pen monoculture of catfish at highly intensive scale	*Pangasius bocourti* (cábasa), *Pangasius hypophthalmus*, *Clarias gariepinus* × *Clarias macrocephalus* (hybrid catfish)	An Giang and Dong Thap

Continued

Table 9.1. Continued.

Production system	Culture species	Extent and location
Other freshwater fish culture systems		
Intensive pond culture of climbing perch	*Anabas testudineus*	Can Tho city
Pond, cage, hapa and tank monoculture of snakehead	*Channa* spp.	An Giang, Dong Thap, Can Tho and Kien Giang
Cage and pond-intensive monoculture of tilapia	*Oreochromis niloticus* (black and red tilapia)	Tien Giang and An Giang
Low- to semi-intensive polyculture of carp	*Cyprinus carpio* (common carp), *Ctenopharyngodon idella* (grass carp), *Chirrhinus molitorella* (mud carp), *Hypophthalmichthys molitrix* (silver carp)	Hau Giang, Can Tho, Vinh Long, Ben Tre, An Giang and Dong Thap
Fish polyculture in rice fields	Tilapia, common carp, silver carp, *Pangasius* catfish, *Helostoma temminckii* (kissing gourami), *Osphronemus goramy* (giant gourami), *Puntius gonionotus* (silver barb)	
Livestock–fish polyculture		
Giant freshwater prawn culture systems		
Rice–prawn rotation (prawn during flood season); integrated rice–prawn; pond and pen monoculture	*Macrobrachium rosenbergii*	An Giang, Can Tho, Ben Tre, Dong Thap
Mariculture systems		
Floating raft culture of oysters	*Crassostrea gigas*	
Cage mariculture of marine finfish	*Panulirus* spp. (spiny lobster), *Epinephelus* spp. (grouper), *Lates calcarifer* (seabass), *Seriola dumerilli* (yellowtail), *Parargyrops edita* (sea bream), *Lutjanus* spp. (snapper), *Hippocampus* sp. (seahorse), *Pinctada maxima* and *P. martensii* (pearl oyster)	Kien Giang

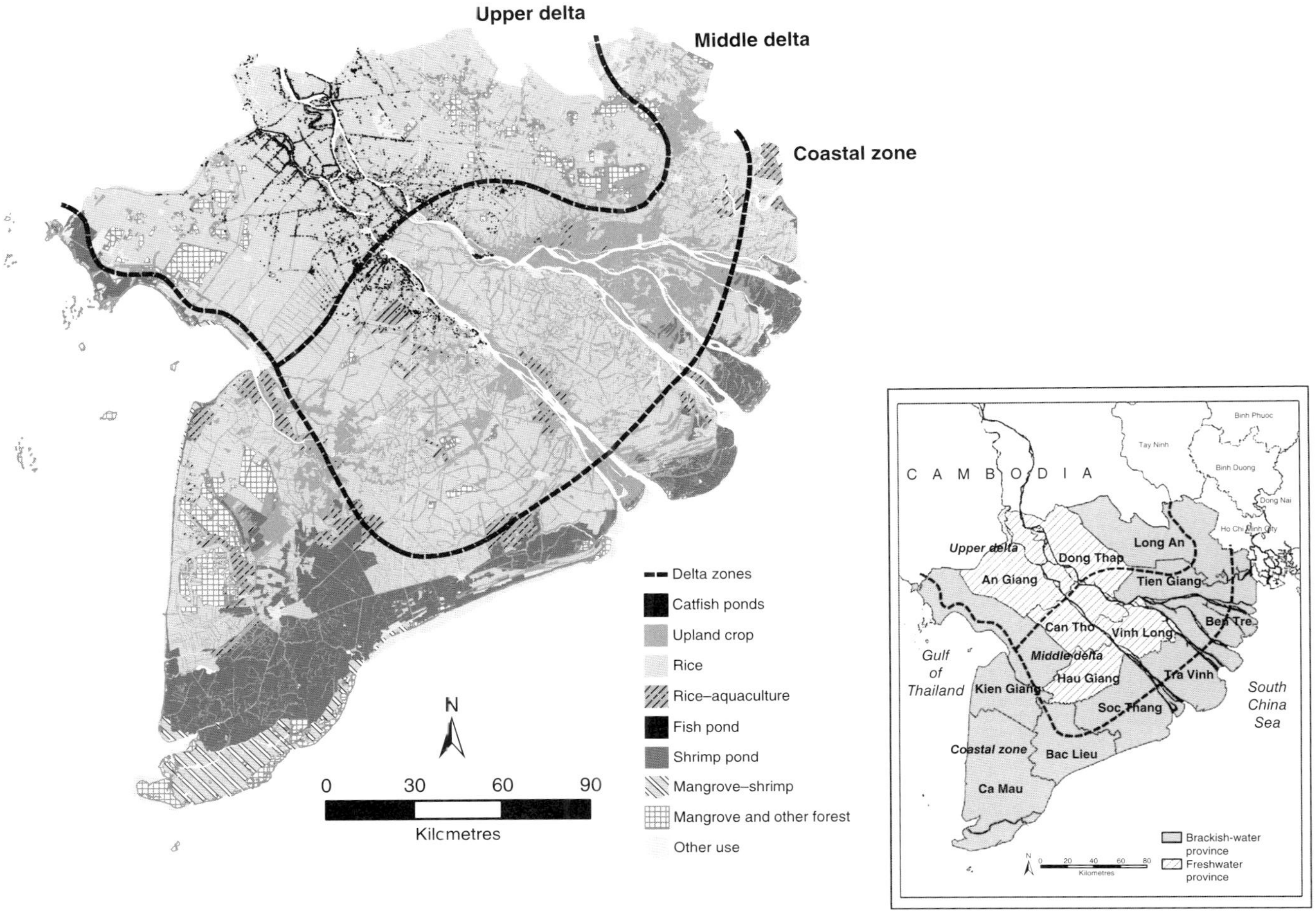

Fig. 9.1. Land use in the Mekong Delta of Vietnam, 20C7 (from The Sub-National Institute for Agricultural Planning and Projection (Sub-NIAPP), Vietnam).

need to be protected by raised dykes. With the Vietnamese government's relaxation of land zoning for catfish production in 2002, farms have expanded towards the coast to take advantage of the stronger tidal movement to lower water-pumping costs. Catfish farming is now found in the coastal provinces of Soc Trang and Ben Tre provinces, as far downstream as the water salinity conditions are tolerable to the riverine catfish.

9.3 Climate Change and its Implications on the Aquaculture Industry in the Mekong Delta

As shown in Table 9.2 for the Mekong Delta, the magnitudes of long-term impacts of anthropogenic climate change on weather and sea events are highly uncertain and are estimated based on plausible scenarios of future socio-economic development that influence greenhouse gas emissions, and consequently the earth's ecophysical responses. Nevertheless, the trends are clear that the Mekong Delta will experience: (i) temperature increases at rates that will appreciate more markedly beyond 2050; (ii) drier dry seasons and wetter rainy seasons; and (iii) SLR that will increase ingress of sea water particularly to parts of the delta that are not already protected.

These climatic and sea-level changes have direct and indirect impacts on aquaculture in the delta. Aquaculture ponds are dependent on water supply from rivers and canals rather than directly from rainfall;

Table 9.2. Comparison of projected changes in environmental variables under moderate and high emission climate change scenarios[a] for the Mekong Delta of Vietnam by 2050 and 2100 (relative to the 1980–1999 period).

Change in impacted variable	Moderate emission (B2)		High emission (A2)		
	By 2050	By 2100	By 2050	By 2100	Remarks
Annual mean temperature (°C)					
Marchand *et al.*, 2011	1.0	2.0	2.0	4.0	
MONRE, 2009	1.0	2.0	1.0	2.6	
Dry season rainfall (%)					
Marchand *et al.*, 2011	0 to –10	–5 to –15	–10 to –20	–20 to –40	Period considered: November–April
MONRE, 2009	–7.5	–14.3	–7.2	–18.2	Period considered: March–May
Wet season rainfall (%)					
Marchand *et al.*, 2011	0–5	5–10	10–20	10–30	Period considered: May–October
MONRE, 2009	0.9	1.6	0.8	2.1	Period considered: June–August
Sea-level rise (cm)					
Marchand *et al.*, 2011	20 to 30	30–50	40–60	100–200	High emission scenario used is
MONRE, 2009	30	75	33	100	A1F1instead of A2
Increase in salinity intrusion					
Marchand *et al.*, 2011	slight	moderate	moderate	dramatic	
Low flow of the Mekong[b] (%)					
Marchand *et al.*, 2011	5 to –5	5 to –15	–10 to –30	–30 to –60	Dry season flow: November–April
Hoanh *et al.*, 2010	40		42		Low flow season: December–May
High flow of the Mekong (%)					
Marchand *et al.*, 2011	no change	10	0–10	20–50	Wet season flow: May–October
Hoanh *et al.*, 2010	–3.3		1.4		High low season: June–August

[a]The emission scenarios identified in the Special Reports on Emission Scenarios (SRES) of the Intergovernmental Panel on Climate Change (IPCC, 2000).

[b]This large discrepancy of projected low flows, from reduced to increased volumes, stems from different model assumptions and reflects the great uncertainty in future scenarios for river flow and discharge into the Mekong Delta, which are influenced by climate change impacts as well as by upstream development of irrigation and hydropower schemes.

hence, changes in hydrological flows and regimes of the delta are expected to be the most significant aspect of climate change impact to affect the aquaculture sector. These changes are influenced not only directly by weather changes, especially rainfall patterns, but also by changes in sea level and by upstream hydropower and irrigation development affecting seasonal flow of the Mekong River into the delta (Hoanh *et al.*, 2010). These changes have large impacts on the availability of water resources, flood events, particularly in the upper delta, and water quality, particularly intrusion of sea water into the coastal zone.

SLR accompanying global warming will result in ingress of sea water further up the branches of the Mekong within the delta, which can exacerbate the accumulation of flood waters from upstream river discharge during the high-flow season and thus aggravate flooding in the upper delta. Figure 9.2 shows the projected increase in maximum flood levels during the rainy season for a 50 cm SLR scenario,[3] superimposed with locations of catfish pond areas in each province.

The greatest increments in flooding depth are projected to occur in An Giang, Dong Thap and Can Tho provinces, which

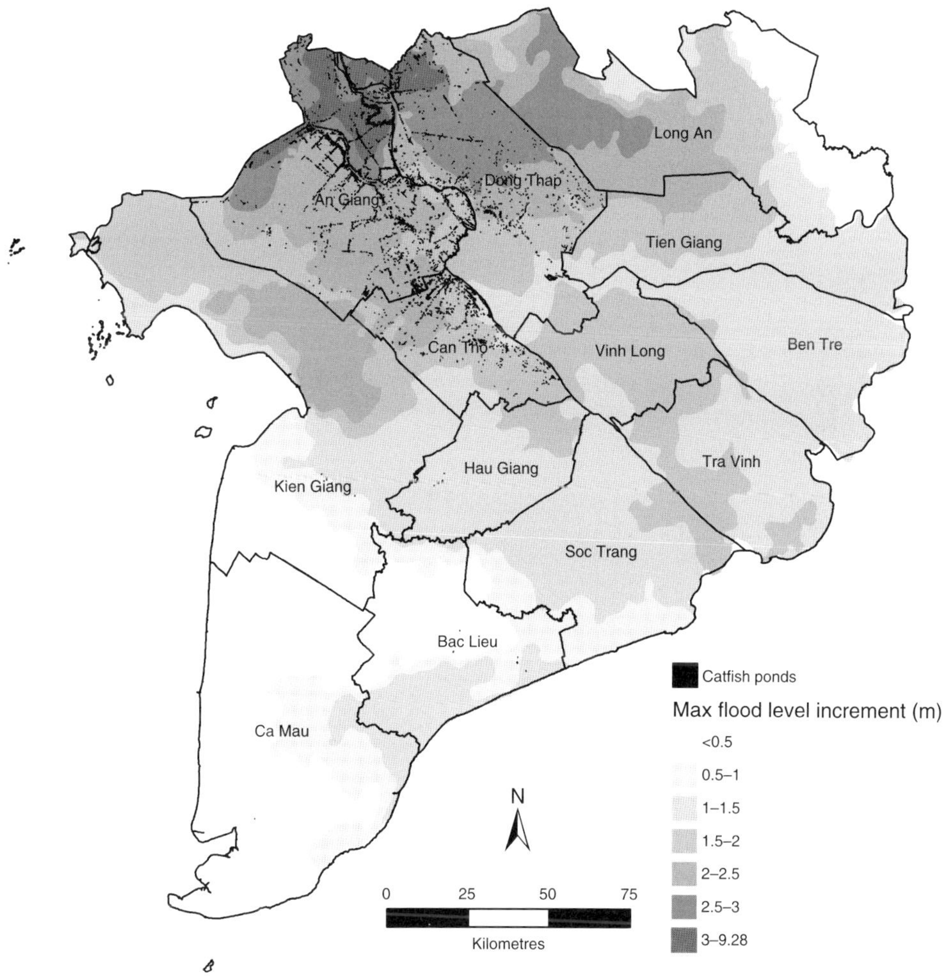

Fig. 9.2. Increment of maximum flood level during the rainy season for a 50-cm SLR scenario, with location of catfish farms in the Mekong Delta of Vietnam (from SIWRP, 2009).

have the largest concentrations of catfish farms in the delta. Catfish ponds would need further protection from seasonal floods. This could be achieved by farmers raising their pond dykes or by benefiting from public initiatives to boost flood-protection measures.

Figure 9.3 shows where increments in maximum water salinity under the 50-cm SLR scenario are projected to occur during the dry season, assuming no additions or enhancements to the existing infrastructure to control salinity intrusion into the delta (SIWRP, 2009). From GIS analysis (Kam *et al.*, 2012), an estimated 180,000 ha (or 55%) of the shrimp-farming area will experience salinity increments of up to 2 parts per thousand (ppt), and another 45,000 ha (or 11%) will be subjected to salinity increments greater than 2 ppt in the dry season. A further 190,000 ha presently in rice and rice–aquaculture areas are likely to experience increased salinity (above the 4 ppt threshold for rice) during the dry season, thereby providing an opportunity for expansion of brackish-water aquaculture in these areas.

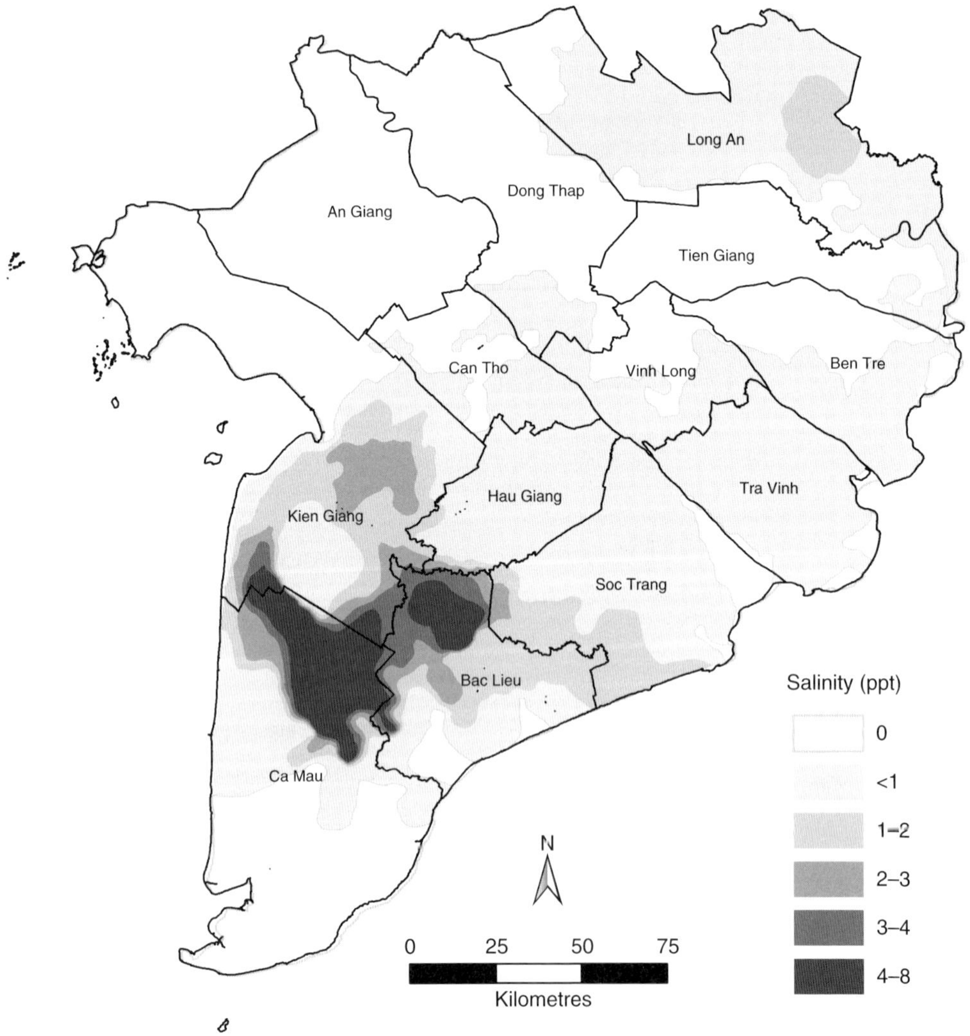

Fig. 9.3. Increment of maximum salinity intrusion during the dry season for the 50-cm SLR scenario in the Mekong Delta of Vietnam (from SIWRP, 2009).

Table 9.3 summarizes the specific impacts of climate change on shrimp and catfish aquaculture in the Mekong Delta, where farmers would have to make adjustments to their farm operations. The associated costs would be taken into account in the economic analysis of autonomous adaptation that is reported in the next section.

9.4 Economic Analysis of Autonomous Adaptation for Shrimp and Catfish Farms

The approach taken in the economic analysis of farm-level autonomous adaptation for shrimp and catfish farms (Kam *et al.*, 2012) involved conducting a farm-level cost–benefit analysis (CBA) for freshwater catfish and brackish-water shrimp in the Mekong River Delta under climate change (CC) with autonomous adaptation and non-climate change (NCC) scenarios. Four pond-based aquaculture production systems were considered: brackish-water shrimp culture at improved extensive and semi-intensive/intensive scales, and freshwater catfish culture in the inland and the coastal provinces. The analysis was done for two time-periods, from 2010 to 2020 (where projections of climate change impacts on input costs and price changes could be made with relative confidence) and from 2021 to 2050 (where projections become more uncertain).

The production budgets for catfish and shrimp operations in the Mekong Delta from past sample surveys conducted by Sinh

Table 9.3. Sensitivity of cultured catfish and shrimp to expected changes in environmental variables due to climate change.

Climate change impacts	Effect on aquaculture organisms
Temperature rise	Expected increased range within tolerance limits;[a] stress and mortality less likely
	Enhanced growth rates and feed conversion (metabolic rate) increases oxygen demand and need for aeration
	Increased decomposition rate of organic detritus in the water reduces oxygen levels and worsens water quality and invasiveness and virulence of bacteria (Dalvia *et al.*, 2009) and increases need for more frequent water exchange, hence greater water demand with adverse effect on water quality of discharge from ponds
	Air-breathing catfish (Browman and Kramer, 1985) are better able to withstand low levels of dissolved oxygen than shrimp, which require more aeration
Drier dry season	Higher evaporation rates from ponds increase the need for replenishing with fresh water, also in shrimp ponds to reduce salinity build-up, hence more pumping and electricity use
Wetter wet season	Tendency for greater prevalence of infection of catfish in the rainy season (Thuy *et al.*, 2010), hence the need and greater costs for disease prevention and control
Increased flooding	Higher risk of loss of fish if ponds overflow, and hence financial loss
Sea-level rise	Increased tidal reach further aggravates inland flooding
Increased salinity intrusion	Increased salinity levels further up the Mekong branches beyond the tolerance limit of river catfish has direct adverse effect on catfish farming in coastal provinces
	Seasonal salinity increases do not affect shrimp survival rates if kept within the range of 10–35 ppt; higher salinity can retard growth
	Increased salinity-affected areas in the coastal zone provide expansion opportunities for brackish-water aquaculture

[a]The range of temperature that supports normal growth of *Penaeus monodon* is between 28°C and 33°C (Duong, 2006); *Pangasius* spp. perform well in water temperatures ranging between 22°C and 26°C (from http://www.fao.org/fishery/culturedspecies/Pangasius_hypophthalmus/en (accessed 31 March 2013)).

(2008) were compiled and used as the basis for estimating the baseline (2010) costs and benefits for farm operations. Benefits from farm operations were measured by gross income, and costs comprised fixed and variable farm operation costs. Given the limited sample sizes of the 2008 survey – 131 inland catfish farms, 60 coastal catfish farms, 50 extensive/improved extensive shrimp farms and 50 intensive/semi-intensive shrimp farms – the production data provided were aggregated and averaged over all farm sizes. The averaged estimates would not represent the diversity of farm characteristics such as location, socio-economic circumstances and access to technology and financing, which would influence economic performance of catfish and shrimp farming.

An expert elicitation approach was used to gather local experts' opinions on the impacts of climate change and other drivers of change on variable costs such as land price, feed use and seed use of the four studied aquaculture systems over the past 10 years and in the next 10 years (2010–2020). The 13 stakeholders consulted comprised shrimp and catfish farmers, provincial aquaculture staff and local university researchers specializing in aquaculture. In consultation with the selected 13 stakeholders, drivers of change affecting shrimp and catfish production costs were grouped into four categories: technical, market, pond environment and climate change. The stakeholders provided their perceptions of how the likely effects of climate change parameters on shrimp and catfish culture (listed in Table 9.3) would impact on operating costs (the focus was on variable inputs such as land, feed and seed uses). While expert elicitation may be considered an acceptable approach in the absence of predictive models on the actual impacts of climate change on fish performance and yields (Moss and Schneider, 2000), estimates based on fish growth models and projected yield changes would reduce the uncertainty surrounding the results.

Farm-level costs and benefits were analysed under scenarios of CC with autonomous adaptation (the CC scenario), and with no climate change (the NCC scenario). Under the CC scenario, the full value of input costs was used. For the NCC scenario, the proportion of costs attributed to CC impacts was omitted. After obtaining the annual net benefits for the analysis period the net present values (NPV) of the CC and NCC scenarios were then compared. The results of the 2010–2020 period are described here as an illustration of economic analysis of the costs of autonomous adaptation at farm level for catfish and shrimp farming.

It is to be noted that the analysis was based on the assumption of linear increase in farm-gate prices for catfish and shrimp. Future market changes were not included in the computation of net benefits, which are highly sensitive to the farm-gate price of fish and shrimp. Similarly, technological advances may impact cost structures, and international trade policies can change demand for Vietnam's aquaculture products. Given the large uncertainty over the interaction of seafood market demand, input prices and costs in the future, these results should be interpreted as the outcome that would exist only if the input price and cost situations assumed in the estimations prevail.

9.4.1 Base production budgets

Tables 9.4 and 9.5 summarize the estimated baseline costs and benefits for catfish and shrimp farming operations, respectively.

Feed constitutes the largest cost in aquaculture production, accounting for 82% and 84% of the variable costs for inland and coastal catfish farms (Table 9.4) and 53% and 66% of variable costs for semi-intensive/intensive and improved extensive shrimp farms (Table 9.5), respectively. Seed and biochemicals account for the next largest costs for both catfish and shrimp production, with biochemicals taking up a high 11% of the variable costs in the semi-intensive/intensive shrimp system. On current trends, catfish farming in general faces a bleak future because gross revenues are not able to keep pace with the past and expected increase in input costs (especially land, feed and seed costs) even in the absence of CC impacts. Only the most efficient and

Table 9.4. Base production budget for inland and coastal catfish farms (costs in Vietnam dollars, VND) (from Sinh, 2008).

Input million ha^{-1} crop^{-1})	Inland (n=131)	Coastal (n=60)
Gross income	4868.9	3738.1
Total costs	4616.8	3644.7
Total fixed costs	20.9	28.3
– Depreciation of ponds	11.6	17.15
– Depreciation of machinery	7.17	8.15
– Land taxes	2.13	3.0
Total variable costs	4596.1	3616.4
– Pond preparation	23.6	27.2
– Seed	329.1	263.7
– Feed	3772.5	3051.2
– Chemicals and drugs	205.4	152.4
– Dyke upgrades	11.0	4.6
– Fuel and electricity	48.7	7.7
– Harvest and transportation	28.8	25.4
– Labour	39.2	44.7
– Interest on loans	127.4	33.9
– Miscellaneous	10.4	5.6
Net income	252.1	93.4

Table 9.5. Base production budget for improved extensive (IE) and semi-intensive/intensive (SII) shrimp farms (from Sinh, 2008).

Input (VND million ha^{-1} crop^{-1})	SII (n=50)	IE (n=50)
Gross income	431.1	65.9
Total costs	193.3	28.8
Total fixed costs	13.53	2.94
– Depreciation of ponds	7.58	1.79
– Depreciation of machinery	4.6	0.85
– Land taxes	1.35	0.30
Total variable costs	179.77	25.86
– Pond preparation	8.09	2.2
– Seed	9.35	3.13
– Feed	119.0	13.7
– Chemicals and drugs	21.0	1.88
– Dyke upgrades	3.05	0.31
– Fuel and electricity	8.63	1.37
– Harvest and transportation	1.61	0.10
– Labour	6.11	1.45
– Interest on loans	1.41	1.14
– Miscellaneous	1.43	0.58
Net income	237.8	37.1

adaptable farmers will survive such a squeeze on farming margins, which are currently in the range of 3–5%. Many catfish farmers, particularly those operating at small scale, might soon find it unprofitable to remain in the sector and will be forced to leave the industry, which will result in industry consolidation in the hands of large-scale stakeholders vertically integrated. This trend has already been observed in the delta in recent years. For example, in An Giang province the number of small catfish farms

(<500 m^2) declined from 3200 in 2004 to only 547 in 2009 (Little and Murray, n.d.).

Current trends are more favourable for shrimp farming, with margins of 123% and 129% for semi-intensive/intensive and improved extensive farms, respectively (Table 9.5).

9.4.2 Costs of autonomous adaptation

The results from Kam *et al.* (2012) for the 2010–2020 period are described here as an illustration of economic analysis of the costs of autonomous adaptation at farm level for catfish and shrimp farming. As shown in Fig. 9.4, net income remains positive until 2018 and only until 2015 for inland catfish and coastal catfish farms, respectively, under the NCC scenario. Under the CC scenario, the additional costs of adapting to climate change will intensify the squeeze on the slim profit margins of catfish farms, thus hastening the onset of net losses. Catfish farms in the coastal provinces will be particularly affected adversely by increased salinity intrusion up the rivers reducing freshwater supply to the ponds. Comparison of NPVs for the NCC and CC scenarios suggests that responding to climate change over this period would result in a decrease in discounted net income of VND4.7bn ha^{-1} for coastal catfish farms over the 2010–2020 period.

In comparison, shrimp farms manage to produce positive net benefits for a longer period than catfish operators due to lower total costs relative to gross income. As shown in Fig. 9.5, net incomes remain positive over the 2010–2020 period but responding to climate change leads to a more rapid decrease of net income. Comparison of NPVs for the NCC and CC scenarios suggest that responding to climate change over this period would result in decreases in discounted net income of VND51.6 million ha^{-1} for improved extensive shrimp farming and VND403.7 million ha^{-1} for semi-intensive and intensive shrimp farming.

The economic analyses show that autonomous adaptation of aquaculture to CC increases farm-level operation costs. Catfish farmers already operate at the brink of economic viability given increases in input costs and reduction in prices of marketed products in recent years. Climate change is just an additional driver to decrease even further the expected profitability of catfish culture. Shrimp farmers overall are able to bear the cost of adaptation over a longer time frame than catfish farmers. This is because outlooks for shrimp markets are in better condition compared to those for catfish. Furthermore, the shrimp industry is more spread out, involving much larger geographical areas and a larger number of stakeholders. It is more mature and less capitalized and so it is projected that the shrimp

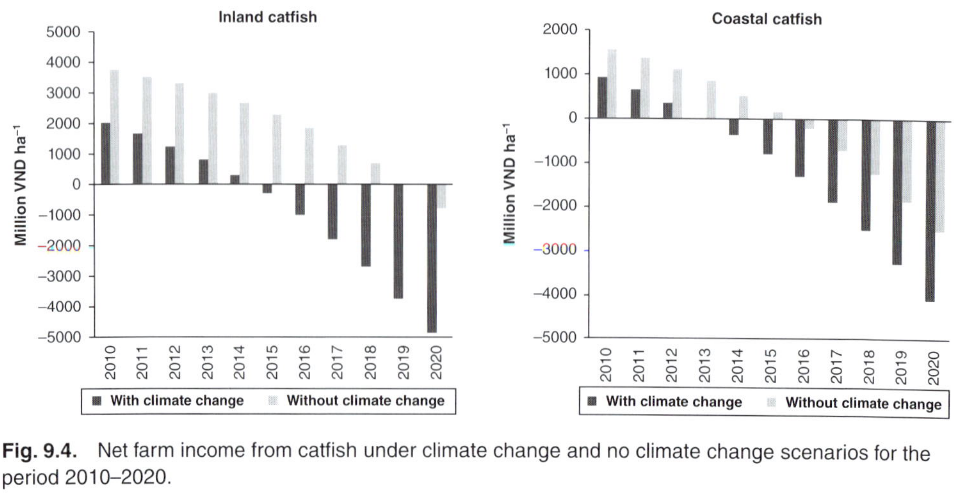

Fig. 9.4. Net farm income from catfish under climate change and no climate change scenarios for the period 2010–2020.

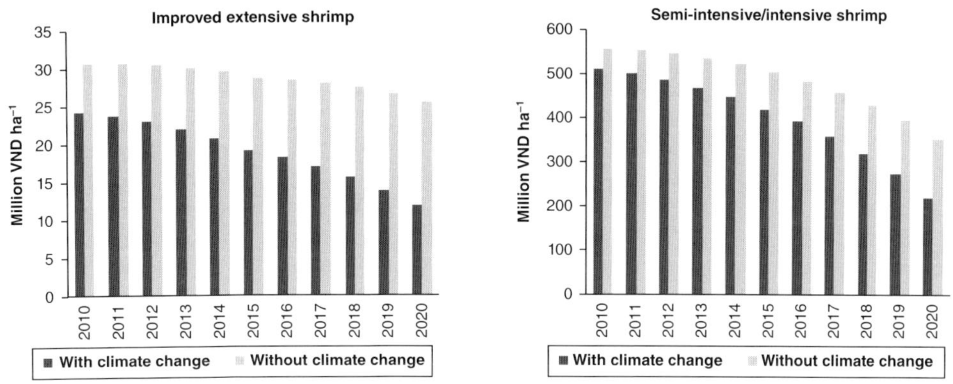

Fig. 9.5. Net farm income from shrimp under climate change and no climate change scenarios for the period 2010–2020.

industry can remain profitable for longer than catfish under CC scenarios. Despite its lower profitability compared with the semi-intensive/intensive system, improved extensive shrimp culture is more sustainable, both environmentally and economically, especially for small-scale farmers.

9.5 Planned Adaptation: The Case of Water Resources Management

Considering the economic importance of the aquaculture industry for Vietnam, government-funded planned adaptation measures that can partially offset certain farm-level costs of autonomous adaptation would bring relief to farmers, especially those already operating with narrow profit margins. In particular, planned adaptation measures relating to water resources management in the delta have direct implications on the aquaculture sector. A recent water resources development plan for the Mekong Delta (Fig. 9.6) takes account of its anticipated agricultural development scenario for this decade and the next, with a vision to 2050 (SIWRP, 2010).

Catfish farms in the inland provinces will mainly benefit from improved flood protection measures that are broadly targeted for rice and other agricultural crops in the upper and middle delta zones. The study by Kam *et al.* (2012) illustrates the case where public investment on constructing flood protection infrastructure in the upper delta would ease the financial burden on catfish farmers to raise their pond dykes. Estimates of the farm-level costs for upgrading catfish pond dykes were scaled up to the industry level by taking account of the prevailing area extents of the two catfish production systems and estimates of the extents to which these areas would be affected by successive increments in flooding depth resulting from SLR, based on a GIS analysis of the map in Fig. 9.2.

Figure 9.7 depicts the escalating costs for raising pond dykes of catfish farms in response to incremental increases in flooding depth for the 2010–2020 period at the industry level for the delta, which amounts to an estimated US$17.6 million of autonomous adaptation cost over this 10-year period. Considering that the catfish export value from the Mekong Delta for just the year 2012 was US$1.7 bn, and that the Mekong Delta accounts for 98% of Vietnam's catfish export value,[4] this cost is nominal but would help defray the operational costs of individual catfish farmers and increase the profitability of farms that are already operating at slim profit margins.

Water management is more complex in the coastal zone where water provides a range of goods in terms of fresh- and brackish-water resources for agriculture, aquaculture, and urban and industrial uses, as well as ecosystem services that support

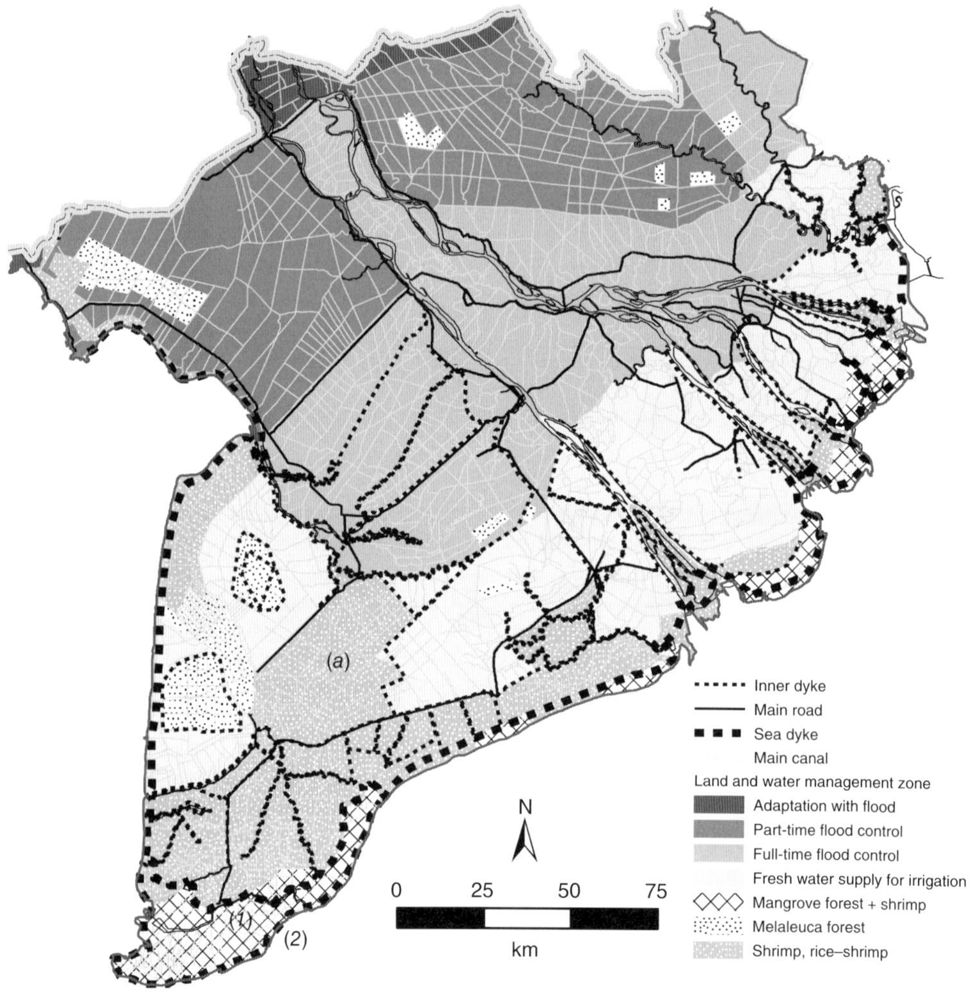

Fig. 9.6. Water resources planning for the Mekong Delta (from SIWRP, 2010).

coastal fisheries and maintain a variety of wetland ecosystems including mangrove forests. Adaptation strategies, including planned adaptation, can bring about potential synergies as well as conflicts among different economic sectors. Achieving a delicate balance that ensures equitable access and efficient sharing of the water resources to meet the needs of the various sectors requires non-structural measures besides structural ones appropriately implemented at field to community and regional levels (Nhan *et al.*, 2007).

The prospects for brackish-water aquaculture in the coastal zone depend on the extent and configuration of water control infrastructure planned for the future. The existing shrimp and shrimp–rice areas inland of the national highway linking Bac Lieu and Ca Mau cities will benefit from plans to continue partial control of salinity intrusion into the coastal zone by judicious operation of sluices and temporary dams that allow intake of sea water in the dry season and flushing out of saline and pollutant-laden water in the rainy season (Hoanh *et al.*, 2009). This operational control of salinity intrusion is itself a nonstructural adaptive mechanism that provides flexibility for managing the water resources to meet

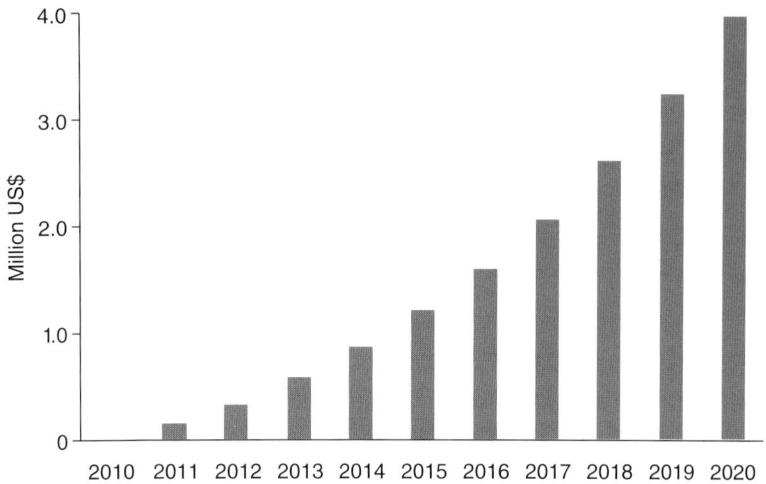

Fig. 9.7. Autonomous adaptation cost for upgrading dykes of inland catfish ponds at industry level for the Mekong Delta, 2010–2020 (from Kam *et al.*, 2012).

agriculture and aquaculture requirements as physical (including climatic) and socio-economic (including markets) conditions change. The projected increase in dry-season salinity in these areas beyond the tolerance limits for rice would, in fact, shift land use in favour of brackish-water aquaculture expansion. On the other hand, structural measures, such as construction of sea embankments for protection of coastal communities and agricultural land from storm surges and full prevention of land salinization, would mean removing the brackish-water aquaculture area marked *(a)* in Fig. 9.6 out of production. Such structural measures will also have negative impacts on fisheries in the coastal zone by preventing the natural movement of food-fish species up the rivers and canal networks.

The water resources planning map (Fig. 9.6) suggests two possible alignments for the sea dyke for coastal protection in the south-western tip of Ca Mau peninsula. The alignment marked *(1)* in Fig. 9.6, favoured by the Ca Mau provincial authorities, is along the shoreline outside of the remaining largest area of mangroves in the delta. The alignment marked *(2)*, as suggested by the Southern Institute for Water Resources Planning (SIWRP), lies inland of this mangrove area. The ecological integrity of the mangrove area is less likely to be compromised with the latter alignment as its ecosystem health is dependent on natural tidal inundation and shoreline accretion. The mangroves and the accreting mudflats constitute wetlands that support a rich diversity of aquatic organisms that have food and economic potential, including molluscs (cockles) and crustaceans (mangrove crabs). Mangrove–aquaculture systems are more environmentally friendly, taking advantage of natural flushing and nutrient supply, and its products can be targeted at elevated prices at niche markets. For example, in 2010 more than 1000 organic shrimp farms with a total of 6200 ha within the mangroves have been certified by Naturland (Samson, 2010). A healthy and expanding mangrove strip also provides a protective buffer against coastal events and storm surges.

9.6 Aquaculture-specific Adaptations

Besides adaptation plans on water resources management that can benefit aquaculture in the Mekong Delta, there are also other possible response strategies to CC impacts that are specific to the aquaculture sector.

A number of such strategies pertain to better farm management practices such as improving feed conversion ratios and the use of local alternatives of feed materials (to be achieved through research) into feed formulation that reduce feed costs as well as the carbon footprint of formulated feeds (Bunting *et al.*, 2009). Technical improvements such as specific pathogen-free shrimp brood-stock technology will help reduce disease risks that might increase with CC impacts (De Silva and Soto, 2009). A combination of selective breeding programmes and changes in farming practices would enable the farming of catfish strains that can better tolerate higher levels of salinity. Many of these strategies would increase farm profit margins, thereby helping underwrite the costs of adaptation. While the adoption of improved stocks and the modification of farming practices will be the task of those responsible for managing aquaculture operations, planned adaptation options such as genetics selection and breeding programmes fall within the mandate of the government. These strategies may be regarded as 'no-regret' strategies in that they contribute towards building a general resilience beyond specific adaptation to CC, and are therefore not overly dependent on detailed quantification of its specific impacts (Heltberg *et al.*, 2009).

Another strategy that applies particularly to export-oriented aquaculture production would be to transfer the cost of adaptation across the value chain, thereby increasing the margins accruing to farmers rather than to export processing companies and retailers in importing countries. This could happen either on the initiative of actors higher up the value chain as a market response to maintain the supply to meet growth in demand, or by government intervention, community- or civil society-driven measures in recognition of the importance of maintaining the welfare of the large number of small farmers and transferring benefits from a globally integrated production system to the rural communities in the delta.

In the long-term, reducing the high dependence on shrimp and catfish culture and diversifying into more ecologically oriented production systems can also hedge the aquaculture industry against the increasing risks and uncertainties brought about by CC. The export-oriented aquaculture industry in the South-east Asian region, based on the industrial monoculture model, is highly dynamic and volatile, and the economic risks are high (Szuster, 2003), not to mention the high environmental and social costs that producer countries have to bear.

Unlike the staple crops and livestock, there is actually a high diversity of aquatic species and production systems that can be, and are being, practised, as indicated in Table 9.1. These range from purely aquaculture systems to integrated production within rice and mangrove environments. The integrated systems among those listed in Table 9.1 are cited by Costa-Pierce (2010) as good examples of the ecological aquaculture paradigm promoted by FAO (Soto *et al.*, 2008), which calls for responsible aquaculture that takes account of its interactions and influences on the surrounding natural and social environments for better sustainability, equity and resilience.

Because these production systems fit into different agro-ecologies, an aquaculture sector that possesses such diversity would have greater adaptability to changes in hydrological conditions that result from CC impacts. On the other hand, water resources management strategies that accommodate the range of saline, brackish water and freshwater environments will provide the opportunities for this diversity of aquaculture systems to be further developed and improved, thereby keeping future options open for viable alternatives to the industrial-style monoculture systems that dominate the sector today. It is also imperative that future aquaculture planning, particularly for integrated systems that are ecologically more sound, should be done in conjunction with irrigation and water resources enhancement and within the broader context of the delta's development and response to climate change.

9.7 Conclusions

Assessing the economic cost of the impacts of CC and adaptation at the farm level in aquaculture remain uncharted research areas. Given the constraints of conventional cost–benefit analysis adopted by Kam *et al.* (2012), a more thorough integrated assessment of the economics of planned adaptation is needed to examine trade-offs in costs and benefits of adaptation options among sectors which include the aquaculture industry. Modelling of the impact of CC on growth, production and yield of cultured species is urgently needed to enable better estimation of projected farm yields under different CC scenarios and management responses.

All economic studies of CC adaptation are subject to substantial uncertainty surrounding the impacts of future CC, changes in input and output of commodity prices, and changes in production technologies and other factors. Indeed, for the Mekong Delta, the future outlook is riddled with uncertainties: in terms of extremities of climate change, of developments that will occur in the upstream Mekong, and of the economic developments that will happen within the delta. These uncertainties impinge upon the aquaculture sector, which is presently dominated by export-oriented monoculture production of shrimp and catfish. Economic analyses of these two major aquaculture production systems suggest that the costs of farmers' autonomous adaptation to CC impacts will further erode their farm profits and render the less-efficient farms economically unviable much sooner, *ceteris paribus*. Part of the autonomous adaptation costs can be offset to some extent by planned adaptation measures undertaken with public investment, as illustrated by the flood mitigation measures defraying catfish farmers' costs for raising pond dykes in the flood-affected upper and middle delta zones.

Structural responses to CC impacts incur high levels of investments into infrastructure development and are less flexible for adjustments in the face of uncertain future outcomes of climate and other global changes. Combinations with non-structural measures for more nimble adaptation strategies would be necessary for long-term resilience of the aquaculture sector. The diversity of aquaculture production systems provides opportunities for fitting into different agro-ecologies and lends itself to a high degree of transformability of the sector in the face of changing hydrological conditions brought about by CC.

The impetus for diversification needs to be driven by policy that is integral to the country's development plan for fisheries and aquaculture, which must include the broader sustainability issues facing the sector. In the same way that Vietnam has a proven capacity to build an aquaculture industry of global significance, the potential exists for the country to take a lead in the transition from an industrial monoculture phase of aquaculture development to a more diversified, integrated and ecologically sensitive phase that promotes innovation and efficiency by incorporating and not externalizing social and environmental costs. The Mekong Delta, with its vast and diverse land and water resources, and faced with the challenges of CC impacts, presents the opportunity to make this transition.

Acknowledgements

This chapter derives from WorldFish research contribution to the World Bank case study, 'Vietnam: Economics of Adaptation to Climate Change' (Kam *et al.*, 2010). Further development of the paper was supported by the CGIAR research program on Climate Change, Agriculture and Food Security (CCAFS).

Notes

1 Based on data obtained from the Vietnam Association of Seafood Exporters and Producers: http://seafood.vasep.com.vn/statistics/50_121/814/1/seafood-export-statistics.htm (accessed 7 November 2015).

2 These areas were originally planned in the 1990s to be fully protected from intrusion of sea water

for expanding rice intensification by constructing a series of sluices. However, rapid expansion of shrimp farming since the early 2000s and soil acidification from exposure of acid sulfate soils in the area rendered these areas less suited for intensive rice cultivation. A government policy reversal enabled these non-structural adjustments to accommodate both shrimp and rice farming.

3 A 50-cm SLR is anticipated by 2100 under the moderate emission scenario and would be surpassed by 2050 under the high emission scenario (Table 8.2).

4 General Statistical Office of Vietnam, http://www.gso.gov.vn/default_en.aspx?tabid=469&idmid=3 (accessed 29 March 2013).

References

Adger, W.N., Agrawala, S., Mirza, M.M.Q., Conde, C., O'Brien, K., Pulhin, J., Pulwarty, R., Smit, B. and Takahashi, K. (2007) Assessment of adaptation practices, options, constraints and capacity. Climate Change 2007: Impacts, Adaptation and Vulnerability. Contribution of Working Group II. In: Parry, M.L., Canziani, O.F., Palutikof, J.P., van der Linden, P.J. and Hanson, C.E. (eds) *Fourth Assessment Report of the Intergovernmental Panel on Climate Change.* Cambridge University Press, Cambridge, UK, pp. 717–743.

Browman, M.W. and Kramer, D.L. (1985) *Pangasius sutchi* (Pangasiidae), an air-breathing catfish that uses the swim bladder as an accessory respiratory organ. *Copeia* 4, 994–998.

Bunting, S.W., Pretty, J. and Muir, J.F. (2009) *Carbon and Energy Consumption Futures for Fisheries and Aquaculture.* Interdisciplinary Centre for Environment and Society, University of Essex, UK and Institute of Aquaculture, University of Stirling, UK.

Costa-Pierce, B.A. (2010) Sustainable ecological aquaculture systems: the need for a new social contract for aquaculture development. *Marine Technology Society Journal* 44(3), 88–112.

Dalvia, R.S., Pala, A.K., Tiwarib, L.R., Dasa, T. and Baruah, K. (2009) Thermal tolerance and oxygen consumption rates of the catfish *Horabagrus brachysoma* (Günther) acclimated to different temperatures. *Aquaculture* 295, 116–119.

De Silva, S.S. and Soto, D. (2009) Climate change and aquaculture: potential impacts, adaptation and mitigation. *Fisheries Technical Paper* 537, FAO, Rome, 137–215.

Duong, N.D. (2006) Applying GAP in shrimp culture in Ca Mau province. Masters thesis, Can Tho University, Can Tho, Vietnam.

FAO (Food and Agriculture Organization of the United Nations) (2012) The State of World Fisheries and Aquaculture 2012. FAO, Rome, 209 pp.

GSO (General Statistical Office) (2008) Statistical yearbook of Vietnam 2008. General Statistical Office of Vietnam, Vietnam. Available at: http://www.gso.gov.vn/default_en.aspx?tabid=515&idmid=5&ItemID=9233 (accessed 15 January 2011).

Heltberg, R., Siegel, P.B. and Jorgensen, S.L. (2009) Addressing human vulnerability to climate change: toward a 'no-regrets' approach. *Global Environmental Change* 19(1), 89–99.

Hoanh, C.T., Phong, N.D., Gowing, J.W., Tuong, T.P., Ngoc, N.V. and Hien, N.X. (2009) Hydraulic and water quality modeling: a tool for managing land use conflicts in inland coastal zones. *Water Policy* 11, 106–120.

Hoanh, C.T., Jirayoot, K., Lacombe, G. and Srinetr, V. (2010) Impacts of climate change and development on Mekong flow regime. First Assessment – 2009. *MRC Technical Paper No. 29.* Mekong River Commission, Vientiane, Lao PDR.

IPCC (Intergovernmental Panel on Climate Change) (2000) IPCC special report emissions scenarios. Summary for policy makers. A Special Report for IPCC Working Group III. Intergovernmental Group for Climate Change. WMO/UNEP, 27 pp. Available at: http://www.ipcc.ch/pdf/special-reports/spm/sres-en.pdf (accessed 23 March 2013).

Kam, S.P., Badjeck, M.-C., Teh, L., Teh, L., Be Nam, V.T., Hien, T.T., Hue, N.T., Phillips, M., Pomeroy, R. and Sinh, L.X. (2010) Economics of adaptation to climate change in Vietnam's aquaculture sector: A case study. Unpublished report to the World Bank.

Kam, S.P., Badjeck, M.-C., Teh, L., Teh, L. and Tran, N. (2012) Autonomous adaptation to climate change by shrimp and catfish farmers in Vietnam's Mekong River delta. *Working Paper 24–2012.* The WorldFish Center, Penang, Malaysia.

Little, D. and Murray, F. (n.d.) Pangasius and Europe: the unparalleled growth of a farmed tropical whitefish in European markets. Sustainable Aquaculture Group, Institute of Aquaculture, University of Stirling, UK. Available at: http://www.seafish.org/media/572923/acig_pangasiusandeurope_universityofstirling.pdf (accessed 24 March 2013).

Marchand, M., Dam, R. and Buck, T. (eds) (2011) *Towards a Mekong Delta Plan: synthesis of water sector assessment.* Netherlands Cooperation, Viet Nam, 58 pp.

MONRE (Ministry of Natural Resources and Environment) (2009) *Climate Change, Sea Level Rise Scenarios for Vietnam*. Vietnam Institute of Meteorology, Hydrology and Environment, Ministry of Natural Resources and Environment, Hanoi, Vietnam, 34 pp. Available at: http://www.preventionweb.net/files/11348_ClimateChangeSeaLevelScenariosforVi.pdf (accessed 15 January 2011).

Moss, R.H. and Schneider, S.H. (2000) Uncertainties in the IPCC TAR: recommendations for lead authors for more consistent reporting. In: Pachauri, R., Taniguchi, T. and Tanaka, K. (eds) *Guidance Papers on the Cross Cutting Issues of the Third Assessment Report of the IPCC*. World Meteorological Organization, Geneva, pp. 33–51.

Nhan, D.K., Be, N.V. and Trung, N.H. (2007) Water use and competition in the Mekong Delta, Vietnam. In: Be, T.T., Sin, B.T. and Miller, F. (eds) *Challenges to Sustainable Development in the Mekong Delta: Regional and National Policy Issues and Research Needs*. The Sustainable Mekong Research Network (Sumernet), Bangkok, Thailand, 56 pp.

Samson, O. (2010) Saving Vietnam's coasts with mangroves and organic shrimps. DW.com. Available at: http://www.dw.de/saving-vietnams-coasts-with-mangroves-and-organic-shrimps/a-5458570 (accessed 6 April 2013).

Sinh, L.X. (2008) *Information for Pangasius Catfish in the Mekong Delta – 2008*. College of Aquaculture & Fisheries, Can Tho University, Can Tho, Vietnam.

SIWRP (Southern Institute for Water Resources Planning) (2009) The effect of sea level rise on the Mekong Delta. Technical Report of the Southern Institute for Water Resources Planning. Ministry of Agriculture and Rural Development. Southern Institute of Water Resources Planning, Ho Chi Minh City, Vietnam.

SIWRP (2010) Draft report on water resources infrastructure assessment study in the Mekong Delta. Vietnam-Netherlands Mekong Delta Masterplan Project. SIWRP, Ho Chi Minh City, Vietnam, 54 pp. Available at: http://wptest.partnersvoorwater.nl/wp-content/uploads/2011/06/WATER-INFRASTRUCTURE-final-draft_1.pdf (accessed 26 March 2013).

Soto, D., Aguilar-Manjarrez, J., Bermúdez, J., Brugère, C., Angel, D., Bailey, C., Black, K., Edwards, P., Costa-Pierce, B., Chopin, T., Deudero, S., Freeman, S., Ambrey, J., Hishamunda, N., Knowler, D., Silvert, W., Marba, N., Mathe, S., Norambuena, R., Simard, F., Tett, F., Troell, M. and Wainberg, A. (2008) Applying an ecosystem-based approach to aquaculture: Principles, scales and some management measures. In: Soto, D., Aguilar-Manjarrez, J. and Hishamunda, N. (eds) *Building an Ecosystem Approach to Aquaculture*. FAO/Universitat de les Illes Balaers Expert Workshop, 7–11 May 2007, Palma de Mallorca, Spain. FAO Fisheries and Aquaculture Proceedings 14. FAO, Rome.

Szuster, B.W. (2003) *Shrimp Farming in Thailand's Chao Phraya River Delta: boom, bust and echo*. IWMI Research Report. Comprehensive Assessment of Water Management in Agriculture. International Water Management Institute, Colombo, Sri Lanka, 57 pp.

Thuy, D.T., Kania, P. and Buchmann, K. (2010) Infection status of zoonotic Trematodemetacercariae in Sutchi catfish (*Pangasianodon hypophthalmus*) in Vietnam: associations with season, management and host age. *Aquaculture* 302, 19–25.

Tuan, L.A., Hoanh, C.T., Miller, F. and Sinh, B.T. (2007) Floods and salinity management in the Mekong Delta, Vietnam. In: Be, T.T., Sinh, B.T. and Miller, F. (eds) *Challenges to Sustainable Development in the Mekong Delta: Regional and National Policy Issues and Research Needs*. The Sustainable Mekong Research Network (Sumernet), Bangkok, Thailand, 54 pp.

10 Groundwater for Food Production and Livelihoods – The Nexus with Climate Change and Transboundary Water Management

Karen Villholth*

International Water Management Institute, Regional Office for Southern Africa, Pretoria, South Africa

Abstract

Groundwater resources in transboundary aquifers may cushion climate change and other impacts of anthropogenic change through irrigation development, which in turn can enhance food security, livelihood benefits and poverty reduction in developing countries. Such resources present significant water reserves that, however, need to undergo critical joint assessment, development and management if they are to provide substantial as well as sustainable scenarios to agricultural and socio-economic development. This chapter explores the nexus between groundwater in shared aquifers, climate change and agricultural growth in the context of Africa, Asia, the Middle East and Latin America and examines the added challenges as well as opportunities that the transboundary setting of these resources may provide in terms of devising lasting solutions. The chapter highlights that both local smaller-scale no-regret as well as larger-scale, strategic adaptation measures that often hinge on integrated surface–groundwater solutions are important. In addition, the socio-economic and institutional aspects, the latter in terms of general international law and specific adapted international agreements as well as bottom-up participatory processes, are critical for attaining success on the ground.

10.1 Introduction

Water management is seen as a key component of climate change adaptation (CCA) in order to enhance societies' resilience against anticipated and emerging impacts (Bates *et al*., 2008). A broadly recognized adaptation strategy involves increasing and managing water storages as a means of offsetting increased variability in precipitation, and consequently in water availability, at various temporal and spatial scales (Taylor, 2009; McCartney and Smakhtin, 2010). Furthermore, improving the extent, performance and sustainability of irrigated agriculture through better agricultural water management is seen as a key adaptation measure in developing regions of the world (Ngigi, 2009).

In this chapter, the role and options for groundwater, and more specifically groundwater in transboundary international settings, are explored in the context of addressing CCA and meeting needs of irrigated agriculture and food security. Groundwater provides reliable, almost ubiquitous, often (albeit not always) renewable water supply, inherent storage and buffering facilities, and hence has been advocated as an

* Corresponding author, e-mail: k.villholth@cgiar.org

important and strategic resource in CCA, if managed properly (Clifton *et al.*, 2010). While transboundary aquifers (TBAs), defined as those groundwater bodies spanning international boundaries, are increasingly recognized as important around the world (Eckstein, 2011), relatively little attention has been paid to these resources for strategic CCA and agricultural water management. Water management has to increasingly transcend the traditional river basin approach, as promoted by the integrated water resources management (IWRM) paradigm (GWP, 2004), in particular when larger transboundary aquifers play a significant role in the hydrological and ecological systems and/or in water provision, which is often the case in semi-arid and arid areas. Vulnerability and resilience of such TBAs towards climate change may also vary substantially, calling for differentiated and prioritized attention in the context of CCA. Hence, though groundwater's role in CCA and transboundary water resource management is increasingly acknowledged, their nexus with agriculture and food production has hitherto received little attention, and this chapter intends to bridge this gap.

The chapter sets out by giving a short inventory of the global extent, diversity and significance of TBAs. Then, the role, options and limits of groundwater in adaptation are briefly discussed, giving a summary of the comparative characteristics and added advantages of groundwater for managed storage, relative to that of, or in combination with, surface water as well as a list of no-regret options for using groundwater as a component of CCA in agriculture. Finally, the major additional considerations and challenges related to groundwater for irrigated agriculture in TBAs under climate change prospects are highlighted, devising a simple typology, including examples of particular TBAs, as a framework for developing best development and management solutions. The chapter focuses geographically on developing continental regions of the world with larger TBAs: Asia, Africa, the Middle East and Latin America.

As in prehistoric times, where early settled agrarian civilizations clustered around rivers, deltas and springs that provided reliable and easily accessible water resources, future human development may increasingly concentrate around significant groundwater resources and aquifer systems, some of them of international dimensions, that provide reliable, replenishable, protected and manageable water reserves linked to harvestable rainwater, surface water, wastewater streams and manufactured (e.g. desalinated) water. A significant premise here is that underground natural or enhanced recharge of fresh water presents a very favourable means of limiting excessive losses to evaporation under a warmer climate while augmenting stable supplies. The positive relation between the value of land and groundwater availability is increasingly acknowledged, especially when the resources decline (Lee and Bagley, 1972) or is in high demand in less-developed regions (Woodhouse and Ganho, 2011).

Groundwater plays and most likely will continue to play an increasingly important role in meeting global water supply and storage demands as global temperatures increase, and climate variability challenges existing dependence on more erratic surface-water resources. Today, more than one-third of the global population is dependent on groundwater for their domestic supply (Morris *et al.*, 2003; Döll *et al.*, 2012), while about 40% of all irrigated land is supplied by groundwater (Döll *et al.*, 2012; Foster and Shah, 2012), most critically in arid and semi-arid regions. These figures have been increasing unprecedentedly over the latter part of the last and into this century as access to, and awareness of, groundwater in development has accelerated, providing profound development benefits in terms of agricultural livelihoods, food production and productivity increases, water security and improved public health (Shah *et al.*, 2007; Carter and Bevan, 2008; Gun, 2012).

Yet, today it is estimated that 18% of global groundwater-based irrigation is unsustainable, i.e. that it derives from depleting aquifers, where groundwater is utilized at rates faster than their replenishment (Wada *et al.*, 2012a). Hence, the challenge presently relates to moderating existing groundwater demands through efficiency gains, enhancing replenishment through purposeful management wherever possible, cautious development of new sources in still undeveloped regions, optimizing and diversifying multiple sources (including surface water), preserving valuable groundwater-dependent terrestrial and aquatic ecosystems, all in the context of larger uncertainty, variability and risks prescribed by climate change as well as in a globally increasingly interdependent community.

10.2 Transboundary Aquifers

Transboundary aquifers are relatively well-defined units of geological and associated groundwater systems that lie partially in more than one sovereign state[1] (Stephan, 2009). They are delineated, based on geological and hydraulic characteristics and boundaries. Because of larger uncertainties related to the characterization and delineation of these underground water-bearing units compared to international river basins, the definition of these systems also depends on a general agreement between aquifer-sharing states of the transboundary nature of the aquifer. While many TBAs may be (partly) hydraulically connected to rivers or other surface water bodies (e.g. lakes) this is not always the case, and more often than not, these systems are not geographically coincident with river basins (see Fig. 10.1 for the example of Africa). This mismatch between geographic extent of the surface and groundwater systems implies complication in relation to defining best management units for integrated and transboundary water resources' management (Schmeier, 2010; Altchenko and Villholth, 2013).

Transboundary groundwater resources are gaining enhanced attention from water developers as well as from the international research community and increasingly also from national and international policy makers due to increasing stress on available water resources (Aureli and Eckstein, 2011). With increasing attention, the knowledge of acknowledged TBAs increases and the number of newly identified and agreed TBAs expands (IGRAC, 2015). At present, a global inventory reveals the existence of 592 TBAs (IGRAC, 2015). This surpasses the present number of international river basins, which stands at 263 (Cooley *et al.*, 2009), documenting that these resources are indeed of global as well as of local significance. Impacts of negligence of recognizing the transboundary nature of these systems partly resemble those for international rivers, in terms of water quantity-sharing aspects as well as potential water quality issues.[2] When the surface water and groundwater systems are linked, these problems become interrelated.[3] However, certain management aspects are particular to the TBAs (and aquifers more broadly) and relate to their invisible, open-source and vulnerable nature (Table 10.1). These aspects need to be given much more attention in sustainable management and protection of TBAs as compared to internationally shared river systems. On the other hand, co-aquifer state cooperation on TBAs and associated systems may provide aggregate shared benefits that outweigh the costs and disadvantages of not cooperating (Box 10.1).

TBAs range from smaller, more local aquifers shared between two nations to larger regional contiguous aquifer systems that partially span up to eight states (e.g. the Lake Chad Aquifer Basin in central-western Africa, occupying a land area of 1.3×10^6 km^2). The largest TBA in the world is the Guaraní Aquifer in South America with a size of 1.9×10^6 km^2 covering parts of Argentina, Brazil, Paraguay and Uruguay. TBAs also vary significantly in geological set-up, depth interval(s) of groundwater occurrence as well as rate and mechanism of replenishment. A complete inventory of global TBAs in terms of these parameters does not exist, but salient data are available from Margat and Gun (2013), IGRAC (2015), and for Africa in Altchenko and Villholth (2013).

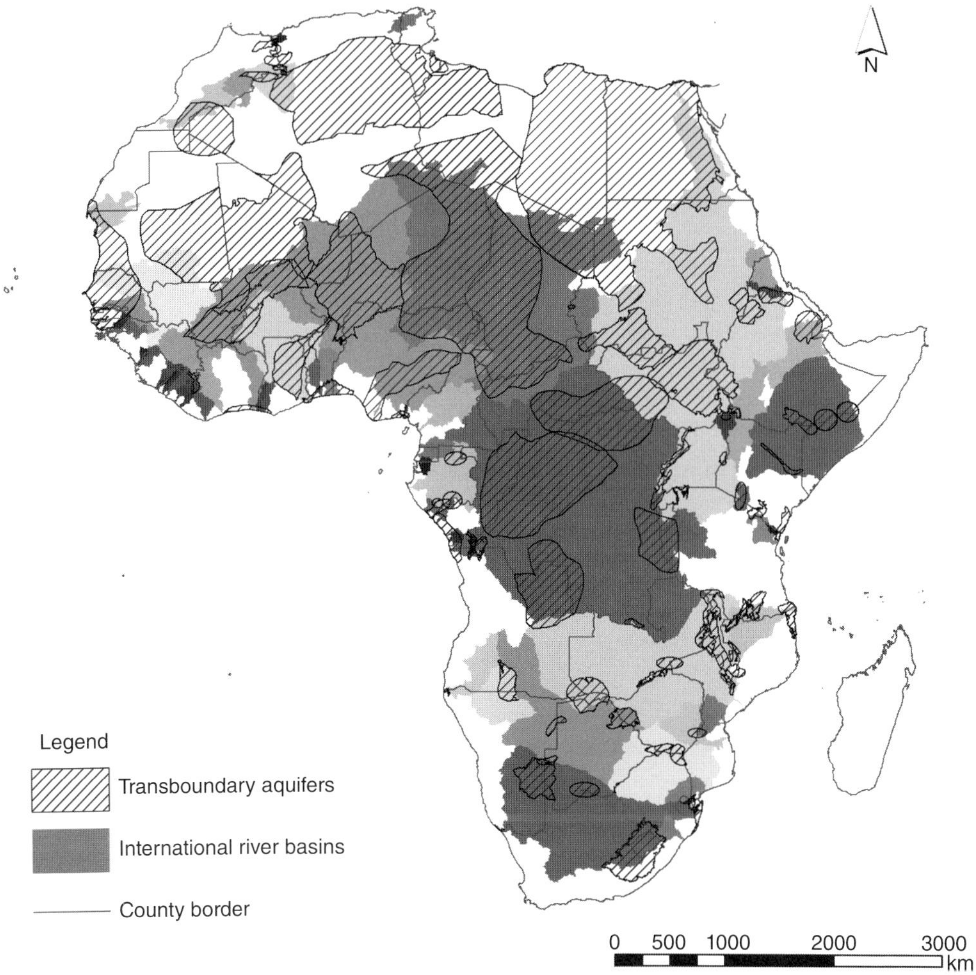

Fig. 10.1. Map of transboundary aquifers of Africa. Depicted are also the international river and lake basins (from IGRAC, 2015). Note that differences in shading weight indicate different river basins.

10.3 Groundwater's Role in Agriculture and Climate Change Adaptation

Groundwater development for agriculture has occurred thanks to the many favourable inherent characteristics of the resource (Table 10.2) and has generally been accompanied by very positive socio-economic transitions (Giordano and Villholth, 2007). Groundwater irrigation has surpassed the role of surface water in terms of acreage, outputs and rural-poverty-alleviating impacts in India (Narayanamoorthy, 2007) and other parts of South Asia, and similar, albeit less pronounced, impacts have been seen in northern China (Foster and Garduño, 2004) and Mexico (World Bank, 2009).

In Africa, particularly south of the Sahara desert, groundwater is presently only contributing minimally to food production, though significant land and hydrological potential exists (Pavelic *et al.*, 2012a). Only

Table 10.1. Particular characteristics of aquifers and implications for management of TBAs.

	Special considerations/provisions needed in TBA management[a]							
Groundwater distinct characteristic	Joint user/use registration, regulation, monitoring and enforcement	Prior notification of development plans to other party	Precautionary principle	Conflict resolution	Stakeholder engagement	Long-term monitoring of resource	Flexibility in conceptual model and clear data-sharing arrangements	Land use and waste regulations	Prioritized protection
Open source	xx				xx				
Invisible and heterogeneous		x	x	x	x	x	x		
Vulnerable to land use impacts					x			xx	x
Slow reacting/delay in response		x	xx	x		xx			
Recharge/discharge is distributed and uneven								x	xx
Boundaries uncertain				x		x	xx		
Climate change impacts uncertain						xx	xx		
Blurred up- and downstream relations			x	x	x	x	xx		

[a]Number of 'x's indicates the degree of importance of considerations/provisions.

Box 10.1. Ten arguments for addressing groundwater in transboundary water management

1. Benefits of groundwater (GW) development and management can be equitably shared across borders to avoid climate-induced distress migration and conflicts.
2. GW development and proper management have a lot to do with achieving the Sustainable Development Goals, poverty alleviation, food security, climate change adaptation, and flood and drought mitigation.
3. An integrated and transboundary approach facilitates enhanced understanding of water flows and water balances within the aquifer basin and supports improved delineation of the aquifer, including active and connected surface water (SW) systems.
4. GW impacts across borders may not be obvious without joint long-term monitoring. Costs and results of monitoring can be shared.
5. Impacts of unilateral GW development and use in one member state may affect another.
6. Developing GW in connection with transboundary SW (conjunctive use) may provide a lot of benefits, e.g. floodwaters may be used to replenish GW in overdrawn aquifers; SW pollution may be reduced through riverbank filtration for better drinking water quality, and managed aquifer recharge (MAR) and recovery may support water banking and salinity control.
7. Many terrestrial ecosystems are GW-dependent and cannot be properly managed without acknowledgement of the GW resources.
8. SW issues involve or even have root in GW-related activities and impacts, e.g. water from the river may be lost through GW abstraction in the vicinity of the river.
9. Lake, river, wetland and estuary water quality may be threatened by GW pollution in adjacent upstream aquifer states (mining, intensive agriculture).
10. No-action and lack of transboundary cooperation may result in significant and long-term risks, e.g. haphazard and chaotic exploitation of aquifers with high remediation costs if at all reversible (like certain types of contamination and land subsidence).

about 4% of the cultivated land in sub-Saharan Africa (SSA) is presently irrigated (NEPAD, 2003), leaving large scope for further expansion aiming at protecting farmers against weather fluctuations at various temporal scales through better control of, and access to, water (Ngigi, 2009), while contributing to closing the gap in food production.

The role of groundwater in CCA relates to its reliable and normally drought-resilient character (Table 10.2) implying that the resource can be reliably drawn upon during dry seasons or times of dry spells or drought, when other sources fail. However, besides the characteristics of the resource itself, the effective buffering capacity of groundwater resources towards drought hinges critically on the resilience and adaptive capacity of populations and agricultural systems dependent on groundwater. In a simple multi-factor assessment, groundwater resources in the regions of South Asia and Africa were considered to be the most vulnerable to climate change, presumably because of the high degree of dependence on groundwater and low adaptive capacity (Clifton *et al.*, 2010). In a similar, but more detailed mapping analysis across the SADC region in southern Africa, Villholth *et al.* (2013) found that the areas underlain by crystalline rock formations, prevalent in SSA, were among the most vulnerable, because of their relatively low water-holding capacity and poor yields combined with high population densities and high drought risk. Furthermore, the vulnerability of drought of these areas was compounded under a projected climate scenario (Villholth *et al.*, 2013). Such interdisciplinary mapping can be invaluable in pro-active planning of best adaptation measures for drought resilience in these vast transboundary settings.

Besides targeting the most vulnerable areas, drought mitigation through groundwater measures should consider the following principles:

1. Access points (wells) should be drought-proof (located in productive parts of the aquifer, intake level not risking dry-out).
2. Access points should be resilient against additional wear and tear during drought,

Table 10.2. Inherent characteristics of groundwater favouring irrigated agriculture and climate change adaptation.

Groundwater property	Irrigated agriculture	Climate change adaptation
Drought-resilient, reliable	Provides year-round on-demand irrigation, encourages intensification	Bridges seasonal and possibly inter-annual variability
Widespread	Supports rural development	Addresses local vulnerability, away from major water infrastructure
Underground	Increases water productivity	Less storage and conveyance losses from evaporation
Amendable to incremental development	Potentially pro-poor. Can be developed by individual farmers or small groups	Requires small investments and lends rapid response
Versatility[a]	Addresses gender needs to diversified water use purposes	Can address several vulnerabilities and water needs
Flexibility[b]	Can help control waterlogging and salinization	Can combine with other sources to optimize storage and increase overall resilience. Can address both drought and floods

[a] Groundwater can be developed for various (multiple) uses in rural areas.
[b] Groundwater can be combined with other sources (conjunctive use) to optimize overall use.

and easily maintainable in distant rural areas.

3. Dedicated drought wells may be reserved for emergency situations.

Importantly, degradation of groundwater resources, due to climate change, prolonged drought or human impacts, will increase the vulnerability of populations dependent on them, ultimately reducing their capacity to shield deficits in other water resources and hence undermining water security for domestic and livelihood purposes. This underpins the critical importance of proper groundwater management.

While basic water supply takes precedence during drought, adapting livelihoods and ensuring availability and accessibility of productive water during projected scenarios of increased rainfall variability are also critical. Since irrigated agriculture serves several development goals, in terms of food security, livelihoods and reduction in rural poverty, developing it wisely as a component of a CCA strategy is further justified. Some pointers as to how to address development and management of groundwater for irrigation and CCA can be put forward. These will be discussed briefly below.

10.4 Ensuring Access to the Poorest

10.4.1 Lowering costs

The major stumbling block for poor farmers to access groundwater in SSA and possibly in other regions of the world, where multidimensional potentials exist, as in parts of the Greater Mekong Region (Johnston *et al.*, 2010), is the requirement for physically drilling holes to the resource and to lift the water to the surface. While progress is seen in terms of declining prices for well drilling and acquiring mechanical pumps along with a developing demand and supply, it is also evident that most smallholder farmers still avail themselves of very rudimentary and labour-intensive means of extracting and lifting groundwater, including manual digging and hand-lifting and watering with buckets, which in prosperous areas limit the accessible resources and hence the level of development (Namara *et al.*, 2011). Where public investment supplies groundwater irrigation facilities, benefits do not necessarily reach the poorest segments. Hence, a critical policy is to support local cost-effective pump and drilling manufacturing, markets and associated services (Abric *et al.*,

2011), ensuring expansion and coverage in rural areas in order to decrease investment and transaction costs. It also entails instituting and supporting small farmers in obtaining access to both import and tax-waiver systems for pumps for irrigation equipment. Finally, well-targeted subsidies, potentially with feasible payback schemes to increase cost recovery and long-term financial viability of public funds and micro-credit services (Nkonya *et al.*, 2010) should be looked into more closely.

10.4.2 Pump and groundwater markets

Previous experience from Asia indicates that the share of farmers, and particularly of the poorest segments benefiting from groundwater, increases due to spontaneous and informal groundwater and pump rental markets (Villholth *et al.*, 2009). Generally, wealthier farmers individually owning pumps and wells sell water (or rather the service of water provision) and/or rent pumps to other farmers without these assets, thereby increasing overall access and income generation. Pump rentals have also been popular in Africa, while groundwater markets have not. While not presently significant, these institutions may spontaneously develop in SSA as groundwater-based irrigation proliferates among smallholder farmers.

10.4.3 Multiple use systems

Rather than dedicated and separate domestic and irrigation infrastructure, multi-purpose systems, catering to the various water needs of rural households need to be further considered. It has been observed (Calow *et al.*, 2009) that domestic water points are often used for productive uses, e.g. in garden irrigation, livestock-rearing, brewing and brick-making, and that this increases the resilience of the households, even though water points are traditionally not designed with multiple uses in mind. Hence, adapting groundwater irrigation structures and their location with these realities in mind better serves the needs of the poorest and female farmers (Koppen *et al.*, 2009). However, the largest challenge confronted in taking this forward is the fact that the sectors for water supply and agriculture remain institutionally and functionally detached. Ensuring basic domestic water supply should, in any case, be first priority in underserved areas, and should not be compromised by poorly planned groundwater development for irrigation.

10.4.4 Energy access

As part of lowering entry barriers for smallholder farmers to groundwater irrigated agriculture in SSA, improving rural electrification is critical. Energy from electricity is generally cheaper in SSA compared to other fossil-fuel based sources (Pavelic *et al.*, 2012b), but the coverage is the world's lowest at only 24% (UNEP, 2012). Experience from South Asia shows that groundwater-irrigating farmers with access to electricity are generally better off than their counterparts who use diesel or other sources (Villholth *et al.*, 2009). While this is partly due to distorting electricity subsidies to the agriculture sector, it shows that groundwater irrigation at a more than subsistence level is linked to better energy provision.

10.5 Ensuring Environmental and Social Sustainability

Globally, irrigation serves about 40% of food production (Wada *et al.*, 2012a); about 40% of water for irrigation derives from groundwater (Foster and Shah, 2012), and of the groundwater extracted approximately 67% goes to irrigation (Gun, 2012). Hence, the way groundwater irrigation is managed significantly influences the resource and any potential negative environmental impacts from irrigation. Today, 18% of irrigation demand is derived from aquifers that are overexploited (Wada *et al.*, 2012a), with serious and growing socio-economic and environmental implications, especially in certain parts of the world where it has proliferated due to a combination of favourable conditions in terms of push-and-pull factors (Moench, 2003; Kajisa *et al.*, 2006; World Bank, 2009). Hence, there is a built-in risk associated with the development of groundwater for irrigation, which is related to some of the same advantageous characteristics of groundwater mentioned in Table 10.2. Management and enforcement of any regulation are hampered by the open-source nature of groundwater, its distributed occurrence with access options for a multitude of dispersed users that are difficult to control in conjunction. A plethora of literature discusses best strategies and options for groundwater management in agriculture (Giordano and Villholth, 2007; Shah, 2009a). Most of this literature focuses on the reactive measures for groundwater management in areas with intensive use and apparent negative impacts. Little attention has been paid to how to pro-actively manage groundwater in areas where the resource is still relatively underdeveloped, ensuring sustainability in terms of lasting poverty alleviation as well as ecosystem and human resilience. In the following sections, a few directions towards this purpose are given realizing that various strategies may serve in the under-developed as well as the over-developed scenario. While these options apply more broadly, they are relevant in the transboundary context in order to optimise use and minimize transboundary effects.

10.5.1 Conjunctive use of groundwater and surface water

Using and managing surface water and groundwater together, or conjunctively, in irrigation, provides options for better overall control, efficiency in use and productivity of both quality and quantity aspects of the resources (Evans and Evans, 2012). Groundwater irrigation may complement canal irrigation in areas where the groundwater table is rising and waterlogging and concomitant salinization is causing decreasing performance. This was evidenced in parts of the transboundary Indus plains (mostly Pakistan) where government drilling schemes helped alleviate waterlogging problems in the 1960s (Scanlon *et al.*, 2007). This may be a viable option in parts of India where areas of waterlogging still persist in the midst of larger areas of groundwater depletion (Foster and van Steenbergen, 2011). Similarly, options exist in existing canal irrigation schemes with waterlogging problems in Africa (Ojo *et al.*, 2011) as well as in suburban areas, where groundwater levels are increasing due to unintentional water leakage or intentional wastewater recharge (Foster *et al.*, 2010). Often, in canal irrigation areas, conjunctive use of groundwater develops spontaneously, partly as a coping mechanism of farmers to get reliable access to water in poorly managed canal irrigation schemes (Foster and van Steenbergen, 2011). However, the challenge is to rather plan, better design and optimize the schemes for conjunctive use, ensuring that headwater as well as tail-water users and areas are reliably served, not compromised by saltwater threats.

Similarly, critically evaluated surface-water transfers may help support irrigation in groundwater-depleted areas, as is seen, for example, in China, where huge interstate transfer schemes supply water from the water-rich south to the relatively water-deficient north, and transferred water is used for irrigation in the form of waste water after serving urban demands (Shu *et al.*, 2012).

In summary, water sources need to be increasingly diversified and integrated, and management needs to reflect this. Similarly, planned integrated conjunctive use may be a better solution than swaying back and forth from primarily depending on one or the other resource as seen in some places around the world (Clifton *et al.*, 2010). Another critical aspect of conjunctive use is that it tends to conserve on overall energy use (Shah, 2009b).

10.5.2 Local management of irrigation

As mentioned, groundwater management is still poorly conceived when it comes to tackling use in agriculture. Basically, an institutional top-down and a bottom-up approach is proposed in addition to more indirect measures (Giordano, 2009) linked, for example, to food, agriculture, energy, health and nature-conservation policies (Sekhri, 2012). In developing countries, the top-down, direct approach to groundwater management, involving user rights and economic instruments like water tariffs, is less tractable and effective, while the more indirect approaches, particularly the link to energy, promises to have some traction (Shah *et al.*, 2012).

The bottom-up approach, where community involvement and mobilization are required to manage local groundwater resources use, deserves further testing, particularly in connection with the more indirect measures at the national and even the international level (Wijnen *et al.*, 2012). These measures, which involve collective rule-setting and enforcement, e.g. in terms of timing of groundwater pumping, well spacing, crop choices, compulsory recharge structures and water-saving irrigation infrastructure, monitoring of the resource, etc. seem to work best if some of the following conditions prevail: existence of strong social capital, relatively homogeneous population groups, strong visionary local leadership to ensure motivation and compliance, the resource being relatively well defined and responding evidently and rapidly to demand management, the regulations are easily monitored, and the number of stakeholders are relatively limited. The challenge is to develop self-motivated and sustainable processes that do not need continued inputs and subsidies (Wijnen *et al.*, 2012), even in transboundary settings.

10.6 Managed Aquifer Recharge and Storage

A continuum of surface and subsurface options for water storage is available, each solution with specific characteristics and options for management (McCartney and Smakhtin, 2010). Managing underground water is a critical component in the context of water scarcity and climate change (Clifton *et al.*, 2010), though traditionally it has received less attention (Taylor, 2009) relative to surface water storage. Though some similarities exist between groundwater storage and surface water storage options (typically dams), groundwater storage presents inherent features that make management significantly different from management of a large impounded reservoir (Table 10.3).

Managing groundwater storage entails the co-management of recharge as well as discharge processes (Dillon *et al.*, 2009), with recharge typically being the most amenable component to manipulate, through various so-called managed aquifer recharge approaches (MAR). In irrigated agriculture, the objectives of MAR typically relate to offsetting abstraction in excess of natural replenishment, the levelling out of seasonal or inter-annual variations in storage, or the expansion of areas of crop production. In most cases, transfer of source water, e.g. from surface water, rainwater or wastewater is involved. This indicates that MAR is just as much a matter of surface water management, or rather integrated management of multiple sources and conjunctive use, as one of only groundwater.

Managing the discharge of groundwater as a part of controlling storage volumes and groundwater levels becomes important in waterlogged surface irrigation schemes, as previously discussed. It is also relevant in planned MAR, where the discharge of groundwater and associated drawdown of water table levels needs to be synchronized relative to the recharge, which is typically governed by seasonal availability of source water.

While MAR and managed groundwater storage present multiple options and potential benefits, numerous considerations need to be taken into account. It needs to be cost-effective and adapted to the local context. It needs to be anchored in institutional set-ups in order to level social inequity in water access rather than exacerbate it. Finally, various environmental impacts (upstream/downstream) need to be closely examined, both in terms of water quantity and quality

Table 10.3. Comparison between groundwater and surface water as manageable water storage and supply, through MAR and large dam reservoirs, respectively.

	Groundwater	Surface water
Storage volume and flux determination	Difficult	Easy
Physical impoundment	Difficult	Easy
Storage regulation	More difficult	Easy
Discharge/abstraction and allocation regulation	More difficult, open source, users ill-defined	Easy, users well-defined
Water source for storage	Needs to be collected and directed to recharge sites, e.g. from storm water, rainwater, floodwater, refuse water	Immediately available from upstream river
Reliability of water source	Depends on the source	Depends on climate and watershed management
Uncertainty in replenishment rate if water source available	High due to clogging phenomena	None
Evaporation losses	Low	High
Drought vulnerability	Lower, due to less evapotranspiration losses and retardation between inflows and outflows	Higher, due to evapotranspiration losses and limited accretion during drought
Drought impacts	Shallow wells and poorest communities hit first	Multiple water uses may be impaired, including hydropower
Drought mitigation options	Can drill/use deeper wells for interim relief	Can temporarily compromise on environmental flow releases
Flood vulnerability	Low if storage managed and located optimally; localized vulnerability in low-lying/discharge zones	Low if storage operated optimally; localized vulnerability along downstream reaches
Flood impacts above storage and detention capacity	Slow-emerging water level rises locally	Can be catastrophic downstream
Flood mitigation options	Floodwater diversions from infiltration/recharge sites, evacuation of vulnerable areas; can prioritize abstraction ahead of crisis	Flood modelling, flood warning, pre-flood releases, evacuation of vulnerable areas
Risk of waterborne diseases	Low	High
Life span	High; depends on clogging control and water quality control	Low; depends on siltation
Water quality	Depends on watershed management, land use and source water for MAR	Depends on watershed management
Catering for environmental flow and storage requirements	More difficult due to releases dependent on level of groundwater	Easy through informed release schemes
Carbon footprint of water use	Depends on depth of pumping and pump efficiency	Depends on need and extent of non-gravity conveyance

issues. Significant experiences with MAR exist from India (Sakthivadivel, 2007) and other developing countries (Dillon *et al.*, 2013), and though the level of documentation of effectiveness and socio-economic impacts is improving, the knowledge of environmental impacts is still limited.

MAR is often brought forward as a panacea for addressing water scarcity and CCA (Clifton *et al.*, 2010), even in the transboundary setting (Puri and Struckmeier, 2010), and also for stabilizing depleting aquifers (Shah, 2009b). However, it needs a lot more detailed examination and actual testing, as well as adaptation to developmental contexts (Dillon *et al.*, 2009). Supporting storage recovery in seriously depleting aquifers through MAR alone seems unrealistic in many cases (Dillon *et al.*, 2013). A multitude of actions, including demand management, soil and catchment management, and even partly curtailing irrigation may be needed (Moench, 2003). However, the role of MAR will become increasingly important as water demand increases and the impacts of climate change and variability become more apparent.

10.7 Climate Change Impact on Groundwater

Despite the fourth assessment report of the Intergovernmental Panel on Climate Change recognizing the deficiency in our understanding of impacts of climate change on groundwater (Bates *et al.*, 2008), recent work is slowly showing progress (Taylor *et al.*, 2013a). The challenges are related to limited long-term monitoring of groundwater resources, as well as imperfect understanding of the climatic changes and impacts on fundamental processes like groundwater recharge. Generally, impacts on groundwater systems are subdued and delayed compared to those on surface water systems.

Precipitation changes may affect recharge in non-linear ways (Clifton *et al.*, 2010), and impacts may depend more on variability of the rainfall and short-term intensity than on longer-term averages (Taylor *et al.*, 2013a). Infiltration and net percolation to groundwater are also governed by soil surface, geological, vegetation and atmospheric conditions and changes therein, whether driven by climatic changes or otherwise. Responses (direction and extent) will depend on the relative strength of these factors in various regions. There seems to be consensus on the projection of increase in number and/or intensity of extreme rainfall events globally (Bates *et al.*, 2008; Gregersen *et al.*, 2013). However, large uncertainty pertains to impacts of this on recharge in various regions. Research indicates that recharge may shift to more episodic events, driven by more intense and extreme rainfall events, especially in the warmer climates. While warming serves to increase evapotranspiration demand thereby limiting excess water for recharge, more intense rainfall would be likely to more than overcome this. Larger infiltration and recharge as a response to higher rainfall intensity in various climates may also be promoted by preferential flow processes, governed by pedological and geological conditions (Villholth *et al.*, 1998). In contrast, other reports suggest increased surface runoff and/or evapotranspiration and, as a net outcome, overall decreased recharge resulting from higher-intensity rainfall (Dourte *et al.*, 2013). In cooler climates, seasonal recharge transitions seem to be shaped by earlier and more intense snowmelt in the spring, which increases recharge, and possibly less recharge due to declining effective rainfall (precipitation minus evapotranspiration) during summer (Okkonen *et al.*, 2010).

Many reports highlight the significance of concurrent human and climate-change-induced impacts on groundwater resources (e.g. Treidel *et al.*, 2012). There is an expected compounded indirect impact of higher overall water demand and increased reliance on groundwater in a warmer and more unpredictable climate (Chen *et al.*, 2004). This indicates a need for addressing both types of forcings, human as well as climate change, which are collectively set to increase in the future. Land-use changes, often associated with new land cultivation, are significant drivers of hydrological change and have also been shown to have comparable or overriding impacts on groundwater resources relative to that of climate change and variability (Scanlon *et al.*, 2006). Increasing irrigation water demand due to higher evapotranspiration rates and the need for overcoming uncertainty and variability in rainfall and hence greater risk in rainfed agriculture, may result in increasing recharge and groundwater levels, due to irrigation return flows (Toews and Allen, 2009; Döll *et al.*, 2012), or conversely in increasing net storage depletion and falling groundwater levels (Shu *et al.*, 2012), depending on the primary source of irrigation water, whether from surface water or groundwater, respectively.

Projecting climate change impacts on groundwater systems is associated with limited confidence, due to uncertainties in global circulation models (GCMs) propagated to uncertainties in nested regional or smaller-scale hydrological models (Zhou *et al.*, 2010), particularly pertaining to precipitation in the former and recharge in the latter. Outcomes are often highly variable and even partly conflicting when ensembles of GCMs and climate scenarios are applied. Furthermore, many larger-scale hydrological models use simplified assumptions related to groundwater-relevant processes (Holman *et al.*, 2011; Taylor *et al.*, 2013b), implying critical uncertainty in quantitatively projecting climate change impacts on groundwater storage and availability. Often-cited global-scale modelling of climate change impact on groundwater (Döll, 2009) assumes groundwater replenishment deriving only from diffuse recharge and disregards focused recharge (from perennial and ephemeral surface water bodies) as well as increased short-term (less than monthly) rainfall variability, the effect of both of which may be critical. Enhancing long-term groundwater monitoring from land and satellite-based sources is an accompanying significant means to improving our knowledge (Taylor *et al.*, 2010).

Impacts of climate change on groundwater quality are generally poorly understood. Adverse groundwater quality impacts of climate change may stem from increased leaching of surface-derived substances in winters in colder climates, similar impacts in warmer climates due to episodic recharge and preferential flow through soils and geological materials, and higher infiltration of contaminated water during flooding events in various climates. Reduced recharge may conversely aggravate groundwater quality through less dilution (Solheim *et al.*, 2010). Great concern relates to projected sea-level rise and increased intrusion of salt water in freshwater coastal areas and on smaller islands (Bates *et al.*, 2008; Villholth, 2013). This will be exacerbated or overruled by intensified groundwater pumping (Ferguson and Gleeson, 2012). It could also, *ceteris paribus*, be counterbalanced by natural groundwater-level lifting processes (Chang *et al.*, 2011). Irrespective, as stated by Custodio (2004), research needs to be careful in attributing salinity increases in coastal areas directly to sea-level rise as salinity may be derived from a complex, sometimes interrelated array of sources, such as innate geological salinity, urban pollution, pumping-induced ingress of salt water, irrigation-derived salinity, etc. Interestingly, groundwater depletion may augment global sea-level rise (Wada *et al.*, 2012b), indicating the complex interrelations between groundwater and climate change.

Transboundary aquifers as a sub-set of all aquifers will be subject to similar climate-change impacts globally, and the study of these and potential CCA options are discussed in the next section.

10.8 The Potential Role of Transboundary Aquifers in Climate Change Adaptation

Transboundary aquifers across the globe are diverse in many respects and their role in CCA will vary accordingly. Multiple criteria can be set up for their potential role, for example, in terms of lateral extent, stored and presently recurrently replenished water volumes, depth of access, water quality, countries sharing, present degree of development and pressures. Some TBAs are still relatively undeveloped, presenting significant opportunities for further joint development and adaptive use, while others are already stressed from existing human development. As an example of the first, the Ohangwena freshwater aquifer between Angola and Namibia (part of the Cuvelai-Etosha Basin) is presently being investigated for potential exploitation in the border region between the two countries, including for small-scale irrigation (Christelis *et al.*, 2012). Development of this aquifer could supplement supply from international surface water transfer from Angola to Namibia and support further economic development. However, the recharge status of the aquifer (whether presently recharged and if so how much) still needs to be assessed. An example of an over-exploited TBA is the Santa Cruz aquifer between Mexico and the USA, which presents similar characteristics as other over-developed aquifers shared between the two countries (Scott *et al.*, 2012). However, other TBAs exhibit partial development in only parts of the aquifer, with huge potential in others, like the great Guaraní aquifer in South America, shared between four nations (OAS, 2009). Similarly, in the Ganges Basin, shared between India, Nepal and Bangladesh (and possibly Bhutan and Burma) groundwater development varies substantially across the basin. Generally, the degree of development and interest in the TBAs increase from humid to arid areas and with level of human development. Large and significant TBAs (like the Nubian Sandstone aquifer and the North Western Sahara aquifer) located in arid regions are non-renewable, giving rise to special challenges in terms of overall sustainable management (Foster and Loucks, 2006).

The arguments for focusing on TBAs in terms of CCA include the following aspects:

1. Climate change is transboundary, and adaptation measures and related water governance structures need to reflect this.
2. There is increasing dependence upon groundwater resources and many exploited aquifers are transboundary with potential transboundary implications (i.e. abstraction in one country affects another).
3. Many transboundary river basins include or intersect with TBAs and the surface water and groundwater resources are hydraulically interlinked.
4. Use and management of groundwater, in combination with surface water, offers more sustainable solutions than the current predominant focus on surface water in transboundary water management.
5. Joint development and management of the TBAs could lead to more equitable and sustainable agricultural development and regional stability and integration.

These aspects are relevant in addition to the more general points brought forward on the role of groundwater in CCA in Section 10.3. Ensuring equitable development of groundwater and food production and sharing of benefits across borders may alleviate impacts of climate change and extremes and prevent mass migration during droughts (Calow and MacDonald, 2009; Namara *et al.*, 2011). Likewise, conjunctive use of groundwater and surface water for agriculture and other purposes may limit the need for large transfer schemes across borders, if water can be drafted from groundwater with its in-built transmission capabilities rather than through relatively costly dam and pipeline/canal systems. Better management of wet-season river flows or floodwaters via MAR in transboundary river basins may benefit from joint management of the storages (in impoundments and aquifers) across the borders, as illustrated in the case from the Fergana Valley in the Syr Darya River Basin, now a transboundary basin spanning Kyrgyzstan, Tajikistan, Uzbekistan and Kazakhstan (Karimov *et al.*, 2009). In this case, storing underground the excessive winter releases from upstream dams necessary for electricity production could provide reliable irrigation water sources in the summer and could prevent waterlogging and salinity problems and flooding risk downstream. Large opportunities for CCA and better aquifer management will be foregone if such transboundary solutions are not sought and optimized. In heavily exploited transboundary aquifers (like the Indus, the US–Mexican aquifers and many aquifers in the Middle East) reducing stress is the key, and the options for expanding irrigation are limited and entail more innovative approaches, like reusing or using wastewater or poorer-quality groundwater infeasible for human consumption, while optimizing not only production with limited resources but use across the borders (Table 10.4).

Small-scale irrigation from TBAs may be considered 'no regret'[4] adaptation measures in many transboundary regions. Though solutions will vary from context to context, some options that may serve to enhance resilience in irrigation and agricultural development are listed in Box 10.2. In fact, many of these 'no regrets' adaptations can be implemented in areas where water

Table 10.4. Best approaches for CCA in TBAs depending on development level and degree of current natural replenishment.

Undeveloped (replenished)	Over-developed (replenished)	Non-renewable (over-developed or not)
Floodwater management, proper drainage, conjunctive use	Conjunctive use to optimize resource use	Equitable use across countries, sectors and users. Protection of domestic users
Simple MAR to capture and control floodwater	MAR from recycled or floodwater	Exit strategies entailing alternative water resources and/or livelihoods
Conjunctive use to avoid waterlogging	Opportunistic small-scale use	Use of renewable energy and efficient pumps for extraction
Intensifying crop-cultivation	Groundwater demand management	Groundwater demand management
Protection of TBA and groundwater-dependent ecosystems	Growing high-value crops, drought-resistant crops and peri-urban cropping	Growing high-value crops, shifting cultivation to non-depleted areas (outside the aquifer) and/or increase imports of food
Drawing down aquifers to enable renewed seasonal storage	Protection of TBA, recharge zones and groundwater-dependent ecosystems Shifting to efficient rainfed agriculture, fallow conditions and urban livelihoods Use of renewable energy and efficient pumps for extraction	
Jointly collect data on shared aquifers	Jointly collect data on shared aquifers	Jointly collect data on shared aquifers
Ensure binding international water agreements between aquifer states	Ensure binding international water agreements between aquifer states	Ensure binding international water agreements between aquifer states
Example: Lower Ganges Basin (India, Nepal, Bangladesh) (Villholth *et al.*, 2009; Sharma *et al.*, 2011) Guaraní Aquifer (Argentina, Brazil, Paraguay and Uruguay) (OAS, 2009)	Example: High Plains aquifer (various states in the USA) (Scanlon *et al.*, 2012) Santa Clara Aquifer (USA, Mexico) (Scott *et al.*, 2012), the transboundary aquifers underlying Euphrates-Tigris rivers (Iran, Iraq, Syria, Turkey) (Voss *et al.*, 2013)	Nubian Sandstone Aquifer (Chad, Egypt, Libya, Sudan) (Foster and Loucks, 2006)

resources are already stressed, regardless of concerns about the uncertainty of climate change projections and assessments of impact on groundwater and surface water resources (Clifton *et al.*, 2010). Though these measures have more general applicability, building resilience of poorer communities in and across border regions will significantly enhance regional stability.

These no-regret, smaller-scale adaptation measures should be combined with larger scale, more strategic measures in the TBA areas, related for example to land use, urban development and water quality protection, which will impact the overall groundwater resource.

In Table 10.4, a simple typology for TBAs in terms of CCA approaches in agriculture is given. The only distinguishing feature is the level of development, though others, such as present climate and human development, could also be critical. Examples of larger TBAs for the various types are also given.

Box 10.2. No-regret options for using transboundary aquifers in climate change adaptation in agriculture

1. Low-cost, low-technology options for MAR from various sources (rainwater, floodwater).
2. Conjunctive use, including capturing and storing surface water underground when excess is available.
3. Opportunistic small-scale groundwater irrigation for smallholders (including peri-urban), dependent on water availability (Clifton *et al.*, 2010).
4. Including groundwater irrigation in or downstream of surface-water irrigation areas and downstream of surface-water impoundments, to increase benefits and to combat waterlogging and salinization.
5. Manage demand for groundwater in irrigation, e.g. through low-cost micro-irrigation.
6. Adapting to current droughts as a surrogate for adapting to future climate change.
7. Protection of groundwater recharge areas.

Box 10.3. Legal aspects of transboundary aquifers

International policies and agreements on TBAs are emerging. Only 15% of 400 international freshwater treaties presently include explicit provisions for groundwater (Jarvis, 2006), albeit often in a rudimentary manner. A set of guiding draft Articles on the Law of Transboundary Aquifers formulated by the United Nations International Law Commission (Stephan, 2009) exist that complement the presently most widely subscribed-to international law on surface waters, the so-called Convention on the Law of the Non-navigational Uses of International Watercourses, adopted by the UN General Assembly in May 1997 (United Nations, 1997). The draft articles have been developed in retrospect, acknowledging that the surface-water-focused convention was inadequate in addressing transboundary groundwater. The result is the coexistence of two guiding documents, only the latter presently ratified, that do not in isolation or in conjunction sufficiently acknowledge the integrated properties and benefits of both resources (McCaffrey, 2009). Another aspect, which has been brought forward in this context, is that the present international agreements do not adequately account for climate change, variability and adaptation (Cooley *et al.*, 2009).

10.9 Challenges and Recommendations for Climate Change Adaptation and Agricultural Growth in Transboundary Aquifers

Traditionally, TBAs have not attained high-level attention from policy makers, international law (Box 10.3) or from relevant water, agriculture and resource management institutions (Eckstein, 2011). Rather, groundwater presently tends to be managed unilaterally with only emerging trends of global and regional emphasis on the transboundary aspects (e.g. SADC, 2011). Even less significance is accorded these resources in terms of potential for CCA and agricultural growth and particularly food security and poverty reduction in developing regions. Evolving work in these areas seems to arise from increasing recognition of the need in water-stressed regions, like in the case of the non-renewable Nubian Sandstone Aquifer, that now counts on a formal data-sharing[5] agreement and results of a joint transboundary diagnostic (Stephan, 2009; Eckstein, 2011), or due to donor-supported impetus, as in the case of the Guaraní Aquifer system (Villar and Ribeiro, 2012). Notwithstanding these, efforts of UNESCO since the 2000s have been instrumental in raising the research and policy attention to TBAs globally through their International Shared Aquifer Resources Management (ISARM) Project (UNESCO, 2010). The disparity in attention given to different TBAs in the continents considered are to be found in variable attentions from the ISARM project and other donor assistance as well as from the different degrees of water stress and water tension between the aquifer-sharing countries.

International tension related to groundwater is emerging, as in the case of the

Middle East (Eckstein and Eckstein, 2003; Voss *et al.*, 2013), demonstrating the significance of these resources for various purposes (including agriculture) and also the need for both bilateral and multilateral negotiations, conflict resolution and agreements on these resources.

Finally, the potential of groundwater, and TBAs as a subset of this, in addressing gaps in food production in prospective areas also needs further attention. The challenge in this respect may lie in the sustainable and equitable development of new resources, ensuring that the resource is not overcommitted, and that development occurs in a fashion that benefits equally all the nations involved and also benefits the poorest segments of the populations. There is a twofold challenge here: (i) the economically stronger nations may drive the development to the detriment of the less developed; and (ii) benefits from agricultural development may accrue mostly to larger-scale commercial farmers.

Africa, and SSA in particular, may epitomize the need and potential for addressing the role of TBAs in climate change and agricultural growth. The region is considered socially vulnerable, a hot-spot for climate change as well as a region with underdeveloped water resources for agriculture, yet a continent with significant sharing of major water resources across borders. In addition, storage capacity of surface water is among the lowest in the world (McCartney and Smakhtin, 2010) and hence focusing more on groundwater and indeed integrated solutions as discussed previously may add to the range of options. In the transboundary sense, groundwater may be more equitable, as aquifer states can share a joint resource without the need for large-scale infrastructure and hence less upstream–downstream controversies. On the other hand, partners will have to come together and create trust and transparency in their individual actions. At the same time, groundwater and also TBA research and advocacy are progressing steadily in SSA, where resources have been mapped to a considerable extent.

However, capacities are limited. To move this agenda forward, there is a need for:

- Addressing the distinct properties of groundwater (Table 10.1) as well as the integrated and conjunctive use potentials of transboundary surface water and groundwater in international water agreements.
- Better understanding of the function, potential, present pressures and vulnerabilities of specific TBAs and how they can sustainably co-benefit the sharing countries.
- Better knowledge of potential climate change impacts on groundwater resources.
- Better knowledge of options and limitations to minimize adverse climate impacts, e.g. MAR.
- Capacity building of institutions and developers on the potentials and limitations of TBAs in enhancing resilience and food productivity.
- Further transboundary cooperation and dialogue with an increased focus on the role of groundwater in agriculture.
- Simultaneous bottom-up and top-down approaches to TBA management.

Proper groundwater management cannot be achieved without the active involvement of users and stakeholders, as access and impacts are often local and/or diffuse and dispersed over the extent of the aquifer area. Conversely, grasping the transboundary significance and seizing the potentials in a long-term sustainable fashion without compromising dependent ecosystems and populations will most often require involvement and commitment at the highest national and international level.

10.10 Conclusions

Groundwater often takes second seat in development policies and strategies, especially related to agriculture and transboundary water collaboration, but it may well be a decisive resource for CCA and social and environmental resilience in many settings. This chapter advocates for increased attention to the role that TBAs can play in addressing climate change resilience, water and food security, and regional integration and cooperation. However, this requires increased emphasis, capacity development and awareness-raising at all levels to harness the options available through integrated solutions. The ability to manage groundwater will affect overall development and adaptation performance and outcomes. While on the one hand groundwater dependence is set to increase partly because of its storage properties and climate variability resilience, on the other, mal-management of groundwater entails loss of exactly that property of the resource.

Notes

1 The definition of TBAs does not imply that groundwater resources in border regions outside of TBAs do not exist or manifest similar properties as TBAs. However, the extent and significance of such resources are presently considered of limited transboundary importance or their transboundary extent has not been identified or acknowledged.

2 Groundwater abstraction in one country may influence groundwater flow and availability in another country. Groundwater pollution in one country may spread through the aquifer to an adjacent country.

3 For example, groundwater pumping in riparian zones of one country may affect river flows downstream in an adjacent country. Likewise, pollution of aquifers in one country may enter an adjacent country, by way of base flow to an international river.

4 Actions that are justifiable from economic, social and environmental perspectives whether climate change takes place or not (Siegel and Jorgensen, 2011).

5 Limited experience with actual and quantitative water sharing of TBAs exists at present (Eckstein, 2011).

References

Abric, S., Sonou, M., Augeard, B., Onimus, F., Durlin, D., Soumaila, A. and Gadelle, F. (2011) *Lessons Learned in the Development of Smallholder Private Irrigation for High-value Crops in West Africa.* World Bank, Washington, DC, 62 pp.

Altchenko, Y. and Villholth, K.G. (2013) Transboundary aquifer mapping and management in Africa: a harmonised approach. *Hydrogeology Journal* 21(7), 1497–1517.

Aureli, A. and Eckstein, G. (2011) Strengthening cooperation on transboundary groundwater resources. *Water International* 36(5), 549–556.

Bates, B.C., Kundzewicz, Z.W., Wu, S. and Palutikof, J.P. (eds) (2008) Climate change and water. *Technical Paper of the Intergovernmental Panel on Climate Change.* IPCC Secretariat, Geneva, 210 pp.

Calow, R. and MacDonald, A. (2009) What will climate change mean for groundwater supply in Africa? *Background Note, 2009.* Overseas Development Institute, UK.

Calow, R.C., MacDonald, A.M., Nicol, A.L. and Robins, N.S. (2009) Groundwater security and drought in Africa: linking availability access and demand. *Ground Water* 48(2), 246–256.

Carter, R.C. and Bevan, J. (2008) Groundwater development for poverty alleviation in sub-Saharan Africa. In: Adelana, S.A. and MacDonald, A.M. (eds) *Applied Groundwater Studies in Africa.* Selected Papers on Hydrogeology, Vol. 13. International Association of Hydrogeologists, CRC Press/Balkema, Leiden, the Netherlands.

Chang, S.W., Clement, T.P., Simpson, M.J. and Lee, K.-K. (2011) Does sea-level rise have an impact on saltwater intrusion? *Advances in Water Resources* 34, 1283–1291.

Chen, Z., Grasby, S.E. and Osadetz, K.G. (2004) Relation between climate variability and groundwater levels in the upper carbonate aquifer, southern Manitoba, Canada. *Journal of Hydrology* 290, 43–62.

Christelis, G., Lohe, C., Quinger, M., Van Wyk, B., Quintino, M. and Villholth, K.G. (2012) Groundwater investigation of the Cuvelai-Etosha Basin – preliminary results and moving towards transboundary cooperation. Poster. *International Waters Science Conference*, 24–26 September, 2012, Bangkok, Thailand.

Clifton, C., Evans, R., Hayes, S., Hirji, R., Puz, G. and Pizarro, C. (2010) Water and climate change: Impacts on groundwater resources and adaptation options. *World Bank Working Note No. 25.* World Bank, Washington, DC.

Cooley, H., Christian-Smith, J., Gleick, P.H., Allen, L. and Cohen, M. (2009) Understanding and reducing the risks of climate change for transboundary waters. *Pacific Institute in Cooperation with the United Nations Environment Programme.* Oakland, California, USA, ISBN: 1-893790-22-3, 38 pp.

Custodio, E. (2004) Myths about seawater intrusion in coastal aquifers. In: Araguás, L., Custodio, E.

and Manzano, M. (eds) *Groundwater and Saline Intrusion*. Selected papers from the 18th SWIM, Cartagena, Spain, pp. 599–608. ISBN: 84-7840-588-7.

Dillon, P., Gale, I., Contreras, S., Pavelic, P., Evans, R. and Ward, J. (2009) Managing aquifer recharge and discharge to sustain irrigation livelihoods under water scarcity and climate change. *Proceedings of Joint IAHS & IAH Convention, Hyderabad, India, 2009*. IAHS Publishing, New Delhi, 12 pp.

Dillon, P., Escalante, E.F. and Tuinhof, A. (2013) Management of aquifer recharge and discharge processes and aquifer storage equilibrium. *Thematic Paper 4. Groundwater Governance: A Global Framework for Country Action*, 49 pp. Available at: http://www.fao.org/fileadmin/user_upload/groundwatergovernance/docs/Thematic_papers/GWG_Thematic_Paper_4.pdf (accessed 19 November 2013).

Döll, P. (2009) Vulnerability to the impact of climate change on renewable groundwater resources: a global-scale assessment. *Environmental Research Letters* 4, 12.

Döll, P., Hoffmann-Dobrev, H., Portmann, F.T., Siebert, S., Eicker, A., Rodell, M., Strassberg, G. and Scanlon, B.R. (2012) Impact of water withdrawals from groundwater and surface water on continental water storage variations. *Journal of Geodynamics* 59–60, 143–156.

Dourte, D., Shukla, S., Singh, P. and Haman, D. (2013) Rainfall intensity-duration-frequency relationships for Andhra Pradesh, India: changing rainfall patterns and implications for runoff and groundwater recharge. *Journal of Hydrologic Engineering* 18(3), 324–330.

Eckstein, G.E. (2011) Managing buried treasure across frontiers: the International Law of Transboundary Aquifers. *Water International* 36(5), 573–583.

Eckstein, Y. and Eckstein, G.E. (2003) Groundwater resources and international law in the Middle East peace process. *Water International* 28(2), 154–161.

Evans, W.R. and Evans, R. (2012) Conjunctive use and management of groundwater and surface water. *Thematic Paper 2. Groundwater Governance: A Global Framework for Country Action*. GEF ID 3726, 40 pp. Available at: http://www.fao.org/fileadmin/user_upload/groundwatergovernance/docs/Thematic_papers/GWG_Thematic_Paper_2.pdf (accessed 19 November 2013).

Ferguson, G. and Gleeson, T. (2012) Vulnerability of coastal aquifers to groundwater use and climate change. *National Climate Change* 2, 342–345.

Foster, S. and Garduño, H. (2004) China: towards sustainable groundwater resource use for irrigated agriculture on the North China Plain. *Case Profile Collection Number 8. GW-MATE*. World Bank, Washington, DC, 16 pp.

Foster, S. and Loucks, D.P. (2006) Non-renewable groundwater resources – a guidebook on socially-sustainable management for water-policy makers. *IHP-VI, Series on Groundwater No. 10*. UNESCO, Paris.

Foster, S. and Shah, T. (2012) Groundwater resources and irrigated agriculture – making a beneficial relation more sustainable. *Global Water Partnership (GWP) Perspectives Papers*. GWP, Stockholm, Sweden, 19 pp.

Foster, S. and van Steenbergen, F. (2011) Conjunctive groundwater use – a 'lost opportunity' for water management in the developing world? *Hydrogeology Journal* 19, 959–962.

Foster, S., Hirata, R., Misra, S. and Garduño, H. (2010) Urban groundwater use policy – balancing the benefits and risks in developing nations. *Strategic Overview Series Number 3*. GW-MATE. World Bank, Washington, DC, 36 pp.

Giordano, M. (2009) Global groundwater? Issues and solutions. *Annual Review of Environment and Resources* 34, 153–178.

Giordano, M. and Villholth, K.G. (eds) (2007) *The Agricultural Groundwater Revolution: opportunities and threats to development*. CAB International in association with IWMI, Wallingford, UK, 419 pp.

Gregersen, I.B., Madsen, H., Rosbjerg, D. and Arnbjerg-Nielsen, K. (2013) A spatial and non-stationary model for the frequency of extreme rainfall events. *Water Resources Research* 49, 127–136.

Gun, J., van der (2012) Groundwater and global change: trends, opportunities and challenges. *United Nations World Water Assessment Programme*. Side Publications Series: 01. UNESCO, Paris, ISBN: 978-92-3-001049-2.

GWP (2004) Integrated water resources management. *TAC Background Papers, No. 4*. Global Water Partnership, Stockholm, 68 pp. ISBN: 91-630-9229-8.

Holman, I.P., Allen, D.M., Cuthbert, M.O. and Goderniaux, P. (2011) Towards best practice for assessing the impacts of climate change on groundwater. *Hydrogeology Journal* 20, 1–4.

IGRAC (2015) Transboundary Aquifers of the World Map 2015. International Groundwater Resources Assessment Centre. Available at: http://www.un-igrac.org/resource/transboundary-aquifers-world-map-2015 (accessed 7 November 2015).

Jarvis, W.T. (2006) Transboundary groundwater: geopolitical consequences, commons sense, and the law of the hidden sea. PhD dissertation, Oregon State University, Oregon, 164 pp.

Johnston, R.M., Hoanh, C.T., Lacombe, G., Noble, A., Smakhtin, V., Suhardiman, D., Kam, S.P. and Choo, P.S. (2010) *Rethinking Agriculture in the Greater Mekong Subregion: how to sustainably meet food needs, enhance ecosystem services and cope with climate change.* International Water Management Institute, Colombo, Sri Lanka, 26 pp.

Kajisa, K., Palanisami, K. and Sakurai, T. (2006) *The Dissemination of Private Wells and Double Tragedies: the overexploitation of groundwater among well users and increased poverty among non-well users in Tamil Nadu, India.* International Association of Agricultural Economics Conference, Gold Coast, Australia, 12–18 August 2006.

Karimov, A., Mavlonov, A., Smakhtin, V., Turral, H. and Gracheva, I. (2009) Groundwater development in Fergana Valley: the adaptation strategy for changed water management in the Syrdarya Basin. In: Bloschl, G., Van De Giesen, N., Muralidharan, D., Ren, L., Seyler, F., Sharma, U. and Vrba, J. (eds) *Improving Integrated Surface and Groundwater Resources Management in a Vulnerable and Changing World.* Proceedings of the JS.3 at the joint IAHS and IAH Convention, Hyderabad, India, 2009. IAHS Publication 330, 2009, ISBN-13: 978-1907161018, CABI, Wallingford, UK, pp. 22–27.

Koppen, B., van Smits, S., Moriarty, P., Penning de Vries, F., Mikhail, M. and Boelee, E. (2009) *Climbing the Water Ladder: multiple-use water services for poverty reduction.* International Water and Sanitation Centre (IRC) and International Water Management Institute (IWMI), The Hague, the Netherlands, 216 pp.

Lee, S.H. and Bagley, E.S. (1972) Ground water and land values in Southwestern Kansas. *Ground Water* 10(6), 27–36.

Margat, J. and van der Gun, J. (2013) *Groundwater Around the World – A Geographic Synopsis.* Taylor & Francis, London, 348 pp.

McCaffrey, S.C. (2009) Current developments – the International Law Commission adopts draft articles on transboundary aquifers. *The American Journal of International Law* 103, 272–293.

McCartney, M. and Smakhtin, V. (2010) Water storage in an era of climate change: Addressing the challenge of increasing rainfall variability. *Blue Paper.* International Water Management Institute, Colombo, Sri Lanka, 14 pp.

Moench, M. (2003) Groundwater and poverty: exploring the connections. In: Llamas R. and Custodio, E. (eds) *Intensive Use of Groundwater: Challenges and Opportunities.* Balkema Publishers, the Netherlands, pp. 441–455.

Morris, B.L., Lawrence, A.R.L., Chilton, P.J.C., Adams, B., Calow, R.C. and Klinck, B.A. (2003) Groundwater and its susceptibility to degradation: a global assessment of the problem and options for management. *Early Warning and Assessment Report Series, RS. 03-3.* United Nations Environment Programme, Nairobi, Kenya, 126 pp.

Namara, R.E., Awuni, J.A., Barry, B., Giordano, M., Hope, L., Owusu, E.S. and Forkuor, G. (2011) Smallholder shallow groundwater irrigation development in the upper east region of Ghana. *IWMI Research Report 143.* International Water Management Institute, Colombo, Sri Lanka, 35 pp.

Narayanamoorthy, A. (2007) Does groundwater irrigation reduce rural poverty? evidence from Indian states. *Irrigation and Drainage* 56, 349–362.

NEPAD (2003) *Comprehensive Africa Agriculture Development Programme.* New Partnership for Africa's Development, Midrand, South Africa, 102 pp, ISBN 0-620-30700-5.

Ngigi, S.N. (2009) Climate Change Adaptation Strategies – Water Resources Management Options for Smallholder Farming Systems in sub-Saharan Africa. *The MDG Centre, East and Southern Africa.* The Earth Institute at Columbia University, New York, 189 pp. ISBN 978-92-9059-264-8.

Nkonya, E., Phillip, D., Mogues, T., Pender, J. and Kato, E. (2010) From the ground up – Impacts of a pro-poor community-driven development project in Nigeria. *Research Monograph.* International Food Policy Research Institute (IFPRI). Washington DC, 77 pp.

OAS (2009) *Guaraní Aquifer: Strategic Action Program.* Organization of American States, USA, 224 pp. ISBN: 978-85-98276-08-3.

Ojo, O.I., Ochieng, G.M. and Otiendo, F.O.A. (2011) Assessment of water logging and salinity problems in South Africa: an overview of Vaal harts irrigation scheme. *WIT Transactions on Ecology and the Environment* 153, 477–484.

Okkonen, J., Jyrkama, M. and Kløve, B. (2010) A conceptual approach for assessing the impact of climate change on groundwater and related surface waters in cold regions (Finland). *Hydrogeology Journal* 18, 429–439.

Pavelic, P., Smakhtin, V., Favreau, G. and Villholth, K.G. (2012a) Water-balance approach for assessing potential for smallholder groundwater irrigation in sub-Saharan Africa. *Water SA* 38(3), 399–406.

Pavelic, P., Giordano, M., Keraita, B., Rao, T. and Ramesh, V. (2012b) *Groundwater Availability and Use in Sub-Saharan Africa. A Review of 15 Countries.* International Water Management Institute (IWMI), Colombo, Sri Lanka. ISBN 978-92-9090-758-9, 255 pp.

Puri, S. and Struckmeier, W. (2010) Aquifer resources in a transboundary context: A hidden resource? – enabling the practitioner to 'see it and bank it' for good use. In: Earle, A., Jägerskog, A. and Öjendal, J. (eds) *Transboundary Water Management – Principles and Practice.* Earthscan, London and Washington, DC, pp. 73–90.

SADC (2011) Regional strategic action plan on integrated water resources development and management. *(2011–2015) RSAP III. Watering Life, Together, Forever.* Southern African Development Community, Gaborone, Botswana, 56 pp. ISBN 978-99912-475-7-1.

Sakthivadivel, R. (2007) The groundwater recharge movement in India. In: Giordano, M. and Villholth, K.G. (eds) *The Agricultural Groundwater Revolution: Opportunities and Threats to Development.* CAB International, Wallingford, UK, pp. 195–210.

Scanlon, B.R., Keese, K.E. and Flint, A.L. (2006) Global synthesis of groundwater recharge in semiarid and arid regions. *Hydrological Processes* 20, 3335–3370.

Scanlon, B.R., Jolly, I., Sophocleous, M.N. and Zhang, L. (2007) Global impacts of conversions from natural to agricultural ecosystems on water resources: quantity versus quality. *Water Resources Research* 43, W03437.

Scanlon, B.R., Faunt, C.C., Longuevergne, L., Reedy, R.C., Alley, W.M., McGuire, V.L. and McMahon, P.B. (2012) Groundwater depletion and sustainability of irrigation in the US High Plains and Central Valley. *PNAS* 109(24), 9320–9325.

Schmeier, S. (2010) Effective governance of transboundary aquifers through institutions – Lessons learned from river basin organizations. *Paper prepared for the International Conference 'Transboundary Aquifers: Challenges and New Directions',* UNESCO, ISARM, 6–8 December 2010, Paris, France.

Scott, C.A., Megdal, S.B., Oroz, L.A., Callegary, J. and Vandervoet, P. (2012) Climate change and population growth impacts on the transboundary Santa Cruz Aquifer. *Climate Research* 51(2), 159–170.

Sekhri, S. (2012) *Sustaining Groundwater: Role of Policy Reforms in Promoting Conservation in India.* India International Centre, New Delhi, India, 31pp.

Shah, T. (2009a) *Taming the Anarchy – Groundwater Governance in South Asia.* RFF Press book. Resources for the Future, Washington DC, 310 pp. ISBN: 978-1-933-115-60-3.

Shah, T. (2009b) Climate change and groundwater: India's opportunities for mitigation and adaptation. *Environmental Research Letters* 4.

Shah, T., Burke, J.J. and Villholth, K.G. (2007) Groundwater: a global assessment of scale and significance. In: Molden, D. (ed.) *Water for Food, Water for Life. Comprehensive Assessment of Water Management in Agriculture Synthesis Report.* Earthscan, London, pp. 395–423. ISBN: 1844073971.

Shah, T., Giordano, M. and Mukherji, A. (2012) Political economy of the energy-groundwater nexus in India: exploring issues and assessing policy options. *Hydrogeology Journal.* DOI: 10.1007/s10040-011-0816-0.

Sharma, B., de Condappa, D. and Bharati, L. (2011) Opportunities for harnessing the increased contribution of glacier and snowmelt flows in the Ganges Basin. *International Conference on Cooperation on the Ganges: Barriers, Myths, and Opportunities.* Institute of Water Policy, LKY School, National University of Singapore, Singapore, 13–14 November 2010, 16 pp.

Shu, Y., Villholth, K.G. Jensen, K.H., Stisen, S. and Lei, Y.P. (2012) Integrated hydrological modeling of the North China Plain: options for sustainable groundwater use in the alluvial plain of Mt. Taihang. *Journal of Hydrology.* DOI: 10.1016/j.jhydrol.2012.06.048.

Siegel, P.B. and Jorgensen, S. (2011) No-regrets approach to increased resilience and climate change justice: toward a risk-adjusted social protection floor. Conference on: 'Social Protection for Social Justice'. Institute for Development Studies (IDS), University of Sussex, Brighton, UK, 13–15 April 2011, 70 pp.

Solheim, A.L., Austnes, K., Eriksen, T.E., Seifert, I. and Holen, S. (2010) Climate change impacts on water quality and biodiversity. *Background Report for EEA European Environment State and Outlook Report 2010.* EEA/ADS/06/001 – *Water,* The European Topic Centre on Water (ETC/W), Prague, 68 pp.

Stephan, R.M. (2009) *Transboundary Aquifers: managing a vital resource, the UNILC Draft Articles on the law of Transboundary Aquifers.* UNESCO, Paray-Vieille-Poste, France, SC-2008/WS/35, 24 pp.

Taylor, R. (2009) Rethinking water scarcity: the role of storage. *Eos* 90(28), 237–248.

Taylor, R., Longuevergne, L., Harding, R., Todd, M., Hewitson, B., Lall, U., Hiscock, K., Treidel, H., Sharma, K.D., Kukuric, N., Struckmeier, W. and Shamsudduha, M. (2010) *Groundwater and Global Hydrological Change – Current*

Challenges and New Insight. Xth Kovacs Colloquium, Paris, 2–3 July 2010, 10 pp.

Taylor, R.G., Scanlon, B., Döll, P., Rodell, M., van Beek, R., Wada, Y., Longuevergne, L., Leblanc, M., Famiglietti, J.S., Edmunds, M., Konikow, L., Green, T.R., Chen, J., Taniguchi, M., Bierkens, M.F.P., MacDonald, A., Fan, Y., Maxwell, R.M., Yechieli, Y., Gurdak, J.J., Allen, D.M., Shamsudduha, M., Hiscock, K., Yeh, P.J.-F., Holman, I. and Treidel, H. (2013a) *Ground Water and Climate Change* 3 (April), 322–329.

Taylor, R.G., Todd, M.C., Kongola, L., Maurice, L., Nahozya, E., Sanga, H. and MacDonald, A.M. (2013b) Evidence of the dependence of groundwater resources on extreme rainfall in East Africa. *Nature Climate Change* 3, 374–378.

Toews, M.W. and Allen, D.M. (2009) Simulated response of groundwater to predicted recharge in a semi-arid region using a scenario of modelled climate change. *Environmental Research Letters* 4, 19.

Treidel, H., Martin-Bordes, J.L. and Gurdak, J.J. (eds) (2012) Climate change effects on groundwater resources – a global synthesis of findings and recommendations. *IAH – International Contributions to Hydrogeology*, Vol. 27. CRC Press/ Balkema, Taylor & Francis, London, 399 pp. ISBN: 978-0-415-68936-6.

UNEP (2012) *Financing Renewable Energy in Developing Countries – drivers and barriers for private finance in sub-Saharan Africa.* United Nations Environmental Programme, Geneva, 53 pp.

UNESCO (2010) Transboundary aquifers: challenges and new directions. *Abstracts, ISARM 2010 International Conference*, Paris, 6–8 December 2010. Unitod Nations Educational, Scientific and Cultural Organization, Paris, 188 pp.

United Nations (1997) *Convention on the Law of Non-Navigational Uses of International Watercourses.* G.A. Res. 51/229, UN GAOR, 51st Sess., UN Doc. A/RES/51/229. Available at: http://legal.un.org/ilc/texts/instruments/english/conventions/8_3_1997.pdf (accessed 28 August 2013).

Villar, P.C. and Ribeiro, W.C. (2012) The agreement on the Guaraní aquifer: A new paradigm for transboundary groundwater management? *Water International* 36(5), 646–660.

Villholth, K.G. (2013) Integrated groundwater use and management in vulnerable coastal zones of Asia-Pacific. In: Wetzelhütter, C. (ed.) *Groundwater in Coastal Zones of Asia-Pacific.* Springer, pp. 317–342. ISBN 978-94-007-5647-2.

Villholth, K.G., Jensen, K.H. and Fredericia, J. (1998) Flow and transport processes in a macroporous subsurface-drained glacial till soil. I: Field investigations. *Journal of Hydrology* 207, 98–120.

Villholth, K.G., Mukherji, A., Sharma, B.R. and Wang, J. (2009) The role of groundwater in agriculture, livelihoods, and rural poverty alleviation in the Indo-Gangetic and Yellow River basins: a review. In: Mukherji, A., Villholth, K.G., Sharma, B.R. and Wang, J. (eds) *Groundwater Governance in the Indo-Gangetic and Yellow River Basins: Realities and Challenges.* Taylor & Francis Group, Wallingford, UK, pp. 3–28.

Villholth, K.G., Tøttrup, C., Stendel, M. and Maherry, A. (2013) Integrated mapping of groundwater drought risk in the Southern African Development Community (SADC) region. *Hydrogeology Journal* 21(4), 863–885.

Voss, K.A., Famiglietti, J.S., Lo, M., de Linage, C., Rodell, M. and Swenson, S.C. (2013) Groundwater depletion in the Middle East from GRACE with implications for transboundary water management in the Tigris-Euphrates-Western Iran region. *Water Resources Research* 49, 904–914.

Wada, Y., van Beek, L.P.H. and Bierkens, M.F.P. (2012a) Nonsustainable groundwater sustaining irrigation: a global assessment. *Water Resources Research* 48, W00L06.

Wada, Y., van Beek, L.P.H., Weiland, F.C.S., Chao, B.F., Wu, Y.-H. and Bierkens, M.F.P. (2012b) Past and future contribution of global groundwater depletion to sea-level rise. *Geophysical Research Letters* 39, L09402.

Wijnen, M., Augeard, B., Hiller, B., Ward, C. and Huntjens, P. (2012) Managing the invisible – understanding and improving groundwater governance. Draft Report. World Bank Water Papers, Washington, DC.

Woodhouse, P. and Ganho, A.S. (2011) Is water the hidden agenda of agricultural land acquisition in sub-Saharan Africa? *International Conference on Global Land Grabbing*, Sussex, UK, 6–8 April 2011.

World Bank (2009) Poverty and Social Impact Analysis of Groundwater Over-Exploitation in Mexico, 73 pp. Available at: http://siteresources.worldbank.org/INTPSIA/Resources/490023-1120841262639/Mexico_groundwater.pdf (accessed 19 November 2013).

Zhou, Y., Zwahlen, F., Wang, Y. and Li, Y. (2010) Impact of climate change on irrigation requirements in terms of groundwater resources. *Hydrogeology Journal* 18, 1571–1582.

11 Irrigated Crop Production in the Syr Darya Basin: Climate Change Rehearsal in the 1990s

Oxana Savoskul[1]* and Elena Shevnina[2]

[1] *International Water Management Institute, Colombo, Sri Lanka;*
[2] *Finnish Meteorological Institute, Helsinki, Finland*

Abstract

Future inflow to the irrigation scheme of the Syr Darya Basin is modelled under two climate scenarios, based on outputs of International Panel on Climate Change (IPCC) core models run under IPCC-SRES A2 emission scenario. Under the GFDL99-R30-based scenario, the mean annual flow (MAF) is likely to increase by 10–20%. Under HadCM3-based scenario, MAF is supposed to decrease by 10–20%. Simulating water allocation in the basin in 2070–2099 shows that 14–21% of water demands in the agriculture sector in a normal hydrological year and 28–51% in a dry year are likely to be unmet. The challenges expected from future climate change can be paralleled to those resulting from the political change due to the collapse of the USSR, which left 18% (normal year) and 46% (dry year) of agricultural water demands unmet in 1992–2001. The study stresses the point that the adaptation measures employed in the post-Soviet transitional period are likely to serve as a basis for the future climate change adaptation strategies, since the development of the agriculture sector under climate change impact will remain handicapped without a more efficient water management at all hierarchical levels.

11.1 Introduction

The Syr Darya (Fig. 11.1), one of the two major river basins belonging to the Aral Sea drainage, is today home to a multi-ethnic population of over 20 million, 73% of whom constitute an impoverished rural population. Livelihoods of these people depend mainly on irrigated crop production. Cotton, one of the principal cash crops, has been produced in the basin since prehistoric times. The first irrigation infrastructures, according to archaeological findings, had been built in this harsh desert and semi-desert environment more than 3000 years ago. Political tensions aimed at gaining control of access to water in this arid land are possibly as old as the first irrigation infrastructures. However, the strains imposed on the basin environment and its inhabitants in the past 50 years have been unparalleled in history.

In the Soviet times, the extensive irrigation schemes and capacious water reservoirs had been constructed in order to intensify the agriculture sector. This measure boosted cotton production, but eventually caused an unprecedented over-exploitation of water resources in the Syr Darya and the adjacent Amu Darya Basin, which resulted in environmental collapse of the aquatic system of the Aral Sea (Raskin *et al.*, 1992). The political collapse of the Soviet Union in 1991 and the emergence of the new post-Soviet states in place of the formerly strongly centralized country created new problems, or rather revived the old ones, since sharing water resources among water users with

* Corresponding author, e-mail: xenia.savoskul@gmail.com

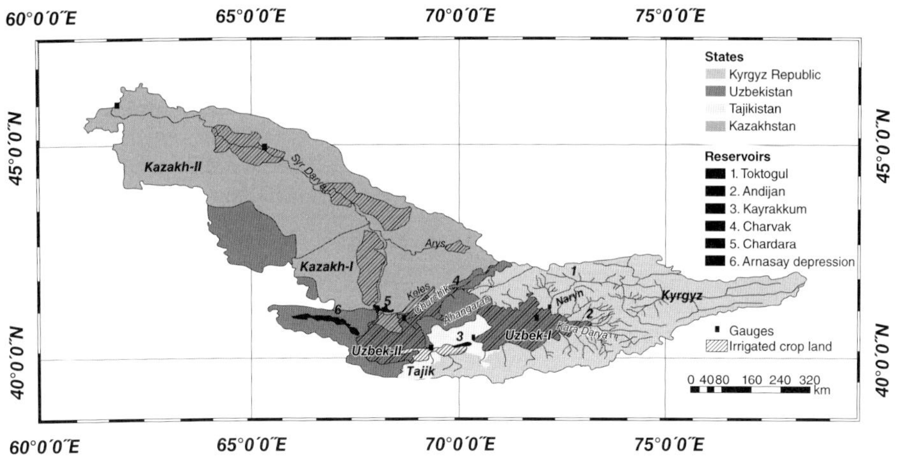

Fig. 11.1. Map of the Syr Darya Basin showing its political divisions, river network, major water reservoirs and location of the irrigated cropland.

competing interests has never been an easy task in the arid environment of Central Asia (Arsel and Spoor, 2009).

The new political and economic mechanisms for the regulation of transboundary water allocation in the basin established by the newly independent Central Asian states of Kyrgyzstan, Uzbekistan, Tajikistan and Kazakhstan in 1992 can hardly be termed very efficient since then (Ul Hassan *et al.*, 2004; Abdullaev *et al.*, 2009; Lerman, 2009); now yet another danger threatens the basin water security. Global climate change (CC) poses additional risks to the sustainable development of the region despite an increase in water availability projected by application of various hydrological models (Malsy *et al.*, 2012; Siegfried *et al.*, 2012; Sutton *et al.*, 2013). The study presented in this chapter examines the potential impacts of future CC in the Syr Darya Basin on the water availability for agriculture, based on modelling of water allocation in the basin under a set of two scenarios derived from General Circulation Model (GCM) outputs.

The key question is to what degree the agriculture sector in the basin is prepared to cope with the CC impacts. The objectives of the study have been: (i) to construct future climate scenarios, which would reflect the most plausible range of CC at the end of the 21st century; (ii) to simulate the changes in streamflow under future climate scenarios; (iii) to simulate water allocation under CC scenarios in order to see how much water will be available for agricultural water use in future; and (iv) to compare the simulated performance of a basin transboundary water allocation system in future with that of the last Soviet decade, 1982–1991, and the first post-Soviet decade, 1992–2001. This decade is of especial interest for this study since it was marked by an initial transitional decline in agricultural production in the Central Asian states, which was reversed, i.e. reached the level of 1991 only in 2003 (Lerman, 2009).

11.2 Constructing Scenarios of Future Climate

The climate scenarios based on the outcomes of GCMs were constructed following the standard protocol suggested by the IPCC guidelines (Parry *et al.*, 2007). The climatological variables used in the scenarios are the 30-year average monthly and annual air temperature and precipitation. The baseline period used as a reference line is 1961–1990.

The future time slice of interest is 2070–2099. GCM runs used here are driven by the SRES A2 emission scenario (Parry *et al.*, 2007).

At the first step, the performance of six core IPCC GCMs described in 2007 IPCC AR4 (IPCC, 2014) over the baseline reference period was examined in order to select the GCMs best suited for simulating the present-day climate in the study area (Table 11.1; Fig. 11.2). The criteria for the selection of the best-performing GCMs were the differences between observed and simulated air temperatures ($T_{mod} - T_{obs}$) and the ratio of simulated and observed precipitation (P_{mod}/P_{obs}). To assess baseline climatological variables in the study area, the re-analysis data set by the Climate Research Unit (CRU) with a resolution of 0.5×0.5 degrees has been used (http://www.cru.uea.ac.uk/data). At this stage, CRU data were up-scaled to match the GCM grid resolution. According to the data presented in Fig. 11.2 and Table 11.1, three GCMs have a good performance in the Syr Darya Basin. Out of the three good-performing GCMs, GFDL99-R30 and HadCM3 have been selected to represent a range of future CC in the study area. GCMs tend to have a poorer performance in the mountains compared to plains, particularly in simulating the precipitation regime. Therefore, an adequate performance of a GCM is of especial importance in this study, since the major part of the river flow in the Syr Darya area originates in the mountains.

The regional baseline climate model (Fig. 11.3) is based on the digital elevation map and data of long-term observations from 238 meteorological stations (De Pauw *et al.*, 2004). It has a resolution of 1×1 km and was designed for the project presented by Savoskul *et al.* (2004). At the final step of the CC scenario construction, the statistical downscaling was done using the change factor method (Parry *et al.*, 2007). The change field based on HadCM3 is presented in Fig. 11.4. Two sets of CC scenarios have been constructed by adding monthly change fields to the baseline regional climate model (Fig. 11.5; Table 11.2).

11.3 Simulation of the Inflow to the Reservoirs and Principal Gauges

With the construction of two new reservoirs in 2010, the total storage capacity of the water reservoirs in the basin increased to 105% of the Syr Darya MAF, of which active storage volume makes 87% of MAF. In the medium and low reaches of the basin, virtually all streamflow is regulated through an immense water storage and irrigation scheme. Because 85% of the flow is formed in the mountains that constitute roughly 20% of the catchment area, the simulation of inflow to the major reservoirs and irrigation schemes can be done using time series of natural flow measurements from the mountain gauges. A semi-distributed streamflow model has been applied for this purpose. A supplementary model block has been designed to account for the changes in contribution from seasonal snow cover and glaciers. Table 11.3 shows the data used for the baseline extent of glaciers and seasonal snow, glacier runoff and seasonal snowmelt contribution to streamflow as well as simulated future values. Snowmelt yields were modelled by the temperature-index approach for the 100-m elevation bands using the method proposed by Mukhin (1991). The glaciological approach described in detail by Savoskul and Smakhtin (2013) was used to model changes in specific glacier elevations, area, volume and glacier runoff. The major uncertainties of the streamflow simulations are inherent in the uncertainties of the CC scenarios, particularly in predicting future precipitation. In this respect, the use of two contrasting scenarios is a commonly recommended option for outlining the potential range of future changes (Parry *et al.*, 2007).

The outputs of the streamflow simulation under the selected CC scenarios are presented in Fig. 11.6. The changes of flow regime under warm and humid GFDL99-R30-based scenario are characterized by a pronounced shift of spring maximum to earlier dates and increase of annual water flow by 27%. Under this scenario, the amount of water in summer months does not show

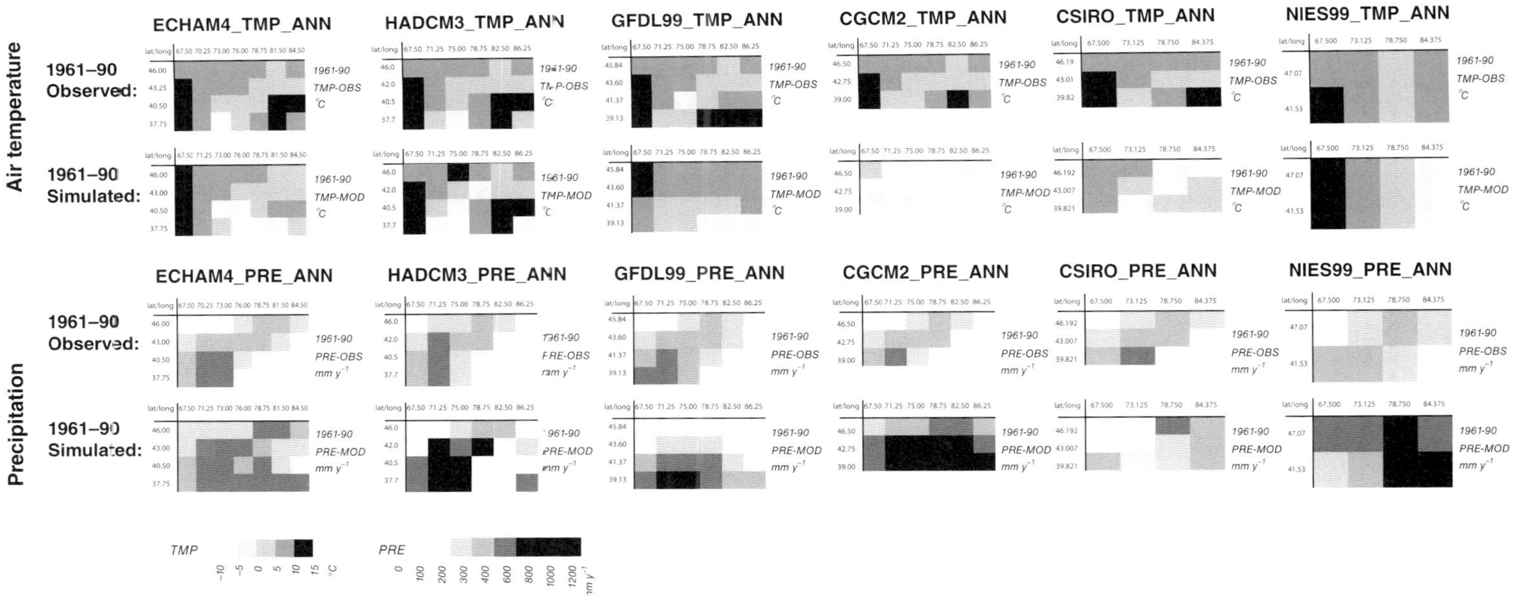

Fig. 11.2. GCM outputs for the baseline period (1961–1990) as compared with CRU observed climatology fields (upscaled to match GCM resolution). (Data sources: observed climatologies: CRU, available at http://www.cru.uea.ac.uk/data, accessed 7 March 2009; simulated climatologies: IPCC Data Distribution Center (DDC), available at http://www.ipcc-data.org/sim/gcm_clim, accessed 9 March 2009. The GCM abbreviations are adopted from the original data sources.)

Table 11.1. Summary of the performance of six core GCMs in simulating present-day climate in the study area. Observed climatologies: CRU, from http://www.cru.uea.ac.uk/data (accessed 7 March 2009); simulated climatologies: IPCC DDC, from http://www.ipcc-data.org/sim/gcm_clim (accessed 9 March 2009). The GCM abbreviations are adopted from the original data sources.

GCM	$T_{mod}-T_{obs}$ (°C)	P_{mod}/P_{obs}	Performance in the study region
ECHAM4	0.1	1.4	Good
HadCM3	–1.0	1.3	Good
GFDL99-R30	–1.3	1.0	Good
CGCM2	–9.8	3.4	Very poor
CSIRO	–4.4	1.0	Poor
NIES99	–1.4	2.8	Poor

T_{mod}, mean annual air temperature simulated (°C); T_{obs}, mean annual air temperature observed (°C); P_{mod}, mean annual precipitation simulated (mm); P_{obs}, mean annual precipitation observed (mm).

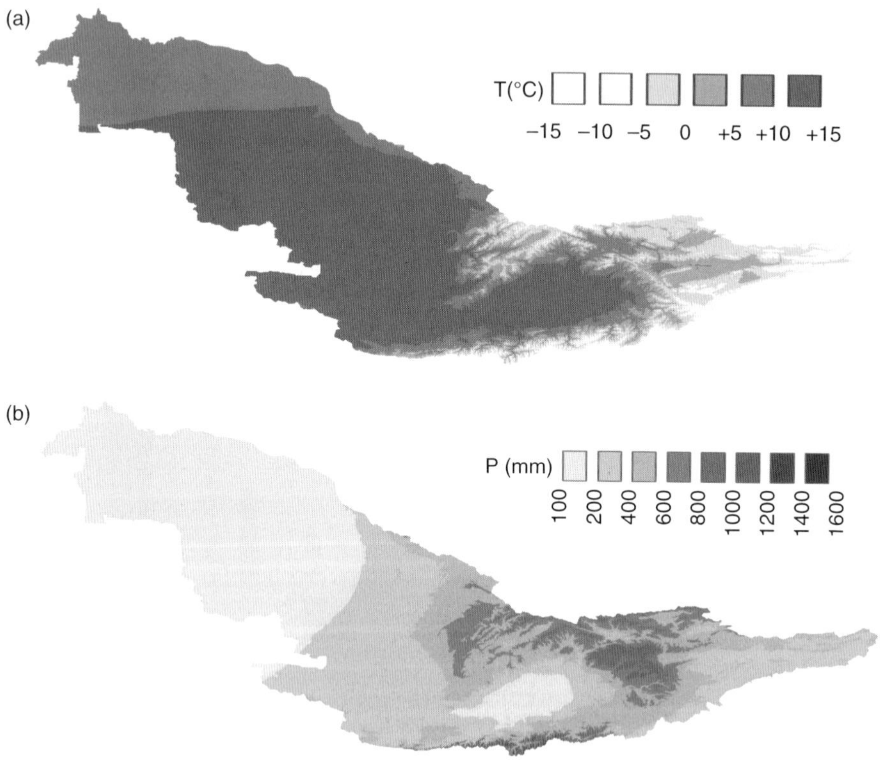

Fig. 11.3. Regional baseline climate model: (a) average annual air temperature in 1961–1990 and (b) average annual precipitation in 1961–1990.

significant changes. The significantly drier and hotter HadCM3-based scenario suggests a decline of annual flow by 18% and a shift of spring high-water season by even earlier dates than under the GFDL99-R30-based scenario. Shift of spring high waters will be due to the earlier onset of seasonal snowmelt, which will significantly reduce under both scenarios, but will still contribute around 10–20% of MAF to the river flow. However, a late summer peak in the discharge of rivers, which is a characteristic feature of baseline (1961–1990) hydrographs and is due to glacier runoff

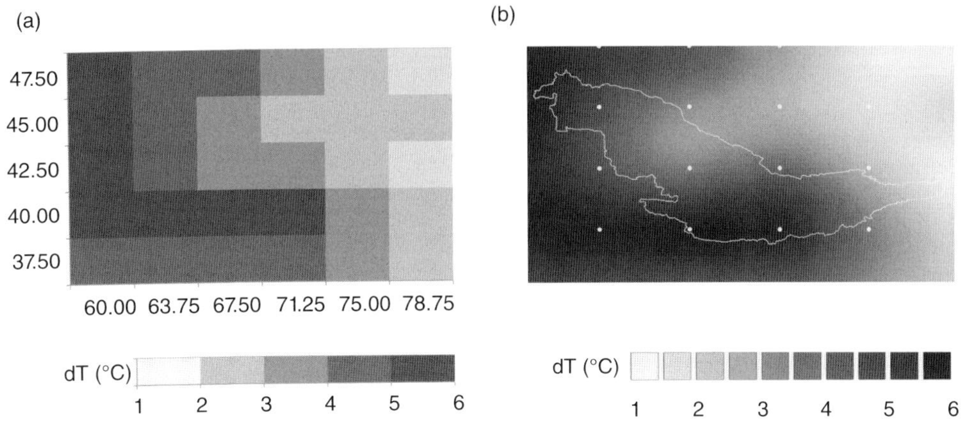

Fig. 11.4. (a) HadCM3 outputs for the air temperature change between baseline 1961–1990 and 2070–2099 periods in the study area (from GCM outputs, IPCC DDC, available at: http://www.ipcc-data.org/sim/gcm_clim/, accessed 9 March 2009). (b) Air temperature change-field based on the HadCM3 data.

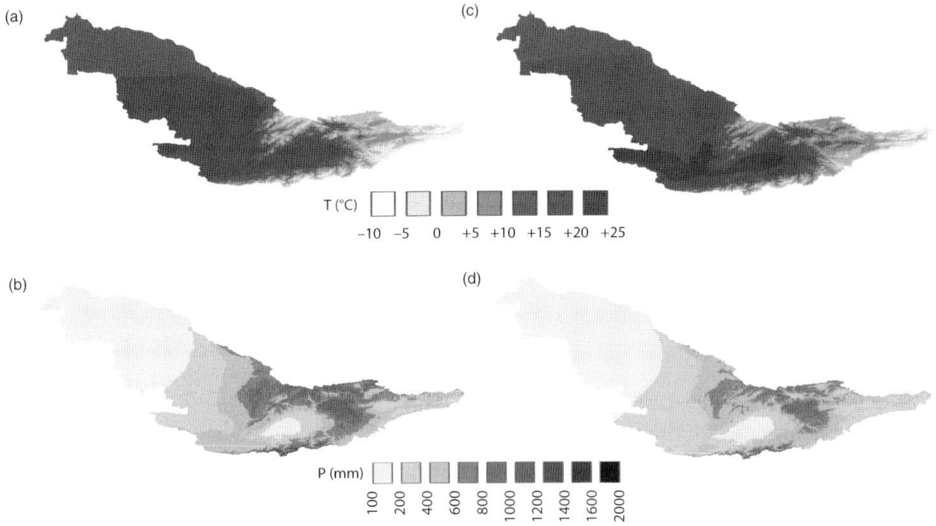

Fig. 11.5. CC scenarios for the period 2070–2099 used in this study: GFDL99-R30-based annual (a) air temperature and (b) precipitation; HadCM3-based (c) air temperature and (d) precipitation.

Table 11.2. Summary of the GCM-based scenarios used in this study. Climatologies: IPCC DDC, from http://www.ipcc-data.org/sim/gcm_clim (accessed 9 March 2009). The GCM abbreviations are adopted from the original data sources.

GCM	$T_{2070-2099}-T_{1961-1990}$ (°C)	$P_{2070-2099}/P_{1961-1990}$	Brief description
GFDL99-R30	3.7	1.34	Warm and humid
HadCM3	4.8	1.07	Hot and dry

$T_{2070-2099}$, future mean annual air temperature simulated (°C); $T_{1961-1990}$, baseline mean annual air temperature simulated (°C); $P_{2070-2099}$, future mean annual precipitation simulated (mm); $P_{1961-1990}$, baseline mean annual precipitation simulated (mm).

Table 11.3. Data on glacier and maximum seasonal extent of snow and contribution to the flow used in the streamflow simulation (from Savoskul and Smakhtin (2013) and authors' estimate).

	Area (km^2) covered by		Contribution to MAF (%)	
Control runs and CC scenarios	Glaciers	Seasonal snow at its maximum	Glacier runoff	Seasonal snowmelt
1961–1990	2,522	413,428	3.4	27.3
2000	1,967	349,358	3.2	19.7
2070–2099 GFDL99-R30	429	139,145	0.7	8.3
2070–2099 HadCM3	101	92,205	0.2	5.5

contribution at the peak of glacier ablation, under both scenarios of future climates, will be insignificant (around 1–2% of MAF under the GFDL99-R30-based scenario and below 1% under the HadCM3-based scenario).

11.4 Setting Water Allocation Model

Simulation of the water allocation was done using Water Evaluation and Planning (WEAP) model (http://www.seib.org/weap/; Savoskul *et al.*, 2004; Chevnina and Savoskul, 2006). WEAP is a basin-scale water allocation model that allows simulating the water budget of a river basin at a reach-by-reach basis. The hydrological linear scheme of the Syr Darya Basin in this application (Fig. 11.7) includes the Syr Darya River and its main tributaries: Naryn and Kara Darya (forming Syr Darya at their confluence), Chirchik, Ahangaran, Keles and Arys. The main types of the WEAP elements are: R, resource; DS, demand site; TL, transmission link; RF, return flow link; O, outflow; and WR, water requirement point. The water resources in the Syr Darya scheme are presented by the five largest reservoirs: Toktogul (total storage capacity 19.5 Bm^3), Andijan (1.9 Bm^3), Kayrakkum (3.5 Bm^3), Charvak (2.0 Bm^3) and Chardara (2.9 Bm^3), which existed in the year of model validation (2000), and two recently constructed water reservoirs, Kambarata (4.7 Bm^3) and Koksarai (3.1 Bm^3), were included in the scheme only for the model runs under future scenarios. The additional local supplies are smaller tributaries, whose summary inflow is introduced at some reaches, the groundwater of Tashkent and Fergana areas and return water flows from agriculture and industrial demand sites.

Among the demand sites, three types are distinguished: agricultural, domestic and industrial. Transmission links, return flow links, outflow and water requirement points are shown in the basin scheme (Fig. 11.7). In this application, the basin is subdivided into six reaches, representing the key political and economic units of the basin (Table 11.4).

The water resources of the basin consist of around 50 Bm^3 per year, of which 39 Bm^3 are river flow, 3 Bm^3 are groundwater and aquifers and 8 Bm^3 are return flow. Approximately 50% of the water resources are consumed, 44% is transmission and return flow losses in the irrigation network and only 6% is the outflow to the Aral Sea, Arnasay and diversions to the desert in the lower Syr Darya reaches. The basin budget in the baseline period (1961–1990) for the dry, normal and wet hydrological years is represented in Table 11.5, along with the budget for year 2000 used for model verification. For the calibration of the WEAP model, simulated baseline (1961–1990) discharge was compared to the observed time series at nine gauges. The model calibration results presented in Fig. 11.8 demonstrate that within its accuracy range of +5%, the model performs satisfactorily in simulating observed flows at the principal gauges of the basin.

The assessment of water demands, consumption, return flow and transmission losses is based on national statistics bulletins and some other sources (Spravochnik, 1981;

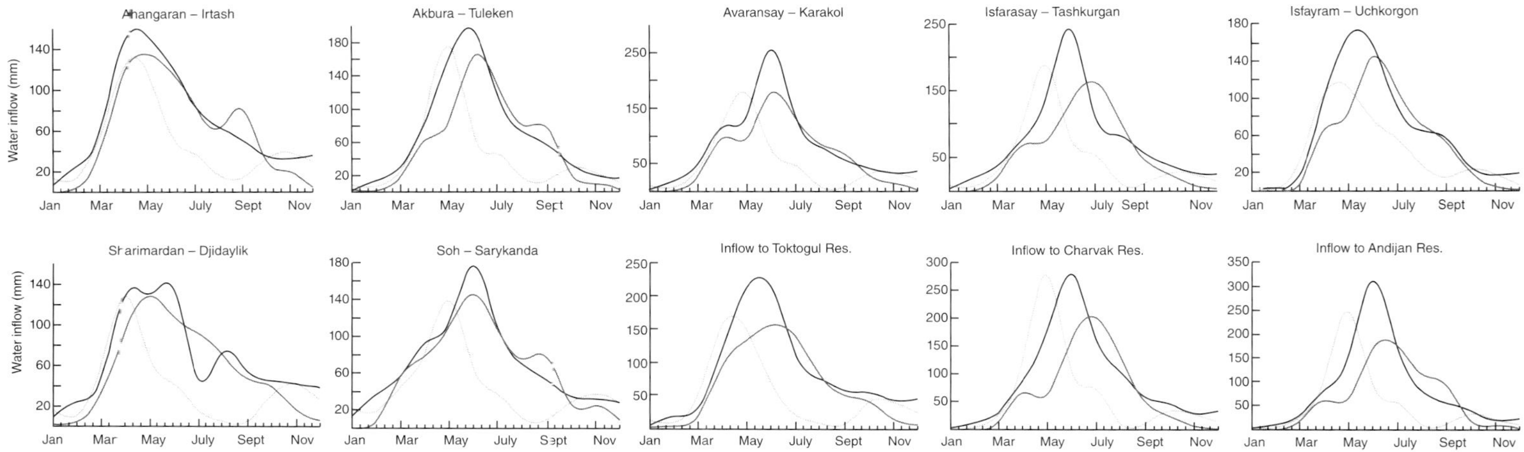

Fig. 11.6. Baseline and future simulated streamflow used to describe the changes of inflow to the reservoirs and irrigation scheme: medium grey line, baseline (1961–1990); dark grey line, scenario based on GFDL99-R30 outputs; light grey line, scenario based on HadCM3 outputs.

Table 11.4. Subdivision of the Syr Darya Basin for WEAP application.

Reach	Hydrological elements	Administrative units (oblasts)	Area, $10^3 \times Bm^2$	Boundary
Kyrgyz (Naryn)	Naryn River headflow, Toktogul and Kambarata reservoirs,[a] Kara Darya River headflow, Andijan Reservoir	Osh, Naryn, Djalal-Abad, Talas	104	Uchkurgan gauge at Naryn Andijan Reservoir water gate at Kara Darya
Uzbek-I (Fergana)	Lower reaches of Naryn and Kara Darya rivers, upper flow of Syr Darya River, tributaries from Fergana Range	Namangan, Fergana, Andijan	18	Akjar gauge at Syr Darya
Tajik (Sogd)	Middle flow of Syr Darya, Kayrakkum Reservoir	Sogd	13	Kzyl Kishlak gauge at Syr Darya
Uzbek-II (CHAKIR)	Middle flow of Syr Darya Tributaries: Ahangaran, Chirchik; Keles Charvak Reservoir (at Chirchik)	Tashkent, Syr Darya, Djizak	39	Chardara Reservoir water gate at Syr Darya
Kazakh-I (ARTUR)	Chardara and Koksarai[a] reservoirs Lower flow of Syr Darya downstream from Chardara Reservoir Tributary: Arys	South Kazakhstan	75	Tumen' Aryk gauge at Syr Darya
Kazakh-II (Kzyl Orda)	Lowest flow of Syr Darya from KzylOrdaup to the delta area	Kzyl Orda	116	Karateren' gauge at Syr Darya

[a]These two reservoirs were constructed in 2010. They were introduced into the basin scheme only in the WEAP runs under future scenarios.

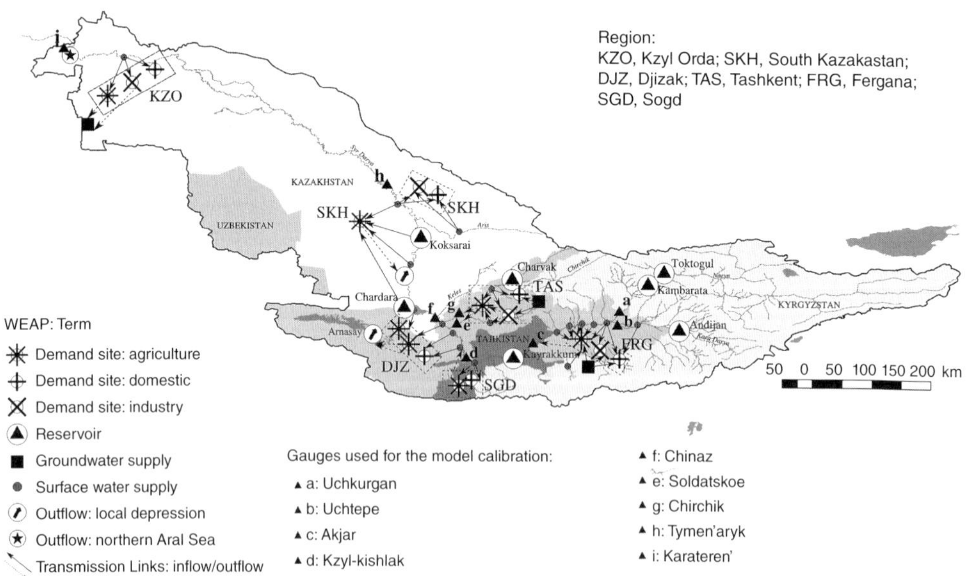

Fig. 11.7. The Syr Darya Basin water use scheme for WEAP application.

Minekonomstat, 2001a, b, c; Ministerstvo, 2001; Abdudjaparov and Toshmatova, 2002; GEF IFAS, 2002; Kipshakaev and Sokolov, 2002; Nurgisaev, 2002; Ryabtsev, 2002). The water demand in agriculture was estimated from crop areas and irrigation norms for the principal crops (Spravochnik, 1981; Minekonomstat, 2001a, b; Ministerstvo, 2001; GEF IFAS, 2002). Domestic water use demands were determined based on population at the demand sites and specific rates of water consumption per capita separately for urban

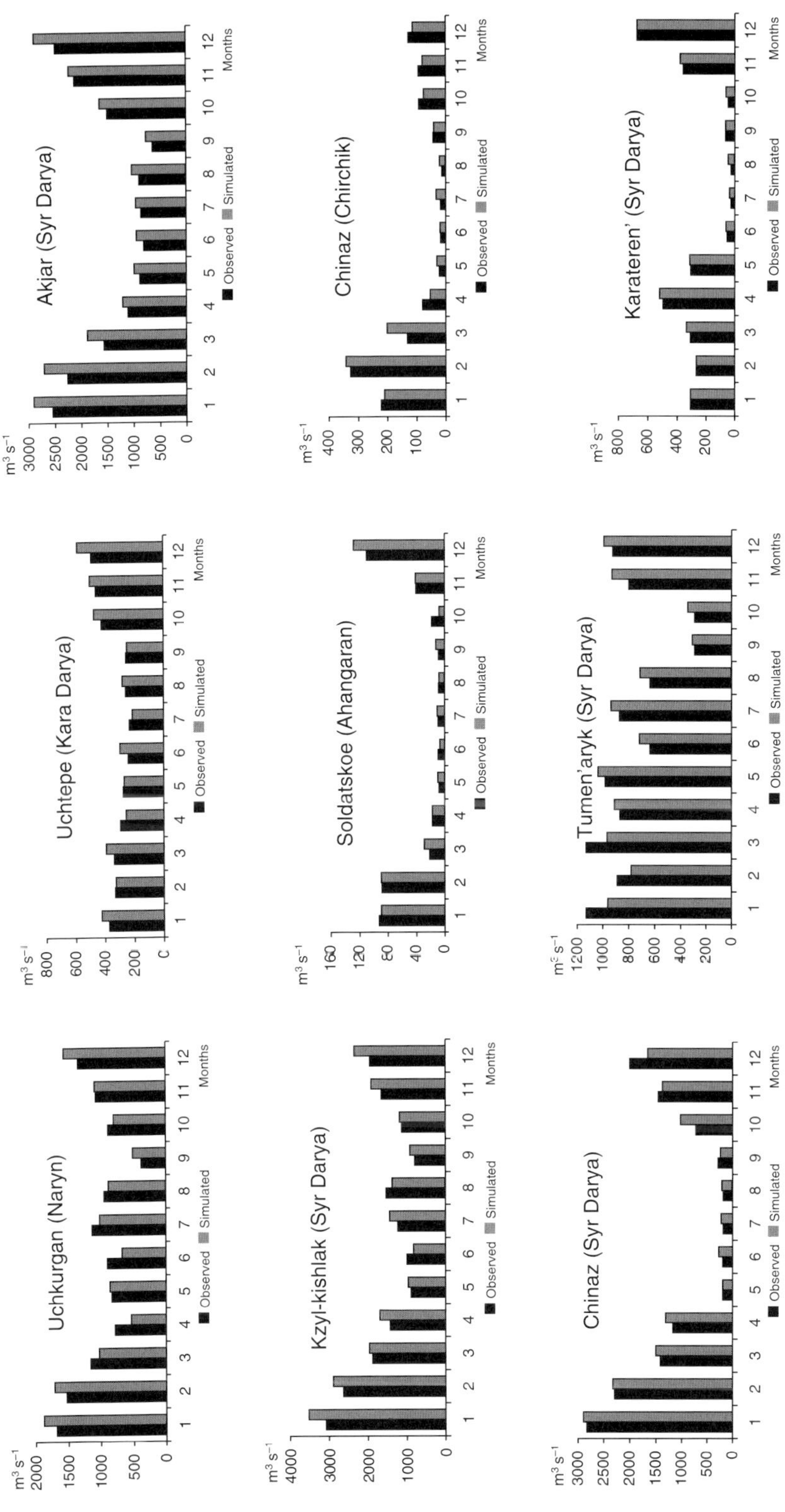

Fig. 11.8. Observed and simulated discharge at the gauges used for the calibration of the WEAP model.

Table 11.5. Basin water-use budget in dry, normal and wet years during the baseline (1961–1990) period, and in year 2000, used for WEAP verification (from Spravochnik, 1981; Minekonomstat, 2001a, b, c; Ministerstvo, 2001; Abdudjaparov and Toshmatova, 2002; GEF IFAS, 2002; Kipshakaev and Sokolov, 2002; Nurgisaev, 2002; Ryabtsev, 2002; Bucknall *et al.*, 2003).

Balance items	Year			
($10^6 \times m^3$)	1975 (dry)	1985 (normal)	1979 (wet)	2000 (used for model verification)
Streamflow	29,750	41,991	53,943	42,076
Groundwater	3,115	3,115	3,115	3,115
Return flow	5,627	8,126	9,931	10,908
Inflow total	38,492	53,232	66,989	56,099
Consumption subtotal	18,257	26,116	31,851	28,751
Agriculture	14,708	22,488	28,151	24,976
Domestic	2,414	2,422	2,426	2,430
Industry	1,135	1,206	1,275	1,345
Losses subtotal	14,312	21,363	26,835	24,457
Transmission losses	11,465	17,109	21,525	19,627
Return flow losses	2,848	4,254	5,310	4,830
Outflow subtotal	7,156	7,344	10,227	7,568
to Northern Aral Lake	3,029	3,049	5,698	3,072
to Arnasay	2,839	2,839	2,839	2,735
Diversion to desert	651	772	882	901
to groundwater	637	684	808	861
Outflow total	39,726	54,823	68,914	60,775
Balance (Bm^3)	–1,234	–1,592	–1,924	–4,677
Balance (%)	–3	–3	–3	–8

and rural populations (Minekonomstat, 2001a, c; Abdudjaparov and Toshmatova, 2002). The estimate of industrial water demands was based on the value of production and annual water use rate (Minekonomstat, 2001). A detailed breakdown of water demands and consumption assessment for each WEAP unit is given in Chevnina and Savoskul (2006).

The principal water user in the basin is the agriculture sector (86% of total water consumption), which by far surpasses other water users: industry (5% of total consumption) and domestic users (9% of total consumption) (Table 11.5). The agriculture sector, apart from the direct consumption, uses water indirectly. It is almost solely responsible for the enormous water losses in the basin, equalling 44% of total water resources, which are due mainly to the transmission seepage in poorly maintained irrigation infrastructure. The water losses in the irrigation network are assessed here based on values provided by GEF IFAS (2002), which are more moderate than the estimate given by Raskin *et al.* (1992).

The highest priority to water use in the model setup runs is given to agriculture to reflect the economic and political reality of the Soviet era. However, since in the post-Soviet period the water use in the basin has switched to the power regime, the highest supply priority in the WEAP runs under future scenarios is given to industrial and domestic demand sites, and the lower one to the agriculture sector. The domestic sector in all WEAP runs has second priority.

WEAP model constraints are of two kinds, related either to the limitations of the model itself or to the data availability. Since the model is designed as a water allocation accounting tool, it is not suited for the evaluation of the changes in crop water requirements due to future rises in air temperature, which will lead to potential increases in evapotranspiration. A glimpse of the potential effects of these factors can be obtained only from application of physically based models (Droogers *et al.*, 2004; Malsy *et al.*, 2012). The study by Sutton *et al.* (2013) indicates that a 10% decline in yields is likely to occur without implementation of adaptation measures. The changes in crop yields of a similar range due to increase in evapotranspiration are projected in Savoskul *et al.* (2004). Malsy *et al.* (2012) suggest that evapotranspirative requirements under CC may increase by 12%. Based on these figures, the constraints of WEAP in accounting for the physically based future increase in water requirements may be roughly estimated as being within a range of 10–15%.

Due to the limitations in data availability for the model calibration, the basin scheme does not include the minor tributary rivers in the middle and low flows of the Syr Darya; those in some instances (e.g. Fergana valley) have been accounted for as an aggregated entry into the unit. Likewise, accounting of the water demands in agriculture is reduced to the major crops (cotton, wheat, potato, fruit and vegetables).

11.5 Simulated Changes in Water Availability for Agriculture in 2070–2099

WEAP simulations of water allocation in the future, under different CC scenarios, are used to quantify the unmet demands of agriculture in order to evaluate the potential range of water deficiency in the future for the three types of hydrological years; dry, normal and wet. Under the GFDL99-R30-based scenario, by 2070–2099 the unmet demands of agriculture are expected to be 28% in a dry year, 14% in a normal year and 2% in a wet year. For comparison, at present, i.e. in 1992–2012, 46% (dry year), 18% (normal year) and 3% (wet year) of agricultural demands are not met. WEAP run under the HadCM3-based scenario suggests a slight increase of unmet demands relative to the present, i.e. 51% (dry year), 22% (normal year) and 5% (wet year) (Fig. 11.9).

Modelling CC impact on future water availability for agriculture is constrained mainly by the uncertainties in future climate projections (Parry *et al.*, 2007). However, these uncertainties with high probability will fall into the range outlined by the two scenarios considered here. Under a more favourable GFDL99-R30-based climate scenario, more water will be available for agriculture in 2070–2099. A second climate scenario based on the HadCM3 model suggests a slight decrease in water availability. However, both scenarios suggest that under business-as-usual water allocation policies and practices, from 14% to 22% of agricultural water demands will remain unmet in a normal year. In a dry year, these values are likely to be between 28% and 51%. A question is 'to what degree the agriculture sector is prepared to cope with the projected changes?' To answer this question, a closer look at recent developments in the water sector might be helpful.

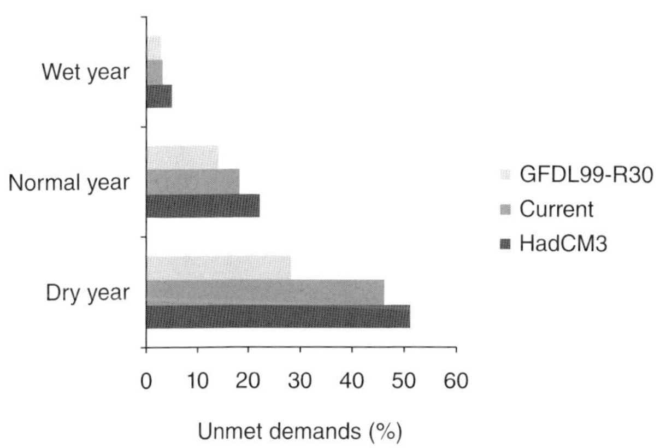

Fig. 11.9. Unmet demands of agriculture at present (1992–2001) and in future (2070–2099) under GFDL99-R30 and HadCM3-based scenarios.

11.6 Climate Change Rehearsal in the 1990s

The total water storage capacity of the water infrastructure in the basin is equal to 105% of MAF. In principle, high intra- and inter-annual regulation capacity of currently existing water reservoirs in the basin allows for the effective optimization of water use in favour of the agriculture sector. Under the centralized Soviet government, the first priority among the water users was given to agriculture. However, after the disintegration of the USSR, the pattern of water allocation in the basin has changed drastically due to the emergence of a conflict of interests among the newly independent states. The focal point of the conflict is Toktogul, the largest water reservoir in the basin, located in Kyrgyzstan, with a total storage capacity of 19.5 Bm^3 of which 14.5 Bm^3 are an active storage. The outflow from Toktogul Reservoir supports a cascade of hydropower plants, which became a vital source of cheap electricity for the Kyrgyz Republic. Under Soviet government, in 1982–1991, 60% of water was released from Toktogul Reservoir during the irrigation season, i.e. between April and August. Starting from 1993, due to high demands of Kyrgyzstan in hydropower in the cold part of the year, from November to March, the outflow from the reservoir in winter months has almost doubled compared to the outflow in the Soviet time, and only 38% of annual flow was released during the irrigation season (Fig. 11.10). Because of high winter water releases the inter-annual water regulation capacity of the Toktogul Reservoir has decreased too.

In the new political situation, every year, starting from 1993, the agriculture sector in downstream countries of Uzbekistan, Tajikistan and Kazakhstan has faced the challenges comparable to what might be expected under the less favourable of the CC scenarios considered here. The water release from Toktogul during the irrigation season in 1992–2003 had declined, on average, by 3 Bm^3 against the release in 1982–1991. In 1992–2001, water demands of the agriculture sector were met only in the wet years, whereas in the dry and normal years, considerable parts of the demands, 46% and 20%, respectively, were left unmet (Fig. 11.11). A closer look at the response of the water sector to this challenge provides an insight into the pathways the CC adaptation strategies are likely to follow in the future.

The new political situation has called for an immediate reorganization of water management in the basin. At the level of the interstate relations, starting from 1992 the transboundary water allocation is regulated via barter arrangements. Downstream Uzbekistan and Kazakhstan provide upstream Kyrgyz Republic with fossil fuel in exchange for guaranteed water releases in summer and restricted water releases in winter. This measure alone, however, could not solve the problem of water deficiency in agriculture. The annual multilateral agreements are not observed fully by all partners. In 2007, the countries entirely failed to sign the multilateral agreement (Abdullaev *et al.*,

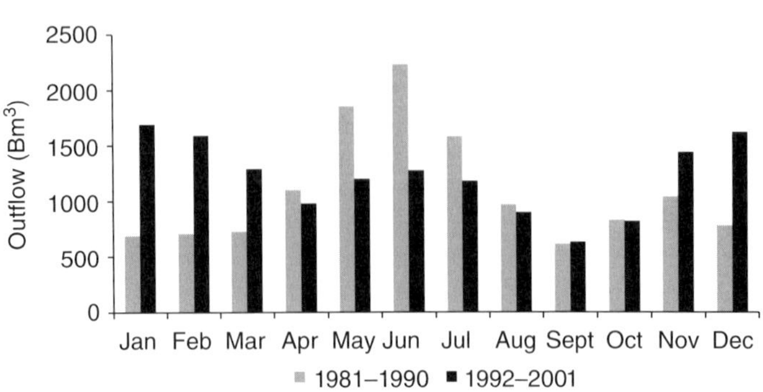

Fig. 11.10. Water releases from the Toktogul Reservoir under centralized Soviet government (1982–1991) and in the first post-Soviet decade (1992–2001).

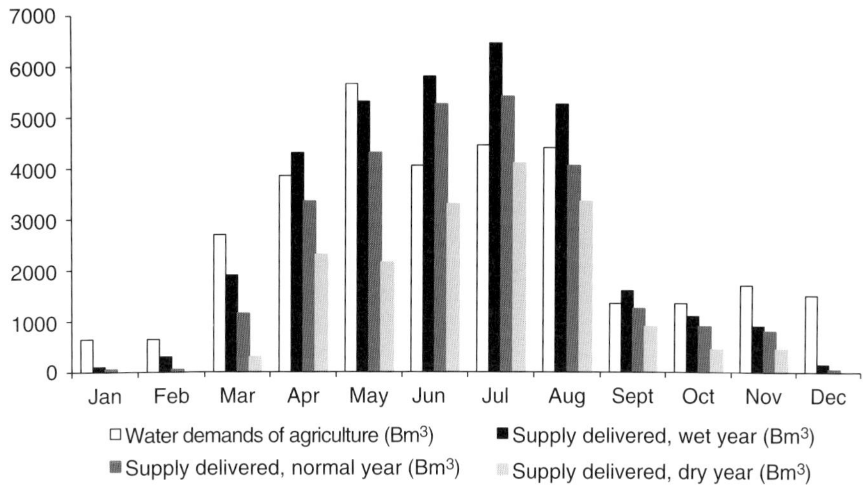

Fig. 11.11. Water demands of the agriculture sector in 1992–2001 and supplies delivered in the dry, normal and wet years. WEAP simulation.

2009). On average, the fossil fuel is delivered with delays, and in response the Kyrgyz Republic releases 2.4 Bm^3 more water in winter than agreed, depleting water storage for summer months (Antipova *et al.*, 2002). There are other challenges the governments had to face in the new economic situation. After disintegration of the Soviet Union, due to drastic reduction of government budgets in the transitional period, maintenance of the water infrastructure in many places came to a standstill, resulting in endangered dam security, silting canals and reservoirs, with gates, barrages and pumping station partly damaged, missing measuring equipment, etc. Other problems in agricultural water management are related to deterioration of water quality and land degradation (Abdullaev *et al.*, 2009).

A number of measures aimed at a more efficient water management had been applied in the first decade of independence (Dukhovny and Sokolov, 2005). A breakthrough initiative was the establishment of the Interstate Commission for Water Coordination in 1992, which is currently the highest water decision-making body in the region. Governments showed interest in the application of high-cost investment measures, such as construction of new storage and hydropower-generation facilities, rehabilitation and better maintenance of the water-allocation infrastructure, which depended mainly on the support from the international donor community (Savoskul *et al.*, 2004; Rakhmatullaev *et al.*, 2010).

At district level, some success was achieved by reorganization of the local water management on an integrated basis, establishment of water user associations (WUAs), installing monitoring facilities, and improving control over water quality, flows and diversions (Sokolov, 2006; Karimov *et al.*, 2012). Responses of the small-scale water users in the agriculture sector to regular water shortages in the 1990s followed three principal pathways: (i) cultivation of cash crops less dependent on irrigation; (ii) application of more efficient water use practices aimed at increasing water productivity; and (iii) reduction of water losses in the irrigation network (Savoskul *et al.*, 2004). All these efforts enhanced the understanding of the importance of improved water management as a principal means to cope with water shortage and related problems (Antipova *et al.*, 2002; Heaven *et al.*, 2002) but, in general, adaptation measures and reforms in the water sector are quite insufficient to solve the current problems of water shortages in the basin (Bucknall *et al.*, 2003; Ul Hassan *et al.*, 2004; Arsel and Spoor, 2009; Sutton *et al.*, 2013).

Application of the models employed in this study is constrained by a range of uncertainties inherent in the business-as-usual type of future scenarios, which do not account for changes in water requirements due to increased crop demands, population growth, industrial development, application of adaptation measures, changes in political environment, etc. Some of these unknown factors will definitely work to increase the capacity of the basin to cope with CC impact. For instance, recent construction of the Koksarai Reservoir significantly increased the capacity of Kazakhstan to accumulate water in the winter to alleviate flooding in the lower reaches of the Syr Darya River and to reduce water shortages for downstream agriculture during the irrigation season. Other factors, like increased evapotranspiration, soil evaporation, increased pressure from future population growth, etc., will impose extra challenges for the transboundary water management (Siegfried *et al.*, 2012; Sutton *et al.*, 2013). However, the use of the models employed in this study is justified by their capacity to help us answer the question asked at the beginning of this chapter, i.e. to what degree is the agriculture sector in the Syr Darya Basin prepared for the challenges imposed by future CC impacts?

Looking in this light at simulated water allocation in 2070–2099 under two different climate scenarios invites some surprising conclusions. First, the CC impacts on the water availability for agriculture in the future will be comparable with the impacts of recent political change in the region due to the disintegration of the USSR. Second, under one of the considered CC scenarios, water deficiency in the basin is likely to be somewhat reduced. Under another scenario, water deficiency in the basin is not likely to significantly exceed that of 1992–2001. Third, the agricultural water users in the basin have already tested some adaptation measures, which might serve as a basis for the development of future CC adaptation strategies.

The study presented here stresses the point that the challenges imposed by CC on the agriculture sector in the Syr Darya Basin will be in many respects comparable to the challenges it faced in the first post-Soviet decade, and their solution will call for a more efficient water management at all hierarchical levels.

Acknowledgements

The authors express their sincere gratitude to Dr Felix Pertziger (URS New Zealand, New Zealand) for letting us use the regional climate model for the Aral Sea Region. Constructive comments of Dr Akmal Karimov (IWMI Central Asian Office, Uzbekistan) and Dr Mariya Glazirina (ICARDA, Uzbekistan) are gratefully acknowledged. The authors would also like to thank Ilhom Babaev (IWMI Central Asian Office, Uzbekistan), Dr Ludmila Vasilina (Swiss Mission for Aral Sea, Uzbekistan Office), Dr Oleg Znay (NRMP, Uzbekistan), Dr Anatoliy Krutov and Dr Dilshod Khidirov (World Bank, Uzbekistan), Dr Lidia Grom (SARNIGMI, Uzbekistan), Dr Norkobil Rahmatov (BVO Syr Darya, Uzbekistan), Dr Alexander Tuchin and Dr Ekaterina Beloglazova (SIC ICWC, Uzbekistan), Dr Bahtiyar Matyakubov and Dr Rafik Murakaev (Geological Survey Service, Uzbekistan), Dr Sirojitdin Aslov (EC IFAS, Tajikistan), Dr Tulegen Sarsembekov (NGO Water Resources, Kazakhstan), Dr Alexander Danshin (Moscow State University, Russia) and Dr Iskandar Abdullaev (ZEF, Germany) for logistical help in data search and acquisition.

References

Abdudjaparov, T.A. and Toshmatova, H.Y. (2002) Osnovnye problemy obespecheniya naseleniya Tadjikistana pit'evoy vodoy i puti ee resheniya. [The basic problems of Tadjikistan domestic water supply and ways for decision.] In: Dukhovny, V.A. (ed.) *Water Resources of the Central Asia.* ICWC, Tashkent, Uzbekistan, pp. 228–233 (in Russian).

Abdullaev, I., Manthritilake, H. and Kazbekov, J. (2009) Water and geopolitics in Central Asia. In: Arsel, M. and Spoor, M. (eds) *Water, Environmental Security and Sustainable Rural Development.* Routledge, London, pp. 125–135.

Antipova, E., Zyryanov, A., McKinney, D.C. and Savitsky, A. (2002) Optimization of Syr Darya water and energy uses. *Water International* 27(4), 504–516.

Arsel, M. and Spoor, M.E. (eds) (2009) *Water, Environmental Security and Sustainable Rural Development: Conflict and Cooperation in Central Eurasia.* Routledge, London.

Bucknall, J., Klytchnikova, I., Lampietti, J., Lundell, M., Scatasta, M. *et al.* (2003) Irrigation in Central Asia: Social, economic and environmental considerations. *Technical Report.* The World Bank, Washington, DC.

Chevnina, E.V. and Savoskul, O.S. (2006) Simulating water allocation pattern in the Syr Darya Basin using WEAP model. *Technical Report.* IWMI, Tashkent, Uzbekistan.

De Pauw, E., Pertziger, F. and Lebed, L. (2004) Agroclimatic characterization for water management in Central Asia. In: Ryan, J., Vlek, P. and Paroda, R. (eds) *Agriculture in Central Asia: Research for development.* International Center for Agricultural Research in the Dry Areas, Aleppo, Syria, pp. 21–43.

Droogers, P., Dam van, J., Hoogeveen, J. and Loeve, R. (2004) Adaptation strategies to climate change to sustain food security. In: Aerts, J.C.J.H. and Droogers, P. (eds) *Climate Change in Contrasting River Basins.* CAB International, Wallingford, UK, pp. 49–71.

Dukhovny, V.A. and Sokolov, V.I. (2005) *Integrated Water Resources Management: Experience and lessons learned from Central Asia towards the Fourth World Water Forum.* Scientific Information Center of Interstate Commission for Water Coordination, Tashkent, Uzbekistan.

GEF IFAS (Global Environmental Facility at International Fund for Aral Sea) (2002) The program of the Aral Sea region: water resource and environment management. Subcomponent A1: Water resource and soil management on regional and national levels. *Technical Report No. 2 (Phase 3 – Regional Needs and Restrictions.).* GEF IFAS, Tashkent, Uzbekistan (in Russian).

Heaven, S., Koloskov, G.B., Lock, A.C. and Tanton, T.W. (2002) Water resources management in the Aral Basin: A river basin management model for the Syr Darya. *Irrigation and Drainage* 51(2), 109–118.

IPPC (2014) Model output described in the 2007 IPCC Fourth Assessment Report (SRES scenarios), multi-year means. Available at: http://www.ipcc-data.org/sim/gcm_clim/SRES_AR4/index.html (accessed 10 July 2015).

Karimov, A., Molden, D., Khamzina, T., Platonov, A. and Ivanov, Y. (2012) A water accounting procedure to determine the water savings potential of the Fergana Valley. *Agricultural Water Management* 108(C), 61–72.

Kipshakaev, I.K. and Sokolov, V.I. (2002) Vodnye resursy basseyna Aralskogo morya – formirovanie, raspredelenie, ispolzovanie. [Water resources of Aral Sea basin – formation, distribution, water use.] In: Dukhovny, V.A. (ed.) *Water Resources of the Central Asia.* ICWC, Tashkent, Uzbekistan, pp. 47–55 (in Russian).

Lerman, Z. (2009) Agricultural development in Uzbekistan: the effect of ongoing reforms. Discussion Paper No. 7.08. The Hebrew University of Jerusalem, The Center for Agricultural Economic Research, Jerusalem, 12 pp.

Malsy, M., BeekAus der, T., Eisner, S. and Flörke, M. (2012) Climate change impacts on Central Asian water resources. *Advances in Geoscience* 32, 77–83.

Minekonomstat Respubliki Uzbekistan (2001a) Itogi okonchatelnogo ucheta posevnyh ploschadey selskohozyaystvennyh kul'tur urogaya 2000 goda (na oroshaemyh zemlyah). [The results of the final accounting of the agricultural areas and harvest of 2000 (for irrigated areas).] Statisticheskiy bulleten Gosudarstvennogo Departamenta Statistiki Minmakroekonomstata [Statistical report of the State Department of Statistics]. Ministry of Economics, State Department of Statistics, Tashkent, Uzbekistan, 134 pp. (in Russian).

Minekonomstat Respubliki Uzbekistan (2001b) Informacionny bulleten ucheta expluatacionnyh zapasov, otbora i urovnya zagryazneniya podzemnyh vod Respubliki Uzbekistan za 2000 god. [Information bulletin: Accounting of the operation capacity, consumption and pollution of the Uzbekistan groundwater in 2000.] Issue 27. Ministry of Economics, State Department of Statistics, Tashkent, Uzbekistan, 32 pp. (in Russian).

Minekonomstat Respubliki Uzbekistan (2001c) Uzbekistan v tsifrah. Statisticheskiy sbornik. [Uzbekistan in numbers. Statistical report.] Ministry of Economics, State Department of Statistics, Tashkent, Uzbekistan, 245 pp. (in Russian).

Ministerstvo Selskogo Xoziaystva Respubliki Uzbekistan (2001) Kratkie rekomendacii po intensivnoy technologii vozdelyvaniyao zimoy pshenitcy v usloviyah orosheniya. [Brief recommendations for intensive technology of the grain production under irrigation.] Ministry of Agriculture, Tashkent, Uzbekistan, 74 pp. (in Russian).

Mukhin, V.M. (1991) Razrabotka metoda dolgosrochnih posledovatelnih prikhodov vodi v Kambaratinskoe vodoxranilische. [Development of a method of long-term forecast of inflow to Kamabarata reservoir.]. Hydrometeorological

Research Institute, Moscow, Russia, 121 pp. (in Russian).

Nurgisaev, S.U. (2002) Socialno-ekonomicheskatya situaciya i vodohozyaystvennay aobstanovka v nizovyah Syr Darya. [Social-economic situation and water demands in downstream Syr Darya.] In: Dukhovny, V.A. (ed.) *Water Resources of the Central Asia*. ICWC, Tashkent, Uzbekistan, pp. 69–72 (in Russian).

Parry, M.L., Canziani, O.F., Palutikof, J.P., van der Linden, P.J. and Hanson, C.E. (eds) (2007) Climate change: impacts, adaptation and vulnerability. *Contribution of Working Group II to the Fourth Assessment Report of the Intergovernmental Panel on Climate Change*. Cambridge University Press, Cambridge, UK.

Rakhmatullaev, S., Huneau, F., Le Coustumer, P., Motelica-Heino, M. and Bakiev, M. (2010) Facts and perspectives of water reservoirs in Central Asia: A special focus on Uzbekistan. *Water* 2, 307–320.

Raskin, P., Hansen, E., Zhu, Z. and Stavisky, D. (1992) Simulation of water supply and demand in the Aral Sea Region. *Water International* 17, 55–67.

Ryabtsev, A.D. (2002) Vodnye resursy Kazahstana: Problemy i perspektivy ih ispolzovaniya. [Water resources of Kazakhstan: problems and perspectives of its uses.] In: Dukhovny, V.A. (ed.) *Water Resources of the Central Asia*. ICWC, Tashkent, Uzbekistan, pp. 7–14 (in Russian).

Savoskul, O.S. and Smakhtin, V. (2013) Glacier systems and seasonal snow cover in six major Asian river basins: water storage properties under changing climate. *IWMI Research Report 149*. International Water Management Institute, Colombo, Sri Lanka, 61 pp.

Savoskul, O.S., Chevnina, E.V., Perziger, F.I., Vasilina, L., Baburin, V.L. (2004) How much water will be available for irrigation in the future? The Syr Darya Basin. In: Aerts, J.C.J.H. and Droogers, P. (eds) *Climate Change in Contrasting River Basins*. CAB International, Wallingford, UK, pp. 93–113.

Siegfried, T., Bernauer, T., Guiennet, R., Sellars, S., Robertson, A.W., Mankin, J., Bauer-Gottwein, P. and Yakolev, A. (2012) Will climate change exacerbate water stress in Central Asia? *Climatic Change* 112 (3–4), 881–899.

Sokolov, V. (2006) Experiences with IWRM in the Central Asia and Caucasus regions. *Water International* 31(1), 59–70.

Spravochnik po hlopkovodstvu (1981) [The reference book of cotton production]. Tashkent, Uzbekistan, 157 pp. (in Russian).

Sutton, W.R., Srivastava, J.P., Neumann, J.E., Droogers, P. and Boehlert, B.B. (2013) *Reducing the Vulnerability of Uzbekistan's Agricultural Systems to Climate Change: impact assessment and adaptation options*. The World Bank, Washington, DC.

Ul Hassan, M., Starkloff, R. and Nizamedinkhodjaeva, N. (2004) Inadequacies in the water reforms in the Kyrgyz Republic: an institutional analysis. *IWMI Research Report 81*. International Water Management Institute, Colombo, Sri Lanka, 44 pp.

12 Mitigating Greenhouse Gas Emissions from Rice Production through Water-saving Techniques: Potential, Adoption and Empirical Evidence

Bjoern Ole Sander,* Reiner Wassmann and Joel D.L.C. Siopongco

Crop and Environmental Sciences Division, International Rice Research Institute (IRRI), Los Baños, Philippines

Abstract

Flooded rice fields are a large anthropogenic source of the greenhouse gas (GHG) methane (CH_4). Aeration of the paddy field can reduce methane emissions and at the same time save water. Different forms of water saving techniques (WST), e.g. alternate wetting and drying (AWD) and midseason drainage (MSD), have been developed and disseminated. This article gives an overview on adoption of AWD in the Philippines and assesses prospects and constraints. It also explains the Clean Development Mechanism (CDM) methodology for rice production and analyses the mitigation potential of WST in the form of a literature review.

The adoption rate of AWD strongly depends on the incentive for the farmer. While direct monetary incentives are limited to areas where saving water is directly linked to reduced costs (e.g. pump irrigation systems), indirect incentives (e.g. improved crop development) have not yet been scientifically assessed. The literature meta-analysis proves the great mitigation potential of WST. Methane emissions can be reduced by an average of 36.5% with a single drainage and by 43% with multiple aerations. Nitrous oxide emissions increase under all WST but this increase does not offset the reduction in CH_4 emissions. This study also shows that the amount of GHG emissions can vary drastically between different regions. This poses a challenge for the transfer of mitigation strategies from one region to another.

12.1 Water-saving Strategies

12.1.1 Principles of alternate wetting and drying and midseason drainage

Producing rice with less irrigation water requirements has been one of the core research objectives of natural resource management at IRRI and other research institutions. Midseason drainage is one strategy that has been widely adopted in China and Japan over the past decades. The principle is to expose the rice field to a dry period of about 7 days towards the end of the vegetative stage. Although water saving might be low under this strategy, grain yield tends to increase (Thompson, 2006) due to suppression of unproductive tillers, which translates to a higher water use efficiency.

While previous in-depth research focused on the identification of thresholds for reducing water use without compromising

* Corresponding author, e-mail: b.sander@irri.org

rice yield, this work has developed into a concise water-saving technology for rice farmers in irrigated lowlands called 'alternate wetting and drying' (AWD) starting in the early 2000s (Bouman *et al.*, 2007). The term AWD has been coined at IRRI and is synonymous with a variety of terms, such as controlled or intermitted irrigation as well as multiple aeration, that are used to describe alternatives to farmers' conventional practice of continuous flooding (CF). The intervals of non-flooded conditions from 1 day to more than 10 days depend on soil type and weather. In this technology, the farmers are taught to monitor the depth of the water table in the field using a perforated water tube that is inserted into the soil (Lampayan *et al.*, 2013). The practice which commences at 1 to 2 weeks after transplanting involves draining the field until the water level reaches 15 cm below the soil surface after which the field is re-flooded to a depth of around 5 cm. This irrigation scheme is done throughout the cropping season except during the flowering stage. The threshold of water at a 15 cm level below soil surface is called 'safe AWD', as this will not cause any yield decline because the roots of the rice plant will still be able to capture water from the saturated soils (Lampayan *et al.*, 2009). The AWD technology can reduce the number of irrigations significantly compared to farmer's practice, thereby lowering irrigation water consumption by 15–30%.

Adoption of alternate wetting and drying

Estimating the number of adopters of AWD is difficult. For the Philippines, the best estimation based on survey responses from national institutions is that around 100,000 farmers have adopted AWD (Lampayan, 2013). This number is based on the number of trainings and demonstration trials in different regions and the level of involvement of farmers and promotion of the technology. However, response may vary from those who practised AWD in the Philippines, e.g. in Canarem (Tarlac Province) the majority of the farmer-cooperators had positive feedback about the effectiveness of AWD as a water-saving technology as follows: (i) no yield difference with farmer's practice of continuous flooding; (ii) saves water; (iii) saves time and labour, thus, less expensive; (iv) heavier and bigger grains, and good shape; (v) more tillers; and (vi) less insect pests and diseases (Palis *et al.*, 2004).

Another example of AWD practised in the Philippines was in Bohol Island. In the face of declining rice production due to insufficient water supply and unequal water distribution, NIA (the National Irrigation Association of the Philippines) established the Bohol Integrated Irrigation System (BIIS) with: (i) the construction of a new dam (Bayongan Dam); and (ii) the implementation of AWD, which was imposed on the whole island by periodic water supply. The adoption of AWD facilitated an optimum use of irrigation water, so that the cropping intensity increased from ca. 119% to ca. 160% (related to the maximum of 200% in these double-cropping systems) (UNFAO, 2010).

The adoption of AWD strongly depends on the incentive for the farmer. In many parts of the Philippines, this incentive is directly linked to the irrigation system. In a pump system where farmers can achieve direct financial savings due to reduced diesel use for pumping under AWD, it is easily adopted and properly implemented. In irrigation systems where farmers pay seasonal fees independent of the actual water usage as currently employed in most of NIA-serviced areas, farmers were found to be reluctant to use water-saving techniques and AWD was not carried out properly.

With the development and improvement of irrigation canals by NIA as part of their nationwide medium-term plan, the use of pumps would become gradually less important. In turn, this – genuinely positive development – may decrease the incentive to adopt AWD as long as there are no policies from the local government units on water savings to support the practice of AWD by other means. As one example, adoption of meter-based (volumetric consumption-based) water rates instead of fixed area-based rates would promote practices of water saving. Volumetric pricing of irrigation water should induce incentive for better collective action toward saving water resources, than does area-based pricing in which marginal cost of using water is zero (Tsusaka *et al.*, 2012).

Potential and constraints

In the perception of farmers, AWD means inadequate soil-water during the dry period, thus carries the risk of drought stress to the crop. However, studies have shown that thoroughly implemented AWD, specifically 'safe AWD', does not lead to any yield declines because the roots of the rice plant can still capture enough water. Flooding of soils over many years triggers the development of a hardpan at 15–30 cm depth which acts as a mechanical barrier for roots and water. Although this sealing may not be complete in terms of percolation losses, it reduces seepage so that roots can acquire enough water even after several days without surface water. It is difficult to convince farmers that the absence of standing water does not automatically imply absence of soil water. Thus, the perforated tube serves a dual purpose: (i) measuring the water table below the soil surface; and (ii) acting as visual assurance to the farmer that the roots still have access to water at the subsurface. One requirement for successful dissemination of AWD, however, is a reliable irrigation source to enable farmers to irrigate whenever it is needed. If irrigation water is scarce sometimes and farmers cannot be sure to have sufficient water, they would prefer to irrigate soon and not wait for a recommended level of drainage to avoid possible drought stress.

Another barrier for adoption of any water-saving strategy by farmers within a wider irrigation system is the physical separation of adopter and benefiter. Farmers near the source of irrigation water ('upstream farmers') who have the potential of saving water have no need to save water. It is the farmers who are far from the irrigation source ('downstream farmers') who would benefit from water saving because those farmers potentially face water scarcity but, as a result, have not much potential to save water themselves.

On the positive side, there is anecdotal evidence through farmers' claims that practising AWD not only saves water but also increases rice yields. This observation may be the exception rather than the rule but it should be followed up for further improving the attractiveness of AWD. Several potential mechanisms have been reported as a means to increase yields under AWD but this needs further investigation:

- lodging resistant culms;
- profuse tillering;
- reduced pests and diseases; and
- better soil conditions at harvest.

Even if the practised AWD management is 'slightly unsafe', i.e. the water level drops below 15 cm below soil surface and yields slightly decrease, the economic yield tends to be higher in AWD (Sibayan *et al.*, 2010) because the cost of irrigation has decreased (in pump systems).

Water savings and greenhouse gas emission

Moreover, AWD technology has a proven potential to mitigate CH_4 emission. Methane is a potent GHG with a global warming potential (GWP) of 25 (IPCC, 2006), which means that it is 25 times more effective in trapping heat inside the Earth's atmosphere than CO_2. Cultivated wetland rice soils emit significant quantities of CH_4 (Smith *et al.*, 2008). Methane is produced anaerobically by methanogenic bacteria, which thrive well in paddy rice fields. Hence, flooded rice fields are a large source of CH_4 emissions contributing about 10–14% of total global anthropogenic CH_4 emissions. Because periodic aeration of the soil inhibits CH_4-producing bacteria, AWD can reduce CH_4 emissions. Various studies on GHG emissions under AWD and other water-saving strategies have been conducted to quantify the mitigation potential of those water management strategies. The results will be further discussed in Section 12.3.

The capability of AWD to reduce CH_4 emissions is also reflected in the IPCC methodology (IPCC, 2006) which is used for computing GHG emissions in the 'National Communications' submitted by countries to the UNFCCC. 'Multiple aeration', the category AWD falls in, is presumed to reduce CH_4 emissions by 48% compared to continuous flooding of rice fields (IPCC, 2006). A single aeration of the field, commonly referred to as 'midseason drainage', reduces CH_4 emissions by 40%, as IPCC guidelines suggest.

However, AWD adoption may also have pitfalls in terms of higher emissions of nitrous oxide (N_2O), a GHG even more potent than CH_4 with a GWP of 298 (IPCC, 2006). Nitrous oxide emissions are generally very low to negligible in continuously flooded systems, so that the IPCC guidelines assign a lower emission factor to rice as compared to non-flooded crops. Under water-saving strategies, N_2O emissions tend to increase due to increased nitrification and denitrification activities with the soil conditions constantly changing between anaerobic and aerobic and related changes in the redox potential. Data on N_2O emissions under different water management regimes is limited and varies drastically as discussed in Section 12.3. The available data, however, suggest that the incremental N_2O emission through AWD is insignificant as long as the N fertilization remains within a reasonable range. Thus, the combination of AWD with efficient fertilization techniques, such as Site-Specific Nutrient Management, is the best way to avoid excessive N levels in the soil and thus, negative trade-offs in terms of mitigation potentials.

12.2 Clean Development Mechanisms

12.2.1 Definition and criteria

The CDM is one of the flexibility mechanisms introduced by the Kyoto Protocol (KP) in 1997. It is a project-based mechanism of emissions trading involving non-Annex 1 parties (developing countries) that do not have any stipulated obligation to reduce their GHG emissions. The idea behind this cooperative mechanism is that reduced GHG emissions will slow global warming – irrespective of the location of the savings. Annex 1 (industrialized) countries can take advantage of a CDM project implemented in a developing country by purchasing Certified Emission Reduction Units (CERs) to meet their targets or emission caps. This mechanism adds more choices and flexibility to comply with the targets and offers economically sound solutions. The non-Annex 1 countries in turn receive capital for investments in projects and clean technologies to reduce their emissions and enhance socioeconomic well-being.

Thus, the CDM has two key goals: (i) to promote sustainable development (SD) objectives in the host country (i.e. non-Annex 1 countries); and (ii) to assist Annex 1 parties to meet their GHG reduction targets. A CDM project activity in a non-Annex 1 country produces certified emission reductions that can be used towards partial compliance of their emission reduction targets.

According to Section 12.5 of the KP, a CDM project has to satisfy the following criteria: (i) parties involved in the project activity do so voluntarily and both approve the project; (ii) the project must produce real, measurable and long-term benefits to the mitigation of climate change; and (iii) the emission reductions should be additional to any that would occur without the project activity (commonly known as the 'additionality' criterion).

Moreover, article 12.2 of the KP states that the purpose of the CDM is to assist non-Annex 1 parties in achieving SD. This is interpreted to suggest that the project activities should be compatible with the SD requirements of the host country. However, neither the KP nor the subsequent Conference of Parties (COPs) have provided guidance on defining sustainability, leaving the decision to the host countries. COP 7 in Marrakech in 2001 stipulated that all participating countries have to establish a 'Designated National Authority' (DNA) to assess if any CDM proposal complies with their own sustainability criteria (Bhattacharyya, 2011). Figure 12.1 gives an overview over the application and approval process of a CDM project.

However, applying CDM projects in rice production faces many challenges. The Bohol case (see 'Adoption of alternate wetting and drying' above) is an example of water

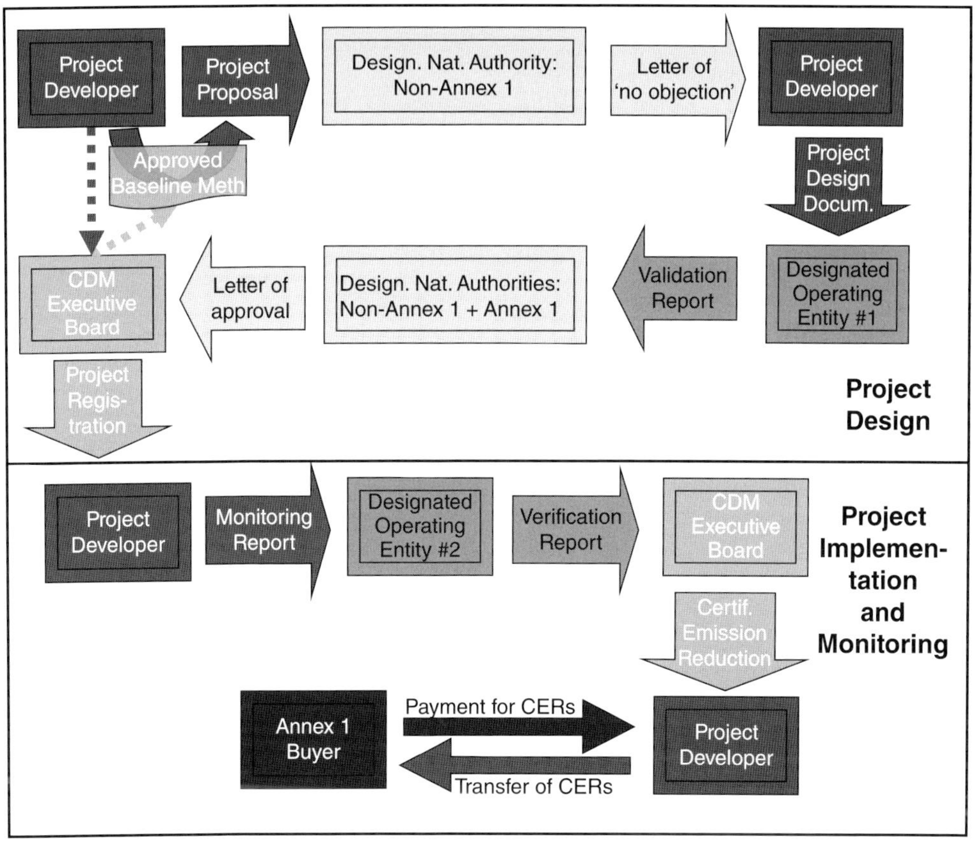

Fig. 12.1. Schematic presentation of the CDM Pipeline (Meth, methodology; Design. Nat. Authority, Designated National Authority; Docum., document).

savings possessing new technologies that increase the income of poor farmers while decreasing GHG emissions. Yet, it is not eligible for CDM because of missing additionality, i.e. AWD was introduced for the purpose of water saving without the incentive of CER generation and would have been introduced even if no GHG saving would have been achieved.

12.2.2 The rice clean development mechanism methodology

The eligibility of projects reducing *in situ* emissions from land use such as CH_4 emissions from rice remains intricate (Wassmann, 2010). However, in 2011 a CDM methodology for 'Methane emission reduction by adjusted water management practice in rice cultivation' was approved by the UNFCCC (2012). The methodology has been modified and is in its third version since August 2012. It now defines default CH_4 emission reduction values for different management practices in rice production. For applying AWD, for example, a reduction of 1.8 kg CH_4 ha^{-1} day^{-1} can be claimed under a certified CDM project. This translates to a saving of 4.5 t CO_2-eq ha^{-1} $season^{-1}$ assuming a 100-day growing period (GWP (CH_4) = 25) or 4.5 CERs.

12.3 Literature Review

For this study we have surveyed peer-reviewed articles on CH_4 and N_2O emissions under different water management techniques in rice fields. The objective was a proof of concept as to what extent water management can be used to mitigate GHG emissions from rice fields. Using an online search engine for scientific literature, ISI Web of Knowledge, we identified 24 articles on field measurements encompassing GHG emission changes as a function of water management of a rice field. The initial number of results of the search was much higher, but many articles on this topic reported mechanistic studies without comparative emission rates under different water management strategies. These 24 articles compiled a total number of 96 experimental comparisons, i.e. one comparison corresponds to one season with adjacent field plots of CF and WST, which can be either multiple aeration (MA) or single aeration (SA). These two WST include AWD and mid-season drainage, respectively, as their most common forms. Moreover, we also included three articles on pot experiments that emulated different water management practices; these articles encompassed four comparisons between CF and WST. For comparing relative emission differences between CF and WST, the pot experiments were included in the analysis. For a comparison of absolute emission differences, however, pot experiments were excluded because of the different environmental effects of 'field' and 'greenhouse'. To assess the effect of different kinds of WST, these 106 comparisons were further classified according to two types of WST: SA and MA.

12.3.1 Results

The emission rates obtained from the different publications are shown in Tables 12.1–12.4 separated by countries/regions (for field measurements) and in Table 12.5 for the pot experiments. In these tables – as well as in the narrative – percentages given are relative GHG emissions of an applied WST as compared to a continuously flooded (CF) field (e.g. a relative emission of 60% shown in these tables translates into a reduction effect of 40%). We recognize that many readers will primarily be interested in the reduction effect, but we felt that the consistent use of relative emission rates will provide a more comprehensive presentation. In some instances in the text, we have given absolute values for reduction in units of kilograms per hectare per day.

These tables list emission rates per day as well as per season. Typically, the articles provided only one of these values, but we computed the corresponding value by using the number of days for one season, which was also obtained from the article. In some articles, emission rates were given as hourly rates and we multiplied it with a factor of 24 for daily emissions (assuming that hourly values provide daily averages).

The articles on field comparisons were sorted according to the location of the experiments into five groups: China (Table 12.1), India (Table 12.2), Japan and South Korea (Table 12.3) and Indonesia and the Philippines (Table 12.4).

As an initial observation, the published studies from South-east Asia are older than 10 years, whereas many studies were conducted in India, China and Japan in more recent years. Emission rates from rice fields in India are much lower than from other parts of Asia, i.e. only 10% of the emission rates observed in field studies in China, Japan and South Korea. One exception is the study by Yue *et al.* (2005) that reports emissions from a CF field in China as low as 24.8 kg CH_4 ha^{-1} $season^{-1}$, but the authors explain the low emission by very low soil temperature in the region of the experiment.

Methane emissions

In total, 19 articles report comparative CH_4 emissions from a continuously flooded field or pot with a field/pot under MA management.

Relative CH_4 emissions in the MA plots as given in these 19 articles (compiling 60 experimental observations) were found in

Table 12.1. Compilation of field studies on GHG emissions as affected by water-saving techniques conducted in China.

		Methane			Nitrous oxide			
		EF under CF	rel. CH_4 emission		EF under CF	rel. N_2O emission		
Citation	Location	kg ha^{-1} $season^{-1}$ (kg ha^{-1} day^{-1})	MA (%)	SA (%)	g N ha^{-1} $season^{-1}$ (g N ha^{-1} day^{-1})	MA (%)	SA (%)	Remarks
Zhang *et al.* (2012)	China, Jiangsu	185 (1.48)		30.8				
Wang *et al.* (2012)	China, Jiangsu	221 (1.77)		33.8	160 (1.28)		137.5	
		278 (2.22)		22.8	550 (4.40)		123.6	
		548 (4.38)		38.9	130 (1.04)		153.8	
		515 (4.12)		52.7	280 (2.24)		121.4	
Qin *et al.* (2010)	China, Jiangsu	127 (1.04)		43.8	180 (1.48)		194.5	
		105 (0.88)		41.0	50 (0.42)		1390	
Jiao *et al.* (2006)	China, Liaoning	230 (1.56)	75.77		296 (2.00)	123.72		4 aerations
Yue *et al.* (2005)	China, Liaoning	24.8 (0.20)	67.74		382 (3.05)	133.33		2 aerations, low soil temperature
Zou *et al.* (2005)	China, Jiangsu	85 (0.72)		35.29	60 (0.51)		2583.3	
		220 (1.86)		64.09	30 (0.25)		4766.7	
Wang *et al.* (2000)	China, Beijing	503 (3.73)	41.2	76.5				Automated system
Lu *et al.* (2000)	China, Zheijang	565 (4.25)	38.9	56.1				Automated system
Wang *et al.* (1999)	China, Beijing	748 (7.48)	41.6					
		145 (1.18)	74.6					Automated system

EF, emission factors given per season and per day, respectively; CF, continuous flooding; MA, multiple aeration; SA, single aeration.

Table 12.2 Emission factors of methane and nitrous oxide under continuous flooding and relative emissions under multiple aeration (MA) from different studies in India.

		Methane		Nitrous oxide		
Citation	Location	EF under CF kg ha^{-1} $season^{-1}$ (kg ha^{-1} day^{-1})	rel. CH_4 emission MA (%)	EF under CF g N ha^{-1} $season^{-1}$ (g N ha^{-1} day^{-1})	rel. N_2O emission MA (%)	Remarks
Khosa *et al.* (2011)	India, Punjab	62.3 (0.53)	46.4			
		36.8 (0.31)	36.2			
Pathak *et al.* (2002, 2003)	India, New Delhi	24.3 (0.27)	34.2	323 (3.63)	95.0	N_2O reported 2002, CH_4 reported 2003, rice/wheat system
		28.1 (0.32)	52.0	735 (8.26)	126.4	
		45.4 (0.51)	61.0	593 (6.66)	120.4	
		20.2 (0.23)	47.5	483 (5.43)	111.8	
Adhya *et al.* (2000)	India, Cuttack	15.7 (0.16)	84.6			Automated system
		30.5 (0.32)	75.0			
Jain *et al.* (2000)	India, New Delhi	39.8 (0.41)	81.4			
		34.8 (0.37)	86.2			
		22.7 (0.23)	42.8			
		23 (0.23)	77.8			
		16.6 (0.17)	78.0			

Table 12.3. Compilation of field studies on CH_4 emissions as affected by water saving techniques conducted in Japan and South Korea (no studies comparing N_2O emissions from this region could be identified; abbreviations, see Table 12.1).

		Methane			
		EF under CF	rel. CH_4 emission		
Citation	Location	kg ha^{-1} $season^{-1}$ (kg ha^{-1} day^{-1})	MA (%)	SA (%)	Remarks
Itoh *et al.* (2011)	Japan, Nagaoka	307 (2.38)	48.1		
		318 (2.50)	25.5		
		662 (5.25)		77.0	
		1044 (8.92)		68.7	
	Japan, Koshi	65 (0.58)		38.0	
		52 (0.44)		102.6	
	Japan, Minamisatsuma	270 (2.48)		129.6	Average of 3 observations
Minamikawa and Sakai (2006)	Japan, Tsukuba	139 (1.03)		47.3	
		142 (1.06)	51.44		
		227 (1.79)	30.77		
		252 (1.98)	25.60		Aeration after EH control
Yagi *et al.* (1996)	Japan, Ryugasaki	148 (1.19)	58.31		
		94.9 (0.65)	54.58		Automated system
Kwun *et al.* (2003)	S. Korea, Milyang	503 (4.70)	85.1		Assumed growth period: 107 days
Park and Yun (2002)	S. Korea, Suwon, Iksan, Milyang	257 (2.40)	62.5		Average of 7 observations, assumed growth period: 107 days
		599 (5.60)	64.3		Average of 5 observations, assumed growth period: 107 days
		396 (3.70)	62.2		Average of 3 observations, assumed growth period: 107 days
		289 (2.70)	63.0		Average of 4 observations, assumed growth period: 107 days
		175 (1.40)	81.8		Average of 4 observations, assumed growth period:125 days

the range between 19.9% and 86.2% of the emissions of the corresponding CF plot. The arithmetic mean is 56.9% (CV: 36%). For single aeration, a total of 40 experiments in 13 articles were identified. One out of four field comparisons in Wassmann *et al.* (2000) was disregarded for this analysis because of non-achievement of the drainage. The relative CH_4 emissions of the remaining 40 experiments varied between 17.9% and 152.6% with an arithmetic mean of 63.5% (CV: 47%) compared to a continuously flooded paddy field/pot.

The absolute CH_4 reduction (in kilograms per hectare per day) has also been assessed for SA and MA. For this assessment, however, only field experiments were considered as explained above. Figures 12.2 and 12.3 show the absolute CH_4 emissions (in CO_2-equivalents) of CF and WST fields for SA and MA, respectively. For further analysis of the absolute mitigation potential, only field experiments with seasonal emissions of 80 kg CH_4 ha^{-1} or more were considered because low-emission fields might not give potential for further emission reduction. For all 42 field experiments on MA, the arithmetic mean of CH_4 reduction was 1.26 kg ha^{-1} day^{-1} with a CV of 69%. For SA the arithmetic mean of reduction of the 26 field experiments was 1.15 kg CH_4 ha^{-1} day^{-1} (CV: 94%).

Table 12.4. Compilation of field studies on GHG emissions as affected by water saving techniques conducted in Indonesia and the Philippines (abbreviations, see Table 12.1).

		Methane			Nitrous oxide			
		EF under CF	rel. CH_4 emission		EF under CF	rel. N_2O emission		
Citation	Location	kg ha^{-1} $season^{-1}$ (kg ha^{-1} day^{-1})	MA (%)	SA (%)	g N ha^{-1} $season^{-1}$ (g N ha^{-1} day^{-1})	MA (%)	SA (%)	Remarks
Suratno *et al.* (1998)	Indonesia, West Java				249 (1.98)	134.94		Water level down to '0 cm' only; assumed growth periods of 91 days and 112 days, respectively
					254 (2.02)	158.10		
					514 (4.08)	95.28		
					622 (4.93)	143.41		
					716 (5.69)	78.69		
					713 (5.66)	117.84		
Husin *et al.* (1995)	Indonesia, West Java	437 (3.06)	43.1					
		381 (2.95)	61.7					
Corton *et al.* (2000)	Philippines, Nueva Ecija	89 (0.91)		57.1				Automated system
		75 (0.73)		63.0				
		348 (3.75)		92.5				
		272 (3.23)		55.1				
Wassmann *et al.* (2000)	Philippines, Laguna	251 (2.51)		17.93				Automated system
		35 (0.35)		31.43				
		10 (0.10)		80.00				
		28 (0.28)		121.43				Rain in SA, no drainage
Bronson *et al.* (1997)	Philippines, Laguna	17.3 (0.20)		38.5	259 (3.05)		246.33	Automated system
		371 (4.36)		57.2	28 (0.33)		589.29	

Table 12.5. Compilation of pot studies on GHG emissions as affected by water saving techniques (abbreviations, see Table 12.1).

		Methane			Nitrous oxide			
		EF under CF	rel. CH_4 emission		EF under CF	rel. N_2O emission		
Citation	Location	kg ha^{-1} $season^{-1}$	MA (%)	SA (%)	g N ha^{-1} $season^{-1}$	MA (%)	SA (%)	Remarks
Katayanagi *et al.* (2012)	Philippines, Laguna	580.7	27.2		10.19	32,476.2		
Minamikawa and Sakai (2005)	Japan, Tsukuba	1353	19.9	55.6				Int. irrigation only after 81 DAT
		1926	29.7	69.5				
Mishra *et al.* (1997)	India, Cuttack		43.80	59.65				BL: 6.393 mg pot^{-1} day^{-1}

Nitrous oxide emissions

Only nine different studies comprising 23 experiments could be identified that measured N_2O emissions from rice fields under different water management practices. N_2O emissions were generally higher under water-saving strategies as compared to continuously flooded fields. However, the

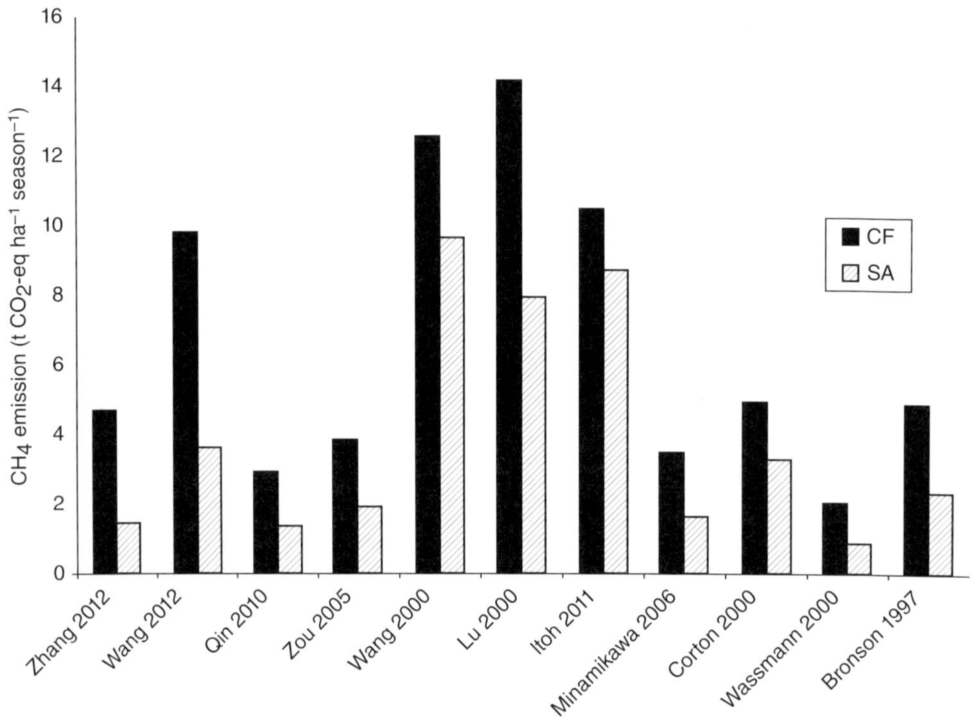

Fig. 12.2. Methane emissions from studies comparing continuous flooding (black) and single aeration (grey). Values are arithmetic means of all experiments in the respective article. GWP (CH_4) = 25.

variation of the results was also higher than for CH_4 results.

For multiple aeration, 12 experiments were analysed and the arithmetic mean of the relative N_2O emissions was found to be 120% (CV: 19%) compared to CF. For SA, 11 relevant experiments were found and the relative N_2O emissions were between 121% and 4767% with an arithmetic mean of 907% (CV: 171%) as compared to a CF reference field. The high coefficient of variation is mainly caused by results of one study (Zou *et al.*, 2005) that reports very high N_2O emission increases for SA. Due to this fact, it might be more meaningful to use another statistical measure, namely the median, which is 176% for relative N_2O emissions under SA. The median for relative N_2O emissions under MA is 122%.

Global warming potential

Only seven field studies were identified measuring both CH_4 and N_2O emissions, as affected by different water management strategies (Fig. 12.4). In all of the studies, CH_4 emissions decrease under WST while N_2O emissions increase. The total GWP, however, decreases in all of them (between 18% and 59%). The contribution of N_2O to the total GWP of continuously flooded fields is between 0.6% and 2.4% for the five studies with a GWP higher than 1 t CO_2-eq ha^{-1} $season^{-1}$. For the other two studies with a very low GWP, Yue *et al.* (2005) and Pathak *et al.* (2002, 2003), contribution of N_2O is 22% and 25%, respectively. In the WST plots, the contribution of N_2O increased from 3.8% to 6.4% for Bronson *et al.* (1997), Jiao *et al.* (2006), Qin *et al.* (2010) and Wang *et al.* (2012), to 25% for Zou *et al.* (2005) and to 36% and 44% for Yue *et al.* (2005) and Pathak *et al.* (2002, 2003), respectively. The increase of N_2O emissions by switching from CF to WST, however, in all the studies (except Zou *et al.*, 2005) is between 17% and 180%, while Zou *et al.* report an N_2O increase of 3300%.

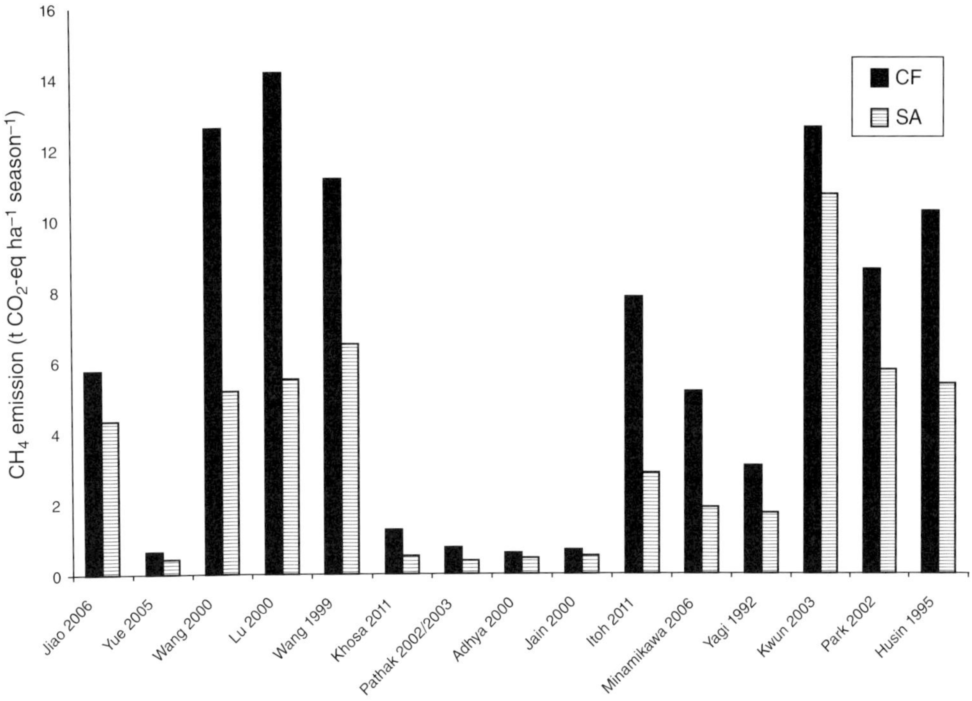

Fig. 12.3. Methane emissions from studies comparing continuous flooding (black) and multiple aeration (grey). Values are arithmetic means of all experiments in the respective article. GWP (CH_4) = 25.

Taking the average of all these seven studies, the GWP decreases from 4.2 t CO_2-eq ha^{-1} $season^{-1}$ under CF to 2.4 t CO_2-eq ha^{-1} $season^{-1}$ under a WST with the contribution of N_2O increasing from 3% to 11%.

12.3.2 Discussions

Derived from this meta-analysis, field drainage in irrigated rice production can be deemed a promising mitigation option with the potential to substantially reduce GHG emissions. Although N_2O emissions increase under WSTs, this increase does not offset the reduction in CH_4 emissions.

The CH_4 reduction potentials of SA and MA are at similar levels – which is a somehow unexpected result. SA was found to reduce CH_4 emissions by 36.5% on average, MA by 43.1%. The explanation for this could be how the drainage is carried out in detail. In studies with only one dry period in the growing season, this drainage might be executed more accurately and maybe even longer (i.e. a lower water level) than the drainages in studies on MA. Hence, this one dry period would have a higher mitigation effect than one dry period in a field managed under MA. Also, the stronger increase of N_2O emissions in SA (median: 176%) than in MA (median: 122%) supports this hypothesis.

Furthermore, CH_4 emissions tend to increase slowly in the beginning of the growth period. The highest flux rates are found towards the middle of the season (Yagi *et al.*, 1996; Hou *et al.*, 2000). Thus, practising a WST in the beginning of the season has a lower mitigation effect than practising it around the middle of the season. After a dry period, CH_4 flux only slowly increases again (Cai *et al.*, 1997).

The relative CH_4 emission levels of plots treated with SA and MA, respectively, as assessed in this literature study are in good agreement with what the IPCC suggests

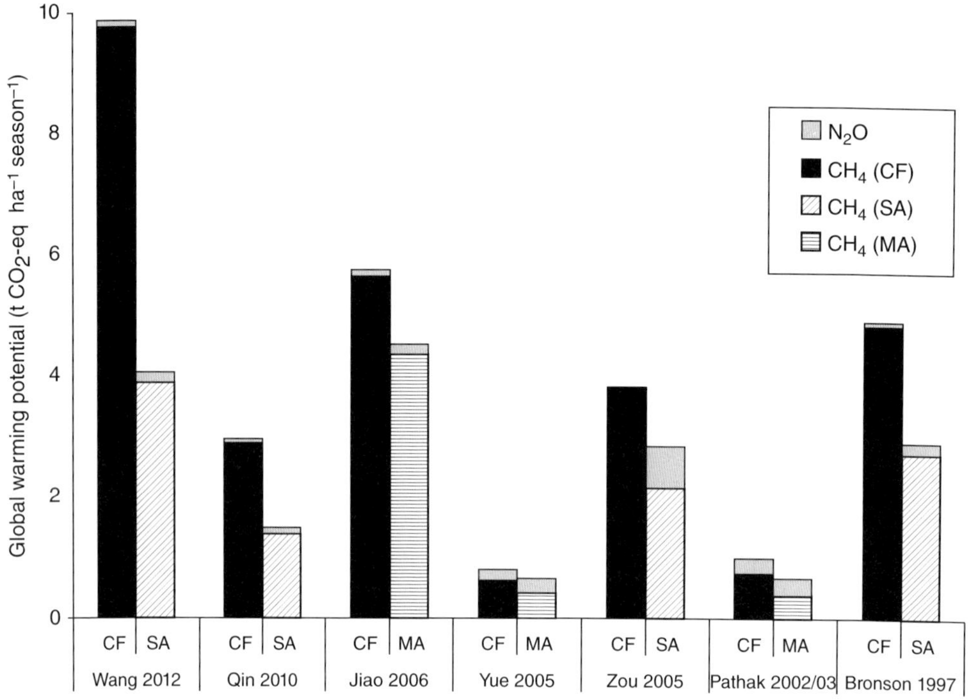

Fig. 12.4. Global warming potential of different water management practices as derived from articles comparing both methane and nitrous oxide emissions. Values are arithmetic means of all experiments in the respective article. GWP (CH_4) = 25, GWP (N_2O) = 298 (CF, continuous flooding; SA, single aeration; MA, multiple aeration).

using the default values given in its 2006 guidelines (IPCC, 2006). For SA the IPCC guidelines recommend a 'scaling factor' of 0.6, i.e. a relative CH_4 emission level of 60% compared to continuous flooding. The average emission level for SA as found in this study is 63.5%. For MA, the IPCC suggests a default scaling factor of 0.52 (i.e. relative CH_4 emissions in a MA field are 52% of those in a CF field) while the average emissions as assessed in this study are 56.9%. It should be noted that the IPCC scaling factors were also founded on a literature survey that probably in large parts is included in this study. But this analysis shows that the IPCC factors still represent good default means even if articles from after 2006 are included in the assessment.

Comparing the absolute values of CH_4 reduction as found in the available literature with what the CDM methodology for rice production gives as standard values, it can be said that for both practices, SA and MA, the CDM standard values are higher than was found in the available literature. The CDM methodology suggest a reduction of 1.8 kg CH_4 ha^{-1} day^{-1} for shifting to intermittent flooding with MA and the arithmetic mean of CH_4 reduction as found in the literature is 1.26 kg ha^{-1} day^{-1}. For midseason drainage, the CDM methodology suggests a reduction of 1.5 kg CH_4 ha^{-1} day^{-1} while the arithmetic mean of all literature findings for SA is 1.15 kg CH_4 ha^{-1} day^{-1}.

The share of N_2O emissions to the total GWP is higher under an applied WST than under continuous flooding. Nitrous oxide contributions under both management strategies, CF and WST, are generally below 10% except when CH_4 emissions are very low as e.g. found in India. Only in one study (Zou *et al.*, 2005) did N_2O emissions exceeded 0.3 t CO_2-eq ha^{-1} season^{-1}.

12.4 Conclusions

AWD and MSD as representative forms of MA and SA, respectively, are potent mitigation options for irrigated rice production systems. The average relative CH_4 emission under SA and MA are at similar levels according to the findings in this literature study. This could have implications on the dissemination of water-saving strategies as mitigation options. Farmers adopt the AWD technology primarily because of the water saved, yet maintained yields. While in areas with pump irrigation AWD is easily adopted because of the direct monetary pay-out, in areas with improved canal irrigation facilities with more than adequate water supply farmers are more reluctant to adopt AWD. Instead of introducing AWD, which might require more effort for a farmer to accurately practise and could be considered as too harsh with its alternating dry phases (thus, has a high adoption barrier), the entry point in those areas could be a single MSD. The mitigation potential of MSD is similar to AWD but it only requires water control during approximately 1 week of the growth period. Thus, farmers might be more willing to adopt this water management strategy and might even practise it more accurately. After adoption of MSD, introduction of AWD could follow. The clean development mechanism may serve as additional incentive if properly coordinated. Aside from this, it is important that other indirect benefits from AWD (e.g. less crop lodging, reduced pest damage, better soil conditions) are further explored and scientifically validated.

This study further shows that the IPCC scaling factors represent good average values according to the articles analysed. However, CH_4 emissions are very low in India compared to other parts of Asia (e.g. China or Japan), which shows that disaggregation for any mitigation strategies is important. Moreover, this finding shows limits for the transfer of any mitigation option from one region to another. Assessment of region-specific characteristics is necessary.

Acknowledgement

B.O. Sander thanks GIZ (Deutsche Gesellschaft fuer Internationale Zusammenarbeit) and CCAFS (the CGIAR Research Program on Climate Change, Agriculture and Food Security) for funding his position at IRRI in 2012 and 2013, respectively.

References

Adhya, T.K., Bharati, K., Mohanty, S.R., Ramakrishnan, B., Rao, V.R., Sethunathan, N. and Wassmann, R. (2000) Methane emission from rice fields at Cuttack, India. *Nutrient Cycling in Agroecosystems* 58, 95–105.

Bhattacharyya, S. (2011) The Clean Development Mechanism. In: Bhattacharyya, S. (ed.) *Energy Economics: concepts, issues, markets and governance*. Springer, London.

Bouman, B.A.M., Lampayan, R.M. and Tuong, T.P. (2007) *Water Management in Irrigated Rice: Coping with Water Scarcity*. International Rice Research Institute, Los Baños, the Philippines.

Bronson, K.F., Neue, H.-U., Singh, U. and Abao, E.B. (1997) Automated chamber measurements of methane and nitrous oxide flux in a flooded rice soil: I. Residue, nitrogen, and water management. *Soil Science Society of America Journal* 61, 981–987.

Cai, Z., Xing, G., Yan, X., Xu, H., Tsuruta, H., Yagi, K. and Minami, K. (1997) Methane and nitrous oxide emissions from rice paddy fields as affected by nitrogen fertilisers and water management. *Plant and Soil* 196, 7–14.

Corton, T.M., Bajita, J.B., Grospe, F.S., Pamplona, R.R., Assis Jr, C.A.A., Wassmann, R., Lantin, R.S. and Buendia, L.V. (2000) Methane emission from irrigated and intensively managed rice fields in Central Luzon (Philippines). *Nutrient Cycling in Agroecosystems* 58, 37–53.

Hou, A.X., Chen, G.X., Wang, Z.P., Cleemput, O.V. and Patrick, W.H. (2000) Methane and nitrous oxide emissions from a rice field in relation to soil redox and microbiological processes. *Soil Science Society of America Journal* 64, 2180–2186.

Husin, Y.A., Murdiyarso, D., Khalil, M.A.K., Rasmussen, R.A., Shearer, M.J., Sabiham, S., Sunar, A. and Adijuwana, H. (1995) Methane Flux from Indonesian Wetland Rice: The Effects of Water Management and Rice Variety. *Chemosphere* 31, 3153–3180.

IPCC (2006) *2006 IPCC Guidelines for National Greenhouse Gas Inventories.* Prepared by the National Greenhouse Gas Inventories Program. IGES, Japan.

Itoh, M., Sudo, S., Mori, S., Saito, H., Yoshida, T., Shiratori, Y., Suga, S., Yoshikawa, N., Suzue, Y., Mizukami, H., Mochida, T. and Yagi, K. (2011) Mitigation of methane emissions from paddy fields by prolonging midseason drainage. *Agriculture, Ecosystems & Environment* 141, 359–372.

Jain, M.C., Kumar, S., Wassmann, R., Mitra, S., Singh, S.D., Singh, J.P., Singh, R., Yadav, A.K. and Gupta, S. (2000) Methane emissions from irrigated rice fields in northern India (New Delhi). *Nutrient Cycling in Agroecosystems* 58, 75–83.

Jiao, Z., Hou, A., Shi, Y., Huang, G., Wang, Y. and Chen, X. (2006) Water management influencing methane and nitrous oxide emissions from rice field in relation to soil redox and microbial community. *Communications in Soil Science and Plant Analysis* 37, 1889–1903.

Katayanagi, N., Furukawa, Y., Fumoto, T. and Hosen, Y. (2012) Validation of the DNDC-Rice model by using CH_4 and N_2O flux data from rice cultivated in pots under alternate wetting and drying irrigation management. *Soil Science and Plant Nutrition* 58, 360–372.

Khosa, M.K., Sidhu, B.S. and Benbi, D.K. (2011) Methane emission from rice fields in relation to management of irrigation water. *Journal of Environmental Biology/Academy of Environmental Biology, India* 32, 169–172.

Kwun, S.-K., Shin, Y.K. and Eom, K. (2003) Estimation of methane emission from rice cultivation in Korea. *Journal of Environmental Science and Health*, Part A 38, 2549–2563.

Lampayan, R. (2013) Smart water technique for rice. Available at: http://www.agritech.tnau.ac.in/agriculture/pdf/csa_pdf/Smart_water_technique_for_rice.pdf (accessed July 2015).

Lampayan, R., Palis, F., Flor, R., Bouman, B., Quicho, E., de Dios, J., Espiritu, A., Sibayan, E., Vicmudo, V., Lactaoen, A. and Soriano, J. (2009) Adoption and dissemination of 'safe alternate wetting and drying' in pump irrigated rice areas in the Philippines. In: *Proceedings of the 60th International Executive Council Meeting and 5th Asian Regional Conference*, 6–11 December 2009, New Delhi, India.

Lampayan, R., Bouman, B.A.M., Flor, R.J. and Palis, F.G. (2013) Developing and disseminating alternate wetting and drying water saving technology in the Philippines. In: Kumar, A. (ed.) *Mitigating Water Shortage Challenges in Rice Cultivation: Aerobic and Alternate Wetting and Drying Rice Water Saving Technologies.* Asian Development Bank, Manila, the Philippines.

Lu, W.F., Chen, W., Duan, B.W., Guo, W.M., Lu, Y., Lantin, R.S., Wassmann, R. and Neue, H.U. (2000) Methane emissions and mitigation options in irrigated rice fields in South-east China. *Nutrient Cycling in Agroecosystems* 58, 65–73.

Minamikawa, K. and Sakai, N. (2005) The effect of water management based on soil redox potential on methane emission from two kinds of paddy soils in Japan. *Agriculture, Ecosystems & Environment* 107, 397–407.

Minamikawa, K. and Sakai, N. (2006) The practical use of water management based on soil redox potential for decreasing methane emission from a paddy field in Japan. *Agriculture, Ecosystems & Environment* 116, 181–188.

Mishra, S., Rath, A.K., Adhya, T.K., Rao, V.R. and Sethunathan, N. (1997) Effect of continuous and alternate water regimes on methane efflux from rice under greenhouse conditions. *Biology and Fertility of Soils* 24, 399–405.

Palis, F.G., Cenas, P.A., Bouman, B.A.M., Lampayan, R.M., Lactaoen, A.T., Norte, T.M., Vicmudo, V.R., Hossain, M. and Castillo, G.T. (2004) A farmer participatory approach in the adaptation and adoption of controlled irrigation for saving water: A case study in Canarem, Victoria, Tarlac, Philippines. *Philippine Journal of Crop Science* 29(3), 3–12.

Park, M.-E. and Yun, S.-H. (2002) Scientific basis for establishing country CH_4 emission estimates for rice-based agriculture: a Korea (South) case study. *Nutrient Cycling in Agroecosystems* 64, 11–17.

Pathak, H., Bhatia, A., Prasad, S., Singh, S., Kumar, S., Jain, M.C. and Kumar, U. (2002) Emission of nitrous oxide from rice-wheat systems of Indo-Gangetic Plains of India. *Environmental Monitoring and Assessment* 77, 163–178.

Pathak, H., Prasad, S., Bhatia, A., Singh, S., Kumar, S., Singh, J. and Jain, M.C. (2003) Methane emission from rice–wheat cropping system in the Indo-Gangetic plain in relation to irrigation, farmyard manure and dicyandiamide application. *Agriculture, Ecosystems & Environment* 97, 309–316.

Qin, Y., Liu, S., Guo, Y., Liu, Q. and Zou, J. (2010) Methane and nitrous oxide emissions from organic and conventional rice cropping systems in South-east China. *Biology and Fertility of Soils* 46, 825–834.

Sibayan, E., de Dios, J. and Lampayan, R. (2010) Outscaling AWD in a public-managed reservoir-type irrigation system: a case study in the Philippines. In: Palis, F.G., Singleton, G.R., Casimero,

M.C. and Hardy, B. (eds) *Research to Impact: Case Studies for Natural Resource Management for Irrigated Rice in Asia.* Los Baños, Philippines.

Smith, P., Nabuurs, G.J., Janssens, I.A., Reis, S., Marland, G., Soussana, J.F., Christensen, T.R., Heath, L., Apps, M., Alexeyev, V., Fang, J.Y., Gattuso, J.P., Guerschman, J.P., Huang, Y., Jobbagy, E., Murdiyarso, D., Ni, J., Nobre, A., Peng, C.H., Walcroft, A., Wang, S.Q., Pan, Y. and Zhou, G.S. (2008) Sectoral approaches to improve regional carbon budgets. *Climate Change* 88(3–4), 209–249.

Suratno, W., Murdiyarsob, D., Suratmoc, F.G., Anas, I., Saenic, M.S. and Rambec, A. (1998) Nitrous oxide flux from irrigated rice fields in West Java. *Environmental Pollution* 102, 159–166.

Thompson, J. (2006) Mid-season 'drainage' of rice – is it worth trialing on your crop? Available at: http://www.irec.org.au/farmer_f/pdf_173/Mid-season drainage of rice.pdf (accessed July 2015).

Tsusaka, T., Kajisa, K., Pede, V. and Aoyagi, K. (2012) Neighborhood effects on social behavior: the case of irrigated and rainfed farmers in Bohol, the Philippines. Paper Presentation at the Agricultural and Applied Economics Association Annual Meeting, August 12–14, 2012, Seattle, Washington.

UNFAO (2010) *'Climate-Smart' Agriculture Policies – Practices and Financing for Food Security, Adaptation and Mitigation.* Rome, Italy.

UNFCCC (2012) Methane emission reduction by adjusted water management practice in rice cultivation Available at: http://cdm.unfccc.int/UserManagement/FileStorage/SLAHVBCKDY2QI86094XZ5UR1OMWEG3 (accessed July 2015).

Wang, B., Xu, Y., Wang, Z., Li, Z., Guo, Y., Shao, K. and Chen, Z. (1999) Methane emissions from ricefields as affected by organic amendment, water regime, crop establishment, and rice cultivar. *Environmental Monitoring and Assessment* 57, 213–228.

Wang, J., Zhang, X., Xiong, Z., Khalil, M.A.K., Zhao, X., Xie, Y. and Xing, G. (2012) Methane emissions from a rice agroecosystem in South China: effects of water regime, straw incorporation and nitrogen fertilizer. *Nutrient Cycling in Agroecosystems* 93, 103–112.

Wang, Z.Y., Xu, Y.C., Li, Z., Guo, Y.X., Wassmann, R., Neue, H.U., Lantin, R.S., Buendia, L.V., Ding, Y.P. and Wang, Z.Z. (2000) A four-year record of methane emissions from irrigated rice fields in the Beijing region of China. *Nutrient Cycling in Agroecosystems* 58, 55–63.

Wassmann, R. (2010) Implementing the clean development mechanism in the land use sector: status and prospects. In: *Climate Change: 'No Regret' Options for Adaptation and Mitigation and Their Potential Uptake. IRRI Limited Proceedings No 16.* International Rice Research Institute, Los Baños, the Philippines, p. 63.

Wassmann, R., Buendia, L.V., Lantin, R.S., Bueno, C.S., Lubigan, L.A., Umali, A., Nocon, N.N., Javellana, A.M. and Neue, H.U. (2000) Mechanisms of crop management impact on methane emissions from rice fields in Los Baños, Philippines. *Nutrient Cycling in Agroecosystems* 58, 107–119.

Yagi, K., Tsuruta, H., Kanda, K.-I. and Minami, K. (1996) Effect of water management on methane emission from a Japanese rice paddy field: automated methane monitoring. *Global Biogeochemical Cycles* 10, 255–267.

Yue, J., Shi, Y., Liang, W., Wu, J., Wang, C. and Huang, G. (2005) Methane and nitrous oxide emissions from rice field and related microorganism in black soil, Northeastern China. *Nutrient Cycling in Agroecosystems* 73, 293–301.

Zhang, G., Ji, Y., Ma, J., Xu, H., Cai, Z. and Yagi, K. (2012) Intermittent irrigation changes production, oxidation, and emission of CH_4 in paddy fields determined with stable carbon isotopc technique. *Soil Biology and Biochemistry* 52, 108–116.

Zou, J., Huang, Y., Jiang, J., Zheng, X. and Sass, R.L. (2005) A 3-year field measurement of methane and nitrous oxide emissions from rice paddies in China: effects of water regime, crop residue, and fertilizer application. *Global Biogeochemical Cycles* 19. DOI: 10.1029/2004GB002401 GB2021.

13 Linking Climate Change Discourse with Climate Change Policy in the Mekong: The Case of Lao PDR

Jana Prosinger,[1] Diana Suhardiman[2]* and Mark Giordano[3]

[1]*School of Oriental and African Studies, London;* [2]*International Water Management Institute, South-east Asia Regional Office, Vientiane, Lao PDR;* [3]*Georgetown's School of Foreign Service, Georgetown University, Washington, DC*

Abstract

Current discourse on climate change highlights the issue of uncertainty, risks and the importance of systems' resilience as a means to cope with impacts of climate change and climate variability. This chapter links the dominant approach of uncertainty as presented in the climate change discourse with policy discussions on climate adaptation strategies in the Lower Mekong Basin. Taking Lao PDR as our case study, we discuss how the idea of uncertainty can be perceived and interpreted differently by policy actors. While these different perceptions and interpretations might lead to multiple problem framings, they also reflect structural impediments and institutional barriers in the overall formulation process of climate change policy and adaptation strategies. The main message of the chapter is that understanding of these different notions of uncertainty is crucial to increase the actual significance of climate change policy. Policy and governance responses to climate change need to be formulated based on a more nuanced, sophisticated understanding of how various policy actors and stakeholders perceive and experience uncertainty.

13.1 Introduction

Climate change is one of the most alarming problems globally. Its effects on nature may range from global warming, glacier melting and rise of seawater level to a high frequency and severity of droughts and floods. Although climate change is widely accepted as a scientific fact, there are various definitions of climate change (Drieschova *et al.*, 2009). These definitions range from climate change being solely human-induced phenomena to one of a natural problem (UNFCCC, 1992). Others, like the International Panel on Climate Change (IPCC), define climate change as combining the impacts of human activity and natural variability or change (IPCC, 2007).

The differences in the very definitions of climate change are a good starting point for discussion on how the idea of climate change uncertainties can be differently perceived and interpreted. Nevertheless, current discourse on climate change approaches the idea of uncertainty primarily within the scope and context of scientific uncertainty in global climate science (Shackley and Wynne, 1996; Pielke, 2007; Weber, 2010). A possible explanation for this might be that a narrow definition of uncertainty is preferred

* Corresponding author, e-mail: d.suhardiman@cgiar.org

in terms of formulating climate change policy (Weber, 2010).

Globally, there is a general tendency to assume that 'policies ideally should rest on reliable, robust and hence certain scientific knowledge' (Shackley and Wynne, 1996, p. 276). Thus, perceiving scientific uncertainty as a hindrance for the formulation and implementation of climate change policies, mitigation and adaptation efforts to cope with climate change are focused on scientific measures to reduce the uncertainty (IPCC, 2007). Policy science, however, highlights that policy formulation is a highly dynamic process, shaped by policy actors' interests and access to resources (Sabatier and Hunter, 1988; Mosse, 2004). In the context of climate policy, this is most apparent from the way policy actors perceive and interpret the idea of uncertainty, not always based on their understanding of climate science.

This chapter highlights the dissonance on how scientists perceive climate change problems as mainly related to uncertainty in global climate science, and how policy makers shape climate change adaptation strategies, not always based on climate science. In addition, it sheds light on how stakeholders (i.e. farmers, dam operators) might perceive climate change from a completely different perspective, not necessarily related to either climate science or climate change adaptation policies.

It argues that policy and governance responses to climate change require more nuanced and sophisticated understanding on how differently policy actors and stakeholders perceive and experience uncertainty. Taking Lao PDR as our case study, we bring to light the need to understand multiple notions of uncertainty and their relevance for climate change policy formulation. We look at how policy actors shape climate change adaptation strategies, based mainly on their perception on how climate change might (not) impact the country's development strategies. The chapter looks at how policies have been formulated and implemented on the ground. It sheds light on the multiple notions of uncertainties and how these have shaped and reshaped the overall process of policy formulation and implementation.[1]

To understand how the idea of uncertainty is approached in climate change discourse, we conducted a literature review on the issue of uncertainty in climate change. In line with this review, we conducted key informant analysis and in-depth interviews with various policy actors and stakeholders to understand how they perceive the idea of uncertainty in climate change. We interviewed staff from the Ministry of Natural Resources and Environment (MoNRE) and Ministry of Agriculture and Forestry (MAF) as two government ministries mainly dealing with climate change policy in Lao PDR. To understand the overall shaping of climate change policy implementation in Lao PDR, we interviewed various stakeholders, working with climate change issues in Lao PDR in general and for the Lower Mekong Basin in particular. These include staff from Australian Aid (AusAid), the Commonwealth Scientific and Industrial Research Organization (CSIRO), German Development Agency (GIZ), Mekong River Commission (MRC), World Bank, United Nations Development Programme (UNDP) and Chiang Mai University.

13.2 Framing Climate Change Uncertainties

The concept of uncertainty has translated into various approaches to climate change throughout the climate change discourse. Politicians, policy actors, scholars, NGOs and diplomats negotiate on emission reductions, responsibilities and strategies for years. Still, one dominant idea has remained: effective policies, strategies and approaches can be formulated to dissolve or minimize the negative impacts of climate change only if the problem of lack of information is addressed (Pielke, 2007). This idea leads to the assumption that coping with climate change requires the collection of all necessary scientific information (Dimitrov, 2003; Tol, 2005). Coping with climate change thus means minimizing uncertainty through statistical information, calculating the probabilities that a certain outcome occurs (statistical approach) (Dessai and van der Sluijs, 2007). It assumes that the data are inaccurately measured and errors might have been occurred through, e.g. a sample error or data inadequacy and data uncertainty, or that more measurement is needed (Heal and Kriström, 2002; Walker *et al.*, 2003). A similar idea has developed to prognosticate future climate scenarios. While there is no precise information, various scenarios can be developed about impacts of climate change potential to plan for the range of possible outcomes (New and Hulme, 2000).

Pielke (2007) coins these approaches as 'factual uncertainty' thinking. While various facts cannot be ascertained, it requires a certain degree of simplification (i.e. through extrapolation) to reduce the overall complexity or uncertainty. In earlier literature, this approach is best described as reducing risks in order to calculate the probability that a certain situation occurs (Wardekker, 2011). The risk management literature defines risk as the 'combination of the probability of an event and its consequences' (ISO/IEC Guide, 73, 2002). Risk is thus the quantitative chance of something occurring in correspondence with the value of possible outcomes (Shackley and Wynne, 1996). In environmental terms, the OECD (2006, p. 21) provides a definition for risk as the 'result of interaction of physically defined hazards with the properties of the exposed systems [...]. Risk can also be considered as the combination of an event, its likelihood, and its consequences'. This has globally translated into mitigation policies, the most prevalent being the reduction in greenhouse-gas (GHG) emissions.

On the other side of discussions are authors like Schneider *et al.* (2002), who recognize that uncertainty might not always be measurable and can also result from disagreement, linguistic imprecision or from other unquantifiable means. It highlights the importance to understand the notion of subjective valuation in climate change science. This thinking on normative uncertainties is often lost in the discussion of climate change uncertainties. With reference to the factual uncertainty thinking, this normative uncertainty is best reflected in the assumption of ignorance of uncertainty. It reaches from substantial uncertainty of knowing too little to make any presumptions about the future or not even knowing that one does not know (Heal and Kriström, 2002; Walker *et al.*, 2003).

While uncertainty is a concept, which is always part of any discourse on climate change, it is seldom defined explicitly. Pielke provides a definition aiming at encompassing various schools of thought: 'Uncertainty means that in a particular situation more than one outcome is consistent with our expectations' (Pielke, 2007, p. 55). Walker *et al.* (2003) demonstrate that there are various dimensions of uncertainties that have to be understood by scientists and policy actors in order to develop a way of how to deal with them. One way to deal with uncertainties is the introduction of the precautionary principle – the idea that action has to be taken in order to minimize, prevent or anticipate the effects of anthropogenic climate change in a cost-effective way – the claim for a way to deal with uncertainties in policies has become more serious (van der Sluijs and Turkenburg, 2006). The UNFCCC (1992, p. 4) states: 'lack of full scientific certainty should not be used as a reason for postponing such measures'.

While various definitions of adaptation exist, the major difference lies in the interpretation of adaptation as an 'outcome' or a 'process'. Adaptation seen as an outcome might require a clear set of goals whereas adaptation regarded as a process is much broader and more open; both interpretations require a diverse set of policies, institutions and financial resources (OECD, 2006). Moreover, Carpenter and Brock (2008) showed that individual or institutional adaptive capacity may not always enhance a system's adaptive capacity. Certain forms of intensive resource management might lead into a rigidity trap, when institutions become inflexible and static, thus maladaptive. Another situation of deadlock is the poverty trap. These are 'situations in which people are impoverished by circumstances beyond their control' (Carpenter and Brock, 2008, p. 2). Carpenter and Brock (2008) thus presented reasons why adaptation and mitigation strategies might be at their best when implemented together so that transformation and persistence of a social-ecological system might exist in parallel. This becomes especially important in situations of great uncertainties, such as in resource management under climate change conditions.

While there is much discussion on various conceptions of uncertainty as shown above, global policy discourse on climate change continues to be governed by the dominant approach towards risk to reduce factual uncertainty.[2] The overstatement of factual uncertainty in various policies is most apparent from the shaping of international policies that aim at reducing GHGs and increase data collection and future modelling of climate change. This has translated into similar policies at national level: management of floods and droughts and exact data collection can be regarded as the main policy focus. In the water sector in particular, climate change adaptation strategies have a strong emphasis on water resources planning, often without taking into account how key policy actors view the overall idea of planning in the first place.

In line with analysis of Walker *et al.* (2003), this chapter argues that understanding of the different notions of uncertainty is crucial to increase the actual significance and effectiveness of climate change policy. Policy and governance responses to climate change need to be formulated based on a more nuanced, sophisticated understanding of how various actors and stakeholders perceive and experience uncertainty. In this chapter we approach uncertainty from the perspective of a perception study, rather than simply treating it in terms of 'objective' data collection and scientific/technical perspective. It highlights how various policy actors can perceive uncertainty differently, and how various interpretations of uncertainty can influence the overall shaping of the design and implementation of a climate change policy.

13.3 The Shaping of Climate Change Policy in Lao PDR

This chapter analyses climate change policies in Lao PDR, looking specifically at the National Adaptation Program of Action (NAPA) and the Climate Change Strategy (CCS). It illustrates how climate change policies are shaped by policy actors' perceptions and interpretations of uncertainty, rather than driven by the need to collect all scientific information, as advocated in the dominant factual uncertainty thinking.

We discuss the actual policy formulation processes and how various notions of uncertainty emerge in and influence the overall process. First, it summarizes the main points described in the NAPA and CCS. Second, it shows how the dominant thinking of factual uncertainty as reflected in global climate change discourse does not match with how national policy actors perceive uncertainties primarily from the angle of economic uncertainty, within the context of the country's development strategies. Third, it highlights how climate change policy in Lao PDR is formulated mainly based on subjective valuation of uncertainty (normative uncertainty) in climate change impacts, in terms of potential disastrous impacts rather than actual incremental impacts.

13.3.1 The National Adaptation Program of Action and Climate Change Strategy

Lao PDR adopted the NAPA in 2008 and 3 years later, the national CCS. The idea to formulate a climate change policy for Lao PDR originated from the United Nations Framework Convention on Climate Change (UNFCCC). The objective of the UNFCCC was to incorporate climate change issues as part of government policies both in developed and developing countries. Like others, the Government of Lao PDR (GoL) is formally obliged to formulate climate change policy in conjunction with its signing of the international agreement on climate change. UNFCCC's objective to ensure the incorporation of climate change issues into government policies is driven primarily by the global climate policy agenda, which is focused on reducing carbon emissions. This agenda came out after the international epistemic community realized the current failure to use the Kyoto protocol as a means to control carbon emissions of countries due, for instance, to USA, China and India's reluctance to sign the agreement. This highlights the hegemonic tendency (Edelman, 1988) in climate change policy formulation, as national climate policies are often imposed through development agendas of international donors. For Lao PDR in particular, the government formulated the NAPA[3] with technical and financial support from international donors (i.e.: World Bank, WB; Asian Development Bank, ADB; donor countries to the Global Environment Facility and the United Nations Development Programme).

The NAPA as well as the CCS are in line with the Millennium Development Goals (MDGs) and aim at reducing climate change effects. Key sectors, which are likely to be negatively affected, are identified such as agriculture, forestry, water resources, energy and transport, industry, urban development and public health (NAPA, 2008/2009; CCS, 2011).

The NAPA focuses on 'reviewing various strategies and measures for managing disasters in the past, present and future, as well as strengthening capacity and assessing alternatives for adaptation to the potential impacts of climate change'. It provides statistical data of adverse effects over the last years and identifies that the severity and frequency of floods and droughts have increased and the temperature has risen. The NAPA also provides predictions of potential future changes in the climate in Lao PDR and its neighbours, relying on climate change models developed by various institutions (e.g. SEA START or the CCAM simulations). Nevertheless, the NAPA lacks a detailed plan on how it can contribute to the overall process of scientific data collection and monitoring, within and beyond the GoL's current focus to improve disaster management strategies.

13.3.2 Scientific uncertainty and national policy actors' perception of uncertainties

As the decision of GoL to formulate the NAPA came originally from its agreement with the UNFCCC, one might assume the central positioning of factual uncertainty thinking in the overall formulation of the NAPA. Driven primarily by the global climate change discourse, one might expect the NAPA to be equipped with policy measures on how to collect necessary scientific information and data on climate change or at least with some proposal on how to address the problem of lack of (scientific) information as means to cope with climate change.

In practice, from our interviews with key policy actors at national level, it was revealed that they do not perceive the current lack of scientific data on climate change as a pertinent issue that needs to be addressed as part of the country's climate change adaptation strategies. Apart from the issue of donors' imposition in the overall formulation of the NAPA, various factors can be brought to light as reasons behind GoL's lack of interest in collecting 'all' necessary information as an integral part of their climate change adaptation strategies. First, policy actors perceive climate change impacts as not self-evident (Weber, 2008). As policy actors framed climate change impacts within the context of potential impacts, they tended to think that climate change impact is not something that they have to think of or deal with immediately. Second, policy actors perceive the country's economic uncertainty as a far greater issue that needs to be addressed, above the need to collect scientific data to make decisions/take actions to adapt to or cope with climate change.

Climate change policy is not in the top priority list of government's policy agenda. Our interviews with national policy actors further indicate that the GoL has various development priorities, which do not necessarily link to climate change or the overall objective of climate change policy to reduce carbon emission and adapt to climate change impacts. The development priority of GoL lies in its attempts to promote the country's rapid economic growth, as means to move up its status from a least developed country (LDC) to a developing country (7th National Socio-Economic Development Plan, 2011–2015). For this purpose, GoL focuses on achieving its development targets in each relevant sector (i.e. energy, agriculture). For the energy sector, for instance, GoL focuses on hydropower development to promote the country's economic growth through revenue generation (Electricity Law, 2010). Similarly, in the agriculture sector, GoL targets irrigation expansion for increased agricultural production (National Growth and Poverty Eradication Strategy, 2010). These defined development goals and targets do not always coincide with climate change policy under the NAPA. From our interviews with relevant stakeholders, we gathered that NAPA has hardly materialized through policy/project implementation on the ground.[4] In practice, existing projects that attempt to tackle climate change issues do not correspond with climate change policy and focus on areas that were listed in the NAPA.

Our Lao PDR case study illustrates the discrepancy between the global notion of scientific uncertainty as reflected in the climate change discourse and how national actors actually perceive uncertainties, primarily from the angle of economic uncertainty. The way factual uncertainty is positioned in global climate change discourse as the main issue needs to be dealt with in coping with climate change and does not correspond to the way policy actors subjectively value scientific uncertainty of climate change impacts, in this case, mainly within the context of the country's economic development. While the factual uncertainty thinkers (global notion) will aim at installing policies to increase scientific knowledge, national actors regard climate change policy formulation as partially important (due to its not self-evident impact) and focus on more urgent issues. These variable perceptions of uncertainties are not reflected in actual policy formulation processes by international development agendas. The discrepancy between national perception and global understanding might lead to ineffective policies as the instruments (or means) provided to cope with climate change would vary. In the case of Lao PDR, this discrepancy is most apparent from the minimal agreement on what role climate change plays and the ponderous implementation of the NAPA.

13.3.3 Scientific understanding and policy actors' perception of climate change impacts

In Lao PDR, the dominant scientific notion of climate change impacts is reduced to the danger of extreme events like floods and droughts. From our interviews with various government staff from both the MoNRE and MAF we found that the GoL perceives climate change potential impacts as closely related to disaster management and the occurrence of extreme events (i.e. floods and droughts). The GoL's focus on disaster management is also apparent in the recent[5] institutional set-up for climate change: climate change is part of the Department for Climate Change and Disaster Management and thus the classification is already visible within the name. The introduction of climate change as part of this department will rather strengthen the focus on disaster management.[6]

Focusing mainly on disaster management as the main policy measure to cope with climate change, key policy actors at national level do not view incremental impacts of climate change as something that needs to be measured and monitored scientifically. While a correlation between climate change and disasters is not untrue, there is a danger of neglecting the incremental impacts of climate change. These incremental impacts include for instance the gradual changes of ecosystems and consequent indirect changes of socio-economic systems or the shift in wet/dry seasons, which if not anticipated can lead to extreme events in the long term.

Our Lao PDR case study illustrates the discrepancy between scientific understanding and policy actors' perception of climate change impacts. While the global notion on scientific uncertainty emphasizes the need for incorporating a plan to collect scientific data and improve technical tools such as modelling and assessing methodology with regard to both potential and actual climate change impacts as part of the NAPA, national policy actors do not view such a plan to be important beyond the context of disaster management. The discrepancy between global and national understanding of climate change (potential) impacts might lead to ineffective discussion on the overall shaping of climate change adaptation strategies, especially taking into account the important role played by international donors in the overall formulation process of the NAPA.

13.4 The Emergence of Institutional and Financial Uncertainty

This section illustrates how the formulation of climate change policy in Lao PDR resulted in the emergence of other types of uncertainty: institutional and financial. It brings to light the issue of institutional uncertainties in the overall shaping of climate change policy and how this can be a hindrance to effective policy implementation. In addition, it illustrates the existing problem of funding uncertainty and its policy implications, which often result in the inconsistency between projects listed in the NAPA and projects actually implemented on the ground. In summary, it illustrates how multiple notions of uncertainties in climate change policy in Lao PDR manifest in a massive disconnect between climate change policy and concrete activities on the ground.

13.4.1 Institutional uncertainty: a hindrance for effective policy implementation

While climate change could be a topic mainstreamed into all ministries and sectors, in Lao PDR there is a lack of institutional clarity on where to locate issues related to climate change. Initially, the Water Resources and Environment Administration (WREA) under the Prime Minister's Office was in charge of dealing with issues related to climate change. In line with donors' recommendation, a climate change office was formed under WREA, and climate change technical working groups (TWG) were formed in each relevant sectoral ministry. The idea was to mainstream climate change policy into their respective sectoral ministry's development plan. In practice, the group has not really functioned, as staff/officials have not seen direct benefits of incorporating climate change policy into existing sectoral development programmes. Similarly, the climate change office at WREA could not deal with the rapid and increasing issues and problems of climate change. In 2011, the GoL decided to move some staff from the Department of Disaster Management under the Ministry of Labour and Social Welfare to strengthen the MoNRE[7] and form the Department of Climate Change and Disaster Management.

Now, the Department of Climate Change and Disaster Management under MoNRE is in charge of the formulation and implementation of climate change policy in Lao PDR. In practice, however, while this department can formulate climate change policy, it cannot ensure the policy incorporation or its implementation by each sectoral ministry.[8] Moreover, there is still much confusion as there are currently two departments of climate change and disaster management: one located under MoNRE and the other under the Ministry of Labour and Social Welfare. This institutional discrepancy leads to a further level of uncertainties in the climate change policy formulation: national actors are bound to secure their working space, practise new ways of communication and act in their new mandate.

Within the MoNRE itself, various uncertainties exist in relation to its status as a newly formed ministry. The communication channel, the decision-making abilities and the role division and mandate for each department have not yet been precisely defined. This lack of certainty is part of the notion of normative uncertainties. However, internationally and also nationally, these uncertainties are not regarded as a hindrance for effective policy implementation. The relevance of these institutional uncertainties is not reflected in international discourse on climate change policy formulation, even though they certainly influence decisions on climate change in every country. The incorporation of institutional uncertainties into policies such as the NAPA could lead to more effective implementation on the ground.

13.4.2 Funding uncertainty and its policy implications

The NAPA identifies 12 high-priority projects, which should be created as part of the GoL's climate change adaptation strategies. As listed in the NAPA, these high-priority projects should address the following needs, to: (i) strengthen the capacity of the national disaster management committees; (ii) promote the secondary professions as mitigation measures; (iii) raise awareness in water resources management; (iv) map flood-prone areas; (v) establish an early warning system; (vi) strengthen institutional capacities in water resources management; (vii) study the possibility of building multiple use reservoirs in drought-prone areas; (viii) improve drinking water and sanitation; (ix) strengthen capacity building; (x) survey underground water sources; (xi) eradicate slash-and-burn practices; and (xii) strengthen forest management.

However, only one project (IRAS) has yet been established[9] to strengthen institutional capacities in water resources management. From our interviews with national policy actors we discovered that project implementation was often halted due to funding uncertainty. Donors' strategies might be mainly to technically and financially support the government in developing its climate change policy and assume that they will implement this policy using the government budget. In practice, the government does not perceive climate change project implementation as a high-priority development agenda.

Donors like ADB and WB structure current climate change project financing by distinguishing between regular and additional climate change activities. Applying and implementing NAPA fall into the regular climate change activity. Additional climate change funds cannot be used to fund project implementation under NAPA. Thus, even if the GoL/MoNRE is intrigued to develop more concrete projects based on NAPA, these donors (who promoted and funded the NAPA formulation) are not able to fund it from additional climate change funds.

Climate change policy implementation takes place through projects funded by various donors in Lao PDR, especially those who are not involved in the climate change policy formulation processes. As the channelling of these additional funds would require the formulation of new climate change activity, not included in the NAPA, this not only makes it difficult to materialize potential projects defined in the NAPA, but also rules out actual implementation of projects dealing with climate change from the NAPA.[10] Sometimes, these projects are materialized as other donors have decided to fund them under their climate change or other development themes. For instance, the project by WB on disaster management and climate risk reduction fits perfectly with the conception of the NAPA. Nevertheless, as many projects seem to be separately implemented, interviews reveal that MoNRE as the agency in charge of climate change policy formulation and implementation often does not know of the existence of certain projects.

The lack of clarity in funding and the lack of information of an overview of existing projects are clearly uncertainties which are not taken into account by existing policies. This leads to a massive disconnect between policy and project implementation.

We argue that synergizing how scientists, policy makers and stakeholders perceive climate change is crucial to increasing the actual significance of climate change policy. In Lao PDR, policy actors and stakeholders do not perceive the lack of scientific certainty as the major problem in coping with climate change (Pielke, 2007). It is more likely that various actors and stakeholders might experience uncertainties primarily in relation to their actual working condition, access to funding and natural resources (predominantly land and water), and livelihood options. Farmers might perceive the actual timing for rice transplanting more important than reducing scientific uncertainty (in terms of rainfall variability) for establishing long-term crop planning; government officials might be more concerned that their department will remain in existence over the next legislation period. Similarly, international donors might have the tendency to focus on certain formalities for funds disbursement, regardless of its role and objectives to clarify long-term future scenarios.

Despite current efforts to reduce scientific uncertainty for instance, shaping of climate change policy remains problematic in terms of its implementation (Termeer *et al.*, 2009). Current practice shows that many climate change policies (especially those in developing countries) are formulated without setting any target in terms of carbon emission or any other indicators to measure climate change impacts.

In the case of Lao PDR, the discrepancy between predominantly scientific uncertainty and the way policy actors perceive uncertainty is most apparent in the overall formulation of the NAPA and the CCS, its focus on disaster risk management of floods and droughts, and the current disconnect between climate change policy and projects implementation. While existing projects include various understandings of uncertainties – depending on the institution implementing it – these projects seem not to be in large part coordinated and harmonized. Some projects aim at enhancing social and environmental resilience or increase farmers' adaptation, others target at modelling scenarios and collecting data in order to reduce the scientific uncertainties (Table 13.1).

13.5 Conclusions

Our Lao PDR case study illustrates the existing discrepancy between global climate change discourse and how climate change policy is formulated and implemented. We argue that this discrepancy is rooted mainly in the current misfit between the dominant factual uncertainty thinking and how policy actors perceive and interpret scientific information.

The dominant scientific notion of uncertainties in the climate change discourse only resembles a fragment of a bigger picture. This is highlighted in the way water scarcity can be viewed differently by various actors and stakeholders at multiple scales, depending on their role in the overall water management. Water resources planners, for instance, perceive water scarcity induced by climate change in close relation to issues of rainfall variability; irrigation system operators tend to look at water scarcity in relation to their ability to increase the flexibility of overall systems operation; farmers would experience water scarcity primarily in relation to shifting seasonal patterns and changes in their surrounding ecosystems.

The actual significance of climate change policy is determined not only by whether it has the 'right scientific' rationale, but also by how policy actors perceive and interpret the overall idea of uncertainty, and how such perceptions shape climate change policy formulation processes. While this highlights the importance to incorporate normative uncertainty thinking in the overall shaping of climate change policy, climate change policy formulation in Lao PDR has also created other types of uncertainty: institutional and funding uncertainty. Current literature on climate change policy highlights this institutional and funding uncertainty by showing for instance how domestic authority is so often overlapping and divided when dealing with climate change issues, and how the mainstream approach in climate policy needs to rely on upscaling models which are envisaged to extract lessons from local adaptation processes.

Understanding of the different notions of uncertainty (normative, institutional and financial) is crucial to increase the actual significance and effectiveness of climate change policy. Our Lao PDR case study shows, for instance, how the overall shaping of climate

Table 13.1. List of stakeholders and climate change projects in Lao PDR.[a]

Name of project/ stakeholder	Implementing agency	Responsible government agency	Funded by	Financial resources	Year	Region	Goal/key activities	Policy impact	Part of existing climate change policies?
IRAS (interview with Manfred Staab)	UNDP	NAFRI, MAF	GEF, LDC Fund, UNDP	n/a	2012–2015	Savannaketh, Xayaboury	Improve the capacity of farmers and government staff in order to deal with the effects of climate change	Little	Part of the NAPA (one of the two projects of agricultural priority)
Lao Local Demonstration Project of the MRC Climate Change and Adaptation Initiative (Interview with Dr Kien)	Lao National Mekong Committee (LNMC) and the MRC Secretariat	Lao National Mekong Committee Authorities of Savannakhet Province and Champone District	For the CCAI: Australia, the EU, Denmark, Germany, Luxembourg, Sweden, Finland	US$10,000	2011–2013	Savannaketh (in Lao PDR) as a demonstration site	Climate change impact assessment and adaptation planning and implementation within the Mekong River Basin Testing of some local adaptation measures (for Lao Demonstration Project)	Lao Demo-Project assisted provincial authority to mainstream CC into development plans	No
Exploring Mekong Region Futures (interview with John Ward)	CSIRO	MoNRE	CSIRO AusAID Research for Development Alliance	n/a	2009–2013	The Mekong Future Program worked in GMS with Lao National component	Improve the sustainability of the MR by investigating the complex relationships between the production, distribution and use of energy, food and water of the region	No	No
Climate Protection through Avoided Deforestation (interview with Dietmar Bräutigam)	KfW/GIZ	MAF, Department of Forestry	KfW/GIZ	€4 million	2009–2012	Luang Prabang Province, Xayaboury	At intermediary level, the necessary policy and institutional framework and initial implementation strategies are in place for effective forest conservation in line with the international debate around REDD+. As well as helping to protect climate and biodiversity, this also improves the conditions under which the rural population live	Yes, including REDD into policy	Not sure?!

Continued

Table 13.1. Continued.

Name of project/ stakeholder	Implementing agency	Responsible government agency	Funded by	Financial resources	Year	Region	Goal/key activities	Policy impact	Part of existing climate change policies?
Clean Air and Climate Change Mitigation for Smaller Cities in the ASEAN Region	GIZ	ASEAN-Secretariat	GIZ	n/a	2009–2012	Vientiane	Smaller and medium-sized cities are increasingly able to develop and implement measures to improve the air quality and contribute to sustainable city development	Not sure	Not sure?
Environmental education to cope with climate change	GIZ	MoNRE	GIZ	n/a	2012–2014	National	Few members of the general public or even decision makers in politics and business are aware of the danger because they know little about the correlation between environment and climate change. The project goal is to educate those	Yes, results shall be included in the action plan and new CC strategy	Not yet, but should be included in the new action plan
Climate Change and Disasters Program	Save the Children		AUD, AusAID	Over US$2 million	Started 2008	Mainly central and southern Lao PDR (Xayaboury)	Child Centred, Child Led and Child Focused DRR, recognizing children as agents of change and encouraging participation of children and communities in the DRR measures which aim to improve the lives of children facing disasters	Through the project 'Establishing Disaster Information Systems', they have influence of policies	A Disaster Information System is part of the NAPA; however, it may not be the programme implemented by Save the Children
Mainstreaming Disaster and Climate Risk Management into Investment Decisions	World Bank	MoNRE	World Bank	US$2.77 million	2011–2015				
Capacity Enhancement for Coping with Climate Change	ADB	MoNRE	Water Resources & Environment Administration (WREA)	US$3.4 million	2010	National	Addresses several capacity barriers by providing support to National Climate Change Office and the institutions responsible	Policy support	Not as a real project, but as part of the underlying aims

[a] This list might not be exhaustive.

change adaptation strategies is influenced by how key policy actors perceive the overall notion of economic uncertainty in relation to the country's development strategy. It illustrates how different notions of uncertainties are linked to various interpretations of climate change impacts (i.e. potential disastrous impacts as opposed to actual incremental impacts). In addition, it highlights the importance of institutional uncertainties (i.e. government agencies' formal mandates, roles, bureaucratic affiliations, and access to resources) and funding uncertainties (donors' funding regulation) in shaping the actual implementation of the NAPA.

The dominant tendency to focus on scientific uncertainty in climate change discourse and policy formulation might in fact decrease the real issue at stake faced by farmers and other local and national actors (i.e. bureaucratic uncertainty, uncertainty of weather and seasonal pattern). Similarly, current lack of recognition of multiple uncertainties in global climate policy formulation might result in a national climate change policy that does not resemble national/local actors' perspective on how to cope with climate change. Understanding the different notions of uncertainty is important to avoid the shaping of climate change policy merely as a blueprint. As stated by Weber (2008, p. 133): 'Too often the debate over climate change is overly simplistic. Given the scale of the problem and the assumption of catastrophic harm, the tendency is to rely on top-down, one-size-fits-all governance solutions'.

Policy and governance responses to climate change need to be formulated based on a more nuanced, sophisticated understanding of how various policy actors and stakeholders perceive and experience uncertainty. Jones *et al.* (2012), for instance, highlight the role of social capital like trust, norms and social networks in influencing how people perceive uncertainty related to climate change impacts. What could be observed on each level of analysis is that various policy actors and stakeholders perceive and experience uncertainties in a very diverse way. While these uncertainties are not translated into policies, they are implied in the way policy actors shape the overall process of climate change policy formulation and in the way stakeholders shape the policy implementation. This can be often seen as the missing link between theory, policy and practice. Hence, it is pertinent to incorporate the diverse notions of uncertainty as an integral part of climate change policy formulation processes.

Notes

1 Globally, climate change is an important issue. The establishment of the IPCC evidenced not only the importance of the issue but also the need to tackle it at global level, primarily with regard to how to cope with climate change impacts. Scientific understanding of climate change impacts shows how climate change can have both long-term (i.e. accumulated incremental impacts) and short-term (i.e. on extreme events) impacts. For Lao PDR in particular these impacts will be both incremental and extreme events.

2 This standardized approach might not translate into effective policy implementation. Even if normative uncertainty is also recognized it might not be perceived as equally important.

3 For an outline of the content of the NAPA and the SCC, see the following section.

4 For more information on projects on the ground, see Section 13.4.2.

5 For more information on recent institutional development, please see Section 13.4.1.

6 This change in institutional set up will be further elaborated in Section 13.4.1.

7 WREA was upgraded into MoNRE in 2011. Unlike earlier, MoNRE has full status as a government ministry and does not function under the official mandate of the Prime Minister's office as did WREA earlier.

8 Here, the issue of institutional discrepancy becomes closely related with the issue of bureaucratic competition and sectoral fragmentation.

9 This information is based on interviews in December 2012 and might already be outdated.

10 Many NGOs, development institutions and the GoL have been involved in climate change projects. The aims of the projects are as diverse as their locations: reaching from capacity development, adaptation enhancement and disaster management to mitigation programmes like Reducing Emissions from Deforestation and Forest degradation (REDD), environmental education and clean-air programmes. The

projects are sometimes located in specific regions (mainly Xayaburi and Savannaketh) or even have a national scope (often based in Vientiane). While the variety of project interests and locations might not intrinsically be wrong, there is limited connection to the official policies.

References

Carpenter, S.R. and Brock, W.A. (2008) Adaptive capacity and traps. Resilience alliance. *Ecology and Society* 13(2), 40.

Climate Change Strategy (CCS)(2011) Government of Lao PDR, Vientiane, Lao People's Democratic Republic.

Dessai, S. and van der Sluijs, J. (2007) *Uncertainty and Climate Change Adaptation – A Scoping Study.* Copernicus Institute. Research Institute for Sustainable Development and Innovation, Utrecht, the Netherlands.

Dimitrov, R.S. (2003) Knowledge, power and interests in environmental regime formation. *International Studies Quarterly* 47(1) (March), 123–150.

Drieschova, A., Fischhendler, I. and Giordano, M. (2009) Climate change, international cooperation and adaptation in transboundary water management. In: Adger, W.N., Lorenzoni, I. and O'Brien, K. (eds) *Adapting to Climate Change: Thresholds, Values, Governance.* Cambridge University Press, Cambridge, UK.

Edelman, M. (1988) *Constructing the Political Spectacle.* University of Illinois Press, Urbana, Illinois.

Heal, G. and Kriström, B. (2002) Uncertainty and climate change. *Environmental and Resource Economics* 22(1–2), 3–39.

International Organisation for Standardisation (ISO)/International Eleoctrotechnical Commission (IEC) Guide 73 (2002) Risk Management – Vocabulary – Guidelines for Use in Standards. Available at: http://www.iso.org/iso/catalogue_detail?csnumber=34998 (accessed 18 December 2013).

IPCC (2007) Climate change: impacts, adaptation and vulnerability. Fourth Assessment Report. Intergovernmental Panel on Climate Change, Geneva, Switzerland.

Jones, N., Clark, J. and Tripidaki, G. (2012) Social risk assessment and social capital: a significant parameter for the formation of climate change policies. *The Social Science Journal* 49(1), 33–41.

Mosse, D. (2004) *Cultivating Development: An Ethnography of Aid Policy and Practice.* Pluto Press, London.

National Adaptation Programme of Action to Climate Change (NAPA) (2008/2009) Water Resources and Environment Administration. Lao People's Democratic Republic.

New, M. and Hulme, M. (2000) Representing uncertainty in climate change scenarios: a Monte-Carlo approach. *Integrated Assessment* 1, 203–213.

OECD (2006) Key adaptation concepts and terms. OECD/IEA Project for the Annex 1 Expert Group on the UNFCCC, Paris, 7 March 2006.

Pielke, R.A. (2007) *The Honest Broker. Making Sense of Science in Policy and Politics.* Cambridge University Press, New York.

Sabatier, P. and Hunter, S. (1988) The incorporation of causal perceptions into models of elite belief systems. *Western Political Quarterly* 42(9), 229–261.

Schneider, S.H., Rosencranz, A. and Niles, J.O. (2002) *Climate Change Policy: a survey.* Island Press, Washington, DC.

Shackley, S. and Wynne, B. (1996) Representing uncertainty in global climate change science and policy: boundary-ordering devices and authority. *Science, Technology, & Human Values* 21(3), 275–302.

Termeer, C.J.A.M., Nieuwaal, K. van, Driessen, P.P.J. and Spit, T.J.M. (2009) *A State of the Art of Governance Literature on adaptation to climate change. Towards a research agenda.* Nationaal Onderzoeksprogramma Kennis voor Klimaat, KvK, Utrecht, the Netherlands.

Tol, R.S.J. (2005) The marginal damage costs of carbon dioxide emissions: An assessment of the uncertainties. *Energy Policy* 33(16), 2064–2074.

UNFCCC (United Nations Framework Convention on Climate Change) (1992) S. *Treaty Doc No. 102–138,* 9 May 1992, 1771 UNTS, 107.

Van der Sluijs, J.P. and Turkenburg, W. (2006) Climate change and the precautionary principle. In: Fisher, E., Jones, J. and Schomberg, R. von (eds) *Implementing the Precautionary Principle, Perspectives and Prospects.* Edward Elgar, Cheltenham, UK, pp. 245–269.

Walker, W.E., Harremoes, P., Rotmans, J., van der Sluijs, J.P., van Asselt, M., Janssen, P. and Krayer von Krauss, M. (2003) Defining uncertainty. A conceptual basis for uncertainty management in model-based decision support. *Integrated Assessment* 4(1), 5–17.

Wardekker, A. (2011) Climate change impact assessment and adaptation under uncertainty. PhD dissertation. Utrecht University, Utrecht, the Netherlands.

Weber, E.P. (2008) Facing and managing climate change: assumptions, science, and governance responses. *Political Science* 60(1), 133–149.

Weber, E.U. (2010) What shapes perceptions of climate change? *Wiley Interdiscplinary Reviews: Climate Change* 1(3), 332–342.

Index

Page numbers for figures and tables are shown in **bold**.